AF755043

Financial Surveillance

STATISTICS IN PRACTICE

Founding Editor

Vic Barnett
Nottingham Trent University, UK

Statistics in Practice is an important international series of texts which provide detailed coverage of statistical concepts, methods and worked case studies in specific fields of investigation and study.

With sound motivation and many worked practical examples, the books show in down-to-earth terms how to select and use an appropriate range of statistical techniques in a particular practical field within each title's special topic area.

The books provide statistical support for professionals and research workers across a range of employment fields and research environments. Subject areas covered include medicine and pharmaceutics; industry, finance and commerce; public services; the earth and environmental sciences, and so on.

The books also provide support to students studying statistical courses applied to the above areas. The demand for graduates to be equipped for the work environment has led to such courses becoming increasingly prevalent at universities and colleges.

It is our aim to present judiciously chosen and well-written workbooks to meet everyday practical needs. Feedback of views from readers will be most valuable to monitor the success of this aim.

A complete list of titles in this series appears at the end of the volume.

Financial Surveillance

Edited by

Marianne Frisén

Göteborg University, Sweden

John Wiley & Sons, Ltd

Telephone (+44) 1243 779777

Email (for orders and customer service enquiries): cs-books@wiley.co.uk
Visit our Home Page on www.wileyeurope.com or www.wiley.com

Other Wiley Editorial Offices

John Wiley & Sons Inc., 111 River Street, Hoboken, NJ 07030, USA

Jossey-Bass, 989 Market Street, San Francisco, CA 94103-1741, USA

Wiley-VCH Verlag GmbH, Boschstr. 12, D-69469 Weinheim, Germany

John Wiley & Sons Australia Ltd, 42 McDougall Street, Milton, Queensland 4064, Australia

John Wiley & Sons (Asia) Pte Ltd, 2 Clementi Loop #02-01, Jin Xing Distripark, Singapore 129809

John Wiley & Sons Canada Ltd, 6045 Freemont Blvd, Mississauga, Ontario, L5R 4J3, Canada

Wiley also publishes its books in a variety of electronic formats. Some content that appears in print may not be available in electronic books.

Library of Congress Cataloging-in-Publication Data

Financial surveillance / edited by Marianne Frisén.
p. cm.
Includes bibliographical references and index.
ISBN 978-0-470-06188-6 (cloth : acid free paper)
1. Econometric models. 2. Mathematical optimization. I. Frisén, Marianne.
HB141.F75 2007
332.01′519 – dc22

2007045546

British Library Cataloguing in Publication Data

A catalogue record for this book is available from the British Library

ISBN 978-0-470-06188-6 (HB)

Typeset in 10.5/13pt Times by Laserwords Private Limited, Chennai, India
Printed and bound in Great Britain by TJ International, Padstow, Cornwall
This book is printed on acid-free paper responsibly manufactured from sustainable forestry in which at least two trees are planted for each one used for paper production.

Contents

List of Contributors

EVA ANDERSSON Statistical Research Unit, Department of Economics, Göteborg University, PO Box 640, SE 405 30 Göteborg, Sweden; and Department of Occupational and Environmental Medicine, Sahlgrenska University Hospital, PO Box 414, SE 405 30 Göteborg, Sweden

DAVID BOCK Statistical Research Unit, Department of Economics, Göteborg University, PO Box 640, SE 405 30 Göteborg, Sweden

MARIANNE FRISÉN Statistical Research Unit, Department of Economics, Göteborg University, PO Box 640, SE 405 30 Göteborg, Sweden

VASYL GOLOSNOY Institute of Statistics and Econometrics University of Kiel, Olshausenstraße 40, D-24118 Kiel, Germany

IRYNA OKHRIN Department of Statistics, Europa-Universität Viadrina, Grosse Scharrnstr. 59, D-15230 Frankfurt (Oder), Germany

YAREMA OKHRIN Department of Statistics, Europa-Universität Viadrina, Grosse Scharrnstr. 59, D-15230 Frankfurt (Oder), Germany

WOLFGANG SCHMID Department of Statistics, Europa-Universität Viadrina, Grosse Scharrnstr. 59, D-15230 Frankfurt (Oder), Germany

HELGI TÓMASSON Faculty of Economics and Business Administrations, University of Iceland, Oddi v/Sturlugotu, IS-101 Reykjavik, Iceland

1

Introduction to financial surveillance

Marianne Frisén
Statistical Research Unit, School of Business, Economics and Law, Göteborg University, PO Box 660, SE 405 30 Göteborg, Sweden

1.1 What is financial surveillance?

In financial surveillance the aim is to signal at the optimal trading time. A systematic decision strategy is used. The information available at each possible decision time is evaluated in order to judge whether or not there is enough information for a decision about an action or if more information is necessary so that the decision should be postponed. Financial surveillance gives timely decisions.

The authors of this book hope that it will serve two purposes. First, we hope that it will stimulate an increased use of surveillance in finance by providing methods which have not been available before. Second, we hope that the statistical community will use the book as a spur to further development of techniques and to research on questions unanswered by the following chapters. Financial surveillance is a new area, and some open problems are discussed in Chapter 9.

Financial decision strategies are based, in one way or another, on continuous observation and analysis of information. This is financial surveillance. Statistical surveillance uses decision theory and statistical inference in order to derive timely decision strategies. Hopefully, this book will serve as a bridge between finance and statistical surveillance.

Financial Surveillance Edited by Marianne Frisén

This book is written by statisticians with an interest in finance. Textbooks describing financial problems and statistical methods are for example Föllmer and Schied (2002), Härdle, Kleinow and Stahl (2002), Gourieroux and Jasiak (2002), Franke, Härdle and Hafner (2004), Cizek, Härdle and Weron (2005) and Scherer and Martin (2005). Many and various statistical techniques are described in these books.

In Section 1.2 statistical methods which are useful for financial decisions will be discussed. In Section 1.3 the area of statistical surveillance is described, and the characteristics of surveillance are compared to other areas in statistics. Evaluations in surveillance are described in Section 1.4. This is an important area, since the choice of evaluation measures will decide which methods are considered appropriate. General methods for aggregating information over time are described in Section 1.5. Special aspects of surveillance for financial decisions are discussed in Section 1.6.

The content of the book will be described in Section 1.7, which deals with the relation between the chapters. This section also provides reading guidelines.

1.2 Statistical methods for financial decision strategies

Statistical methods use observations of financial data to give information about the financial process which produces the data. This is in contrast to probability theory, where assumptions about the financial process are used to derive which observations will be generated.

1.2.1 Transaction strategies based on financial data

In finance, the relation between observations and decisions is often informal. Statisticians have taken on the role of presenting statistical summaries of quantitative data. In many areas, including finance, this means providing point and interval estimates for the quantities of interest. Methods for providing such summaries are highly formalized and constantly evolving. The discipline of statistics uses observations to make deductions about the real world. It has its own set of axioms and theorems besides those of probability theory. While decision making is the incentive for much statistical analysis, the process that transforms statistical summaries into decisions usually remains informal and ad hoc.

In finance the timeliness of transactions is important to yield a large return and a low risk. The concept of an efficient arbitrage-free market is of great interest. One central question is whether the history of the price of an asset contains information which can be used to increase the future return. A natural aim is to maximize the return. The theory of stochastic finance has been based on an assumption of an efficient market where the financial markets are arbitrage-free and there is no point in trying to increase the return. Even though this view is generally accepted today, there are some doubts that it is generally applicable. When the information about the process is incomplete, as for example when a change point could occur, there may be an arbitrage opportunity, as demonstrated by Shiryaev (2002). In Chapter 3 it will be discussed how technical analysis relies on the possibility of using history to increase future returns. The support for the efficient market hypothesis depends on the knowledge about the model, as is discussed below in Section 1.2.2.

1.2.2 Modelling

In finance, advanced stochastic models are necessary to capture all empirical features. The expected value could depend on time in a complicated nonlinear way. Parameters other than the expected value are often of great interest, and the risk (measured by variance) is often of great concern. Complicated dependencies are common, which means that complicated measures of variance are necessary. Multivariate data streams are of interest for example when choosing portfolio. The models may be described in continuous or discrete time. Chapter 2 gives an overview of models of interest in finance. Statistical methods for estimation and model choice will also be briefly described in the chapter. The use of the models should be robust to errors in the model specification.

1.2.2.1 Stochastic model assumed known

When the stochastic model is assumed to be completely known there is no expected return to be gained. We will have an arbitrage-free market. We can use probability theory to calculate the optimal transaction conditions. Important contributions are found in the book by Shiryaev (1999) or in articles in the scientific journal *Finance and Stochastics*. Also the proceedings of the conference *Stochastic Finance* in 2004 and 2007 are informative on how to handle financial decisions when the model is completely known.

1.2.2.2 Incomplete knowledge about the stochastic model

When the model is not completely known, the efficient and arbitrage-free market assumptions are violated. Changes at unknown times are possible. One has to evaluate the information continuously to decide whether a transaction at that time is profitable. Statistical inference is needed for the decision (Shiryaev 2002).

1.2.3 Evaluation of information

Statistical inference theory gives guidelines on how to draw conclusions about the real world from data. Statistical hypothesis testing is suitable for testing a single hypothesis but not a decision strategy including repeated decisions, as will be further described in Section 1.4.1. Statistical surveillance is an important branch of inference. The relatively new area of statistical surveillance deals with the sequential evaluation of the amount of information at hand. It provides a theory for deciding at what time the amount of information is enough to make a decision and take action. This bridges the gap between statistical analysis and decisions.

In this book, we concentrate on the methodology of statistical surveillance. This methodology is of special interest for financial decision strategies, but it is also relatively new in finance. The ambition here is to give a comprehensive description of such aspects of statistical surveillance that may be of interest in finance. Thus the next sections of this chapter will give a short review on statistical surveillance.

1.3 What is statistical surveillance?

1.3.1 General description

Statistical surveillance means that a time series is observed with the aim of detecting an important change in the underlying process as soon as possible after the change has occurred. Statistical methods are necessary to separate important changes in the process from stochastic variation. The inferential problems involved are important for the applications and interesting from a theoretical viewpoint, since they bring different areas of statistical theory together.

Broad surveys and bibliographies on statistical surveillance are given by Lai, who concentrates on the minimax properties of stopping rules, by Woodall and Montgomery (1999) and Ryan (2000), who concentrate on control charts, and by Frisén (2003), who concentrates on the optimality properties of various methods.

The theory of statistical surveillance has developed independently in different statistical subcultures. Thus, the terminology is diverse. Different terms are used to refer to 'statistical surveillance' as described here. However, there are some differences in how the terms are used. 'Optimal stopping rules' are most often used in probability theory, especially in connection with financial problems. However, this does not always include the statistical inference from the observations to the model. Literature on 'change-point problems' does not always treat the case of continuous observations but often considers the case of a retrospective analysis of a fixed number of observations. The term 'early warning system' is most often used in the economic literature. 'Monitoring' is most often used in medical literature and as a nonspecific term. Timeliness, which is important in surveillance, is considered in the vast literature on quality control charts, and here also the simplicity of procedures is stressed. The notations 'statistical process control' and 'quality control' are used in the literature on industrial production and sometimes also include other aspects than the statistical ones.

The statistical methods suitable for surveillance differ from the standard hypothesis testing methods. In the prospective surveillance situation, data accumulated over time is analysed repeatedly. A decision concerning whether, for example, the variance of the price of a stock has increased or not has to be made sequentially, based on the data collected so far. Each new possibility demands a new decision. Thus, there is no fixed data set but an increasing number of observations. In sequential analysis we have repeated decisions, but the hypotheses are fixed. In contrast, there are no fixed hypotheses in surveillance. The statistics derived for a fixed sample may be of great value also in the case of surveillance, but there are great differences between the systems for decision. The difference between hypotheses and on-line surveillance is best seen by studying the difference in evaluation measures (see Section 1.4.1).

In complicated surveillance problems, a stepwise reduction of the problem may be useful. Then, the statistics derived to be optimal for the fixed sample problem can be a component in the construction of the prospective surveillance system. This applies, for example, to the multivariate problems described in Section 1.6.7 in this chapter and in Chapters 5–7.

1.3.2 History

The first modern control charts were developed in the 1920s, by Walter A. Shewhart and coworkers at Bell Telephone Laboratories. In 1931 the famous book *Economic Control of Quality of Manufactured Product* (Shewhart 1931) was published. The same year Shewhart gave a presentation of the new technique to the Royal Statistical Society. This stimulated interest in the UK. The technique was used extensively during World War II both in the UK and in the US. In the 1950s, W. E. Deming introduced the technique in Japan. The success in Japan spurred the interest in the West, and further development started.

In the Shewhart method each observation is judged separately. The next important step was taken when Page suggested the CUSUM method for aggregating information over time. Shortly afterwards, Roberts (1959) suggested another method for aggregating information – the EWMA method. A method based on likelihood which fulfils important optimality conditions was suggested by Shiryaev (1963).

In recent years there have been a growing number of papers in economics, medicine, environmental control and other areas, dealing with the need of methods for surveillance. The threat of bioterrorism and new contagious diseases has been an important reason behind the increased research activity in the theory of surveillance. Hopefully, the time is now ripe for finance to benefit from all these results.

1.3.3 Specifications of the statistical surveillance problem

The general situation of a change in distribution at a certain change-point time τ will now be specified. The variable under surveillance could be the observation itself or an estimator of a variance or some other derived statistic, depending on the specific situation. We denote the process by $X = \{X(t) : t = 1, 2, \ldots\}$, where $X(t)$ is the observation made at time t. The purpose of the monitoring is to detect a possible change. The time for the change is denoted by τ (see Figure 1.1). This can be regarded either as a random variable or as a deterministic but unknown value, depending on what is most suitable for the application.

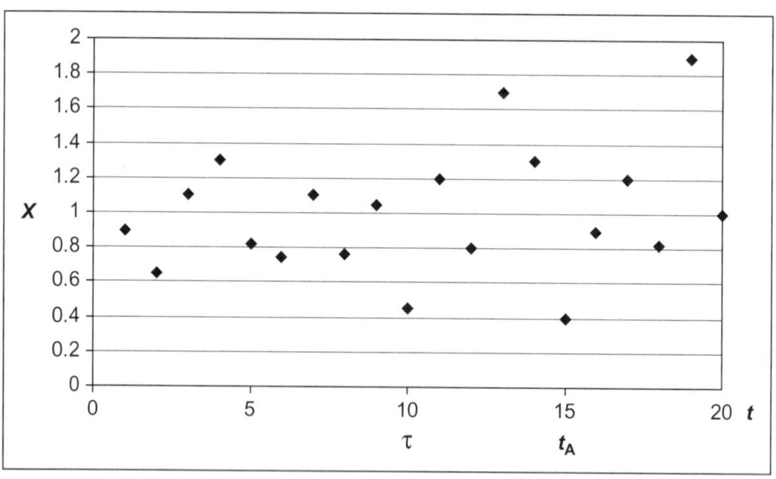

Figure 1.1 The first $\tau - 1$ observations $X_{\tau-1} = \{X(t) : t \leq \tau - 1\}$ are 'in-control' with a small variance. The subsequent observations (from $t = \tau$ (here 10) and onwards) have a larger variance. The alarm time is t_A, which happens to be 15. Thus the delay is $t_A - \tau = 5$.

The properties of the process change at time τ. In many cases we can describe this as

$$X(t) = \begin{cases} Y(t), & t < \tau \\ Y(t) + \Delta, & t \geq \tau \end{cases} \tag{1.1}$$

where Y is the 'in-control' or 'target' process and Δ denotes the change. More generally we can denote the 'in-control' state by D and the state which we want to detect by C. The (possibly random) process that determines the state of the system is here denoted by $\mu(t)$. This could be an expected value, a variance or some other time-dependent characteristic of the distribution. Different types of states between which the process changes are of interest for different applications. Descriptions of the details of the states are made within each chapter.

The change to be detected differs depending on the application. Most studies in literature concern a step change, where a parameter changes from one constant level, say, $\mu(t) = \mu^0$ to another constant level, $\mu(t) = \mu^1$. The case $\mu > 0$ is described here. We have $\mu(t) = \mu^0$ for $t = 1, \ldots, \tau-1$ and $\mu(t) = \mu^1$ for $t = \tau, \tau + 1$.

Even though autocorrelated time series are studied for example by Schmid and Schöne (1997), Petzold, Sonesson, Bergman and Kieler (2004), and in Chapters 5–7, processes which are independent given τ are the most studied and used also in Chapters 3 and 4. This simple situation will be used to introduce general concepts of evaluations, optimality and standard methods.

Some cases of special interest in financial surveillance are discussed in Section 1.6.

1.4 Evaluations

Quick detection and few false alarms are desired properties of methods for surveillance. Knowledge about the properties of the method in question is important. If a method calls an alarm, it is important to know whether this alarm is a strong indication of a change or just a weak indication. The same methods can be derived by Bayesian or frequentistic inference. However, evaluations differ. Here we present measures suitable for frequentistic inference.

1.4.1 The difference between evaluations for hypothesis testing and on-line surveillance

Measures for a fixed sample situation can be adopted for surveillance, but some important differences will be pointed out. In Table 1.1 the measures

Table 1.1 Evaluation measures for hypothesis testing and the corresponding measures for on-line surveillance.

	Test	*Surveillance*
False alarms	Size α, Specificity	ARL^0, MRL^0, PFA
Detection ability	Power, Sensitivity	ARL^1, MRL^1. CED, ED, maxCED, PSD, SADT

conventionally used in hypothesis testing and some measures for surveillance are given. These measures will be described and discussed below.

Different error rates and their implications for a decision system were discussed by Frisén and de Maré (1991). Using a constant probability of exceeding the alarm limit for each decision time means that we have a system of repeated significance tests. This may work well also as a system of surveillance and is often used. The Shewhart method described in Section 1.5.2 has this property. This is probably also the motive for using the limits with the exact variance in the EWMA method described in Section 1.5.4.

Evaluation by significance level, power, specificity and sensitivity, which is useful for a fixed sample, is not appropriate without modification in a surveillance situation since these measures do not have unique values in a surveillance system. One problem with evaluation measures originally suggested for the study of a fixed sample of, say, n observations is that the measures depend on n. For example, the specificity will tend to zero for most methods and the size of the test will tend to one when n increases.

Chu, Stinchcombe and White (1996) and others have suggested methods with a size less than one:

$$\lim_{n\to\infty} P(t_A \leq n|D) < 1.$$

This is convenient since ordinary statements of hypothesis testing can be made. However, Frisén (2003) demonstrated that the detection ability of methods with this property declines rapidly with the value of time τ of the change. Important consequences were illustrated by Bock (2008).

The performance of a method for surveillance depends on the time τ of the change. Generally, the sensitivity will not be the same for early changes as for late ones. It also depends on the length of time for which the evaluation is made. Thus, there is not one unique sensitivity value in surveillance, but other measures may be more useful. Accordingly, conventional measures for fixed samples should be supplemented by other measures designed for statistical surveillance, as will be discussed in the following.

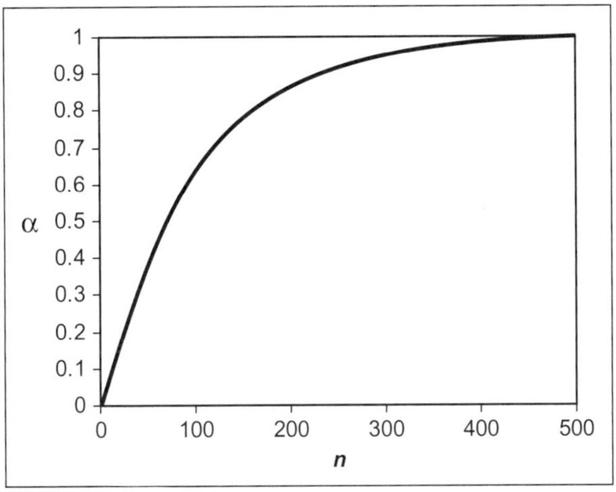

Figure 1.2 The size, α, of a surveillance system which is pursued for n time units, when the probability of a false alarm is 1% at each time point.

1.4.2 Measures of the false alarm rate

The false alarm tendency is more complicated to control in surveillance than in hypothesis testing, as was seen above (for example in Figure 1.2). There are special measures of the false alarm properties which are suitable for surveillance. The most commonly used measure is the Average Run Length when there is no change in the system under surveillance, $ARL^0 = E(t_A | D)$. A variant of the ARL is the Median Run Length, MRL.

A measure commonly used in theoretical work is the false alarm probability, $PFA = P(t_A < \tau)$. This is the probability that the alarm occurs before the change.

1.4.3 Delay of the alarm

The delay time of the detection of a change should be as short as possible. The most commonly used measure of the delay is the average run length until the detection of a true change (that occurred at the same time as the surveillance started), which is denoted by ARL^1. The part of the definition within parentheses is seldom spelled out but generally used in the literature (see, for example, Page 1954 and Ryan 2000). Instead of the average, Gan (1993) advocates that the median run length should be used on the grounds that it may be more easily interpreted. However, also here only a change occurring at the same time as the surveillance started is considered.

In most practical situations it is important to minimize the expected delay of detection whenever the change occurs. Shiryaev (1963) suggested measures of the expected value of the delay. The expected delay from the time of change, $\tau = t$, to the time of alarm, t_A, is denoted by

$$\mathrm{ED}(t) = \mathrm{E}[\max(0, t_A - t)|\tau = t].$$

Note that $\mathrm{ARL}^1 = \mathrm{ED}(1) + 1$. The $\mathrm{ED}(t)$ will typically tend to zero as t increases. Thus, it is easier to evaluate the conditional expected delay

$$\mathrm{CED}(t) = \mathrm{E}[t_A - \tau | t_A \geq \tau = t] = \mathrm{ED}(t)/\mathrm{P}(t_A \geq t).$$

$\mathrm{CED}(\tau)$ is the expected delay for a specific change point τ. The expected delay is generally not the same for early changes as for late ones. For most methods, the CED will converge to a constant value. This value is sometimes named the 'steady state average delay time' or SADT. It is, in a sense, the opposite to ARL^1 since only a very large value of τ is considered. SADT has been advocated for example by Srivastava and Wu (1993), Srivastava (1994) and Knoth (2006).

For some situations and methods the properties are about the same regardless of when the change occurs. However, this is not always true, as illustrated by Frisén and Wessman (1999). Then, it is important to consider more and other cases than just $\tau = 1$. The values of CED can be summarized in different ways. One is the maximal value over τ. Another approach is to regard τ as a random variable with the probabilities $\pi(t) = P(\tau = t)$. These probabilities can also be regarded as priors. The intensity of a change is defined as $\nu(t) = P(\tau = t|\tau \geq t)$, which is usually assumed to be constant over time. Shiryaev (1963) suggested a summarized measure of the expected delay

$$\mathrm{ED} = \mathrm{E}[\mathrm{ED}(\tau)].$$

Sometimes the time available for action is limited. The Probability of Successful Detection suggested by Frisén (1992) measures the probability of detection with a delay time no longer than d

$$\mathrm{PSD}(d, t) = P(t_A - \tau < d | t_A \geq \tau = t).$$

This measure is a function of both the time of the change and the length of the interval in which the detection is defined as successful. Also when there is no absolute limit to the detection time it is often useful to describe the ability to detect the change within a certain time. In such cases it may be useful to calculate the PSD for different time limits d. This has been done by Marshall, Best, Bottle and Aylin (2004). The ability to make a very quick detection (small d) is important in surveillance of sudden major changes, while the long-term detection ability (large d) is more important for ongoing surveillance where smaller changes are expected.

1.4.4 Predictive value

When an alarm is called, one needs to know whether to act as if the change is certain or just plausible. To obtain this, both the risk of false alarms and the risk of delay must be considered. If τ is regarded as a random variable this can be done by one summarizing measure. The probability that a change has occurred when the surveillance method signals was suggested by Frisén (1992) as a time-dependent predictive value

$$PV(t) = P(\tau \leq t_A | t_A = t).$$

When there is an alarm ($t_A = t$), PV indicates whether there is a large probability or not that the change has occurred ($t_A \geq \tau$). Some methods have a constant PV. Others have a low PV at early alarms but a higher one later. In such cases, the early alarms will not prompt the same serious action as later ones.

1.4.5 Optimality

1.4.5.1 Minimal expected delay

Shiryaev (1963) suggested a highly general utility function, in which the expected delay of an alarm plays an important role. Shiryaev treated the case where the gain of an alarm is a linear function of the value of the delay, $t_A - \tau$, and the intensity of the change is constant. The loss associated with a false alarm is a function of the same difference. This utility can be expressed as $U = \mathrm{E}\{u(\tau, t_A)\}$, where

$$u(\tau, t_A) = \begin{cases} h(t_A - \tau) & \text{if } t_A < \tau \\ a_1(t_A - \tau) + a_2 & \text{else.} \end{cases}$$

The function $h(t_A - \tau)$ is usually a constant (say, b), since the false alarm causes the same cost of alerts and investigations irrespectively of how early the false alarm is given. In this case, we have

$$U = b\ \mathrm{P}(t_A < \tau) + a_1\ \mathrm{ED} + a_2.$$

We would have a maximal utility if there is a minimal (a_1 is typically negative) expected delay from the change point for a fixed probability of a false alarm (see Section 4.3). This is termed the ED criterion. Variants of the utility function leading to different optimal weighting of the observations are suggested for example by Poor (1998) and Beibel (2000).

1.4.5.2 Minimax optimality

The minimum of the maximal expected delay after a change considers several possible change times, just like the ED criterion. However, instead of an expected value, which requires a distribution of the time of change, the least favourable value of CED(t) is used. This criterion is used in Chapter 5.

Moustakides (1986) uses an even more pessimistic criterion, the 'worst possible case', by using not only the least favourable value of the change time, but also the least favourable outcome of $X_{\tau-1}$ before the change occurs. This criterion is very pessimistic. The CUSUM method, described in Section 1.5.3, provides a solution to the criterion proposed by Moustakides. The merits of the studies of this criterion have been thoroughly discussed for example by Yashchin (1993) and Lai (1995). Much theoretical research is based on this criterion.

1.4.5.3 ARL optimality

Optimality is often stated as a minimal ARL^1 for a fixed ARL^0. ARL^1 is the expected value under the assumption that all observations belong to the 'out-of-control' distribution, whereas ARL^0 is the expected value given that all observations belong to the 'in-control' distribution. Efficient methods for surveillance (see Section 1.5) will put most weight on the most recent observations. Statistical inference with the aim of discriminating between the two alternatives that all observations come from either of the two specified distributions should, by the ancillarity principle, put the same weight on all observations. To use efficient methods and evaluate them by the ARL criterion is thus in conflict with this inference principle.

Pollak and Siegmund (1985) argue that for many methods, the maximal value of CED(t) is equal to CED(1), and with a minimax perspective this can be an argument for using ARL^1 since $\text{CED}(1) = ARL^1 - 1$. However, this argument is not relevant for all methods. In particular, it is demonstrated by Frisén and Sonesson (2006) that the maximal CED-value is not CED(1) for the EWMA method in Section 1.5.5. In the case of this method, there is no similarity between the optimal parameter values according to the ARL criterion and the minimax criterion, while the optimal parameter values by the criterion of expected delay and the minimax criterion agree well.

The dominating position of the ARL criterion was questioned by Frisén (2003), since methods useless in practice are ARL optimal. The ARL can be used as a descriptive measure and gives a rough impression, but it is questionable as a formal optimality criterion.

1.4.6 Comments on evaluation measures

Computer illustrations of the interpretation of some of the measures mentioned below are made by Frisén and Gottlow (2003). Formulas for numerical approximations of some of the measures are available in literature.

1.5 General methods for aggregating information

In surveillance it is important to aggregate the available information in order to benefit from all information. This aggregation can be carried out in accordance with some general inference principles. Specific methods are then derived from the general ones. Different principles of aggregation have different properties and are thus suitable for different problems.

Some methods are highly flexible and have several parameters. The parameters can be chosen to make the method optimal for the specific conditions of the application (for example, the size of the change or the intensity of changes). Many methods for surveillance are based, in one way or another, on likelihood ratios. Thus, we will start by describing the likelihood ratio component. The likelihood ratio for a fixed value of τ is

$$L(s, t) = f_{X_s}(X_s|\tau = t)/f_{X_s}(X_s|D).$$

Most commonly used methods can be described as different combinations of these components.

1.5.1 The Shiryaev–Roberts method

The simplest way to aggregate the likelihood components is just to add them. This means that all possible times for the change up to the decision time s are given equal weight. Shiryaev (1963) and Roberts (1966) suggested the method, now called the Shiryaev–Roberts method, in which an alarm is triggered at the first time s, so that

$$\sum_{t=1}^{s} L(s, t) > G,$$

where G is a constant alarm limit. This method can also be given a natural interpretation if the time of the change τ is regarded as a random variable. This method can in that case be regarded as a special case of the full likelihood ratio method. This will be further discussed in Section 1.5.6.

1.5.2 The Shewhart method

The Shewhart method (Shewhart 1931; Ryan 2000) is simple and certainly the most commonly used method for surveillance. It can be regarded as performing repeated significance tests. An alarm is triggered as soon as an observation deviates too much from the target. Thus, only the last observation is considered in the Shewhart method. An alarm is triggered at

$$t_{\mathrm{A}} = \min\{s; X(s) > L\},$$

where L is a constant. The alarm criterion for independent observations can be expressed by the condition $L(s, s) > G$ where G is a constant. The alarm statistic of the LR method reduces to that of the Shewhart method when $C(s) = \{\tau = s\}$ and $D(s) = \{\tau > s\}$. This is the case when we want to discriminate between a change at the current time point and the case that no change has happened yet. In this situation, we are only interested to see whether something has happened 'now' or not. Thus, the Shewhart method has optimal error probabilities for these alternatives for each decision time s. For large shifts, the LR method of Section 1.5.6 and the CUSUM method of Section 1.5.3 converge to the Shewhart method (Frisén and Wessman 1999 and Chapter 4). By several criteria, the Shewhart method performs poorly for small and moderate shifts. By the minimax criterion, however, it works nearly as well as the LR method for some situations.

1.5.3 The CUSUM method

The CUSUM method, first suggested by Page (1954), is closely related to the minimax criterion. Yashchin (1993), Siegmund and Venkatraman (1995) and Hawkins and Olwell (1998) give reviews of the CUSUM method. The alarm condition of the method can be expressed by the partial likelihood ratios as

$$t_{\mathrm{A}} = \min\{s; \max(L(s, t); t = 1, 2, \ldots, s) > G\}$$

where G is a constant. The method is sometimes called *the* likelihood ratio method, but this combination of likelihood ratios should not be confused with the full likelihood ratio method, LR.

The most commonly described application of the CUSUM method concerns the case of independent normally distributed variables. In this case, the CUSUM statistic reduces to a function of the cumulative sums

$$C_r = \sum_{t=1}^{r} (X(t) - \mu(t)).$$

There is an alarm for the first time s for which

$$C_s - C_{s-i} > h + ki \text{ for some } i = 1, 2, \ldots, s,$$

where $C_0 = 0$ and h and k are chosen constants. In the case of a step change, the value of the parameter k is usually $k = (\mu^0 + \mu^1)/2$.

Closely related to the CUSUM method are the Generalized Likelihood Ratio (GLR) and Mixture Likelihood Ratio (MLR) methods. For the MLR method suggested by Pollak and Siegmund (1975), a prior for the shift size is used in the CUSUM method. For the GLR method, the alarm statistic is formed by maximizing over possible values of the shift (besides the maximum over possible times of the shift). Lai (1998) describes both GLR and MLR and proves a minimax result for a variant of GRL suitable for autocorrelated data.

The CUSUM method satisfies the minimax criterion of optimality described in Section 1.4.5.2. Other good qualities of the method have been confirmed for example by Srivastava and Wu (1993) and Frisén and Wessman (1999). With respect to the expected delay, the CUSUM method works almost as well as the LR and Shiryaev–Roberts methods.

1.5.4 Moving average and window-based methods

The Moving average method can be expressed by the likelihood ratios as

$$L(s, s-d) > G$$

where G is a constant and d is a fixed window width. In the standard case of normally distributed variables this will be a moving average. It will have the optimal error probabilities of the LR method when we want to detect a change which occurred at time $s - d$ (i.e. for $C = \{\tau = s - d\}$) and will thus have optimal detection abilities for changes which occurred d time points earlier.

Sometimes, as in Lai (1998), advanced methods such as the GLR method are combined with a window technique in order to ease the computational burden.

1.5.5 Exponentially weighted moving average methods

The EWMA method is a variant of a moving average method which does utilize all information. The alarm statistic is based on exponentially weighted moving averages,

$$Z_s = (1 - \lambda) Z_{s-1} + \lambda Y(s), s = 1, 2, \ldots$$

where $0 < \lambda < 1$ and Z_0 is the target value, which is normalized to zero. The EWMA statistic gives the largest weight to the most recent observation and geometrically decreasing weights to all previous ones. If λ is near zero, all observations have approximately the same weight. Note that if $\lambda = 1$ is used, the EWMA method reduces to the Shewhart method. The asymptotic variant, EWMAa, will give an alarm at

$$t_A = \min\{s : Z_s > L\sigma_Z\},$$

where L is a constant. In another variant of the method, EWMAe, the exact standard deviation (which is increasing with s) is used instead of the asymptotic one in the alarm limit. Sonesson (2003) found that the EWMAa version is preferable for most cases.

The EWMA method was described by Roberts (1959). Positive reports of the quality of the method are given for example by Crowder (1989), Lucas and Saccucci (1990), Domangue and Patch (1991) and Knoth and Schmid (2002). The choice of λ is important, and the search for the optimal value of λ has been of great interest in the literature. Small values of λ result in a good ability to detect early changes while larger values are necessary for changes that occur later.

Most reports on optimal values of the parameter λ refer to the ARL criterion. Frisén (2003) demonstrated that by this criterion, λ should approach zero. Methods which allocate the power to the first time points will have good ARL properties but less ability to detect a change that happens later. In fact, and wisely enough, no one seems to have suggested that λ should be chosen to zero, even though this should fulfill the ARL criterion.

The EWMA method can be seen as a linear approximation of the full LR method (see Section 1.5.6). When a change from N(0, σ) to N(μ, σ) occurs with the intensity ν, the parameter λ that gives the optimality properties of the full LR method is

$$\lambda* = 1 - \exp(-\mu^2/2)/(1 - \nu),$$

This was shown by Frisén (2003) and confirmed by large-scale simulation studies by Frisén and Sonesson (2006).

1.5.6 The full likelihood ratio method

When the time of the shift is regarded as a random variable, we can utilize this property. The full likelihood ratio method (LR) is optimal with respect to the criterion of minimal expected delay and also to a wider class of utility functions (Frisén and de Maré 1991). The full likelihood is a weighted sum of the partial likelihoods

$$L(s, t) = f_{X_s}(X_s|\tau = t)/f_{X_s}(X_s|D(s)).$$

The alarm set consists of those values of X for which the full likelihood ratio exceeds a limit. The following notation can be used: At decision time s we want to discriminate between the event $C(s) = \{\tau < s\}$ and the event $D(s) = \{\tau > s\}$. The time of an alarm for the LR method is

$$t_A = \min\left\{s; \frac{f_{X_s}(x_s|C(s))}{f_{X_s}(x_s|D(s))} > \frac{P(\tau > s)}{P(\tau \leq s)} \cdot \frac{K}{1-K}\right\}$$
$$= \min\left\{s; \sum_{t=1}^{s} w(s,t) \cdot L(s,t) > G(s)\right\}$$

where K is a constant and $G(s)$ is the alarm limit. The time of an alarm can equivalently be written as the first time the posterior probability of a change into state C exceeds a fixed level:

$$t_{\mathrm{A}} = \min\{s; P(C(s)|X_s = x_s) > K\}.$$

The posterior probability of a change has been suggested as an alarm criterion for example by Smith and West (1983). When there are only two states, C and D, this criterion leads to the LR method (Frisén and de Maré 1991). In cases where several changes may follow after each other, the process may be characterised as a hidden Markov chain and the posterior probability for a certain state may be determined (for example Harrison and Stevens 1976 and Hamilton 1989). Sometimes the use of the posterior distribution, or equivalently the likelihood ratio, is named 'the Bayes method'. However, it depends on the situation whether the distribution of τ should be considered as a 'prior', as an observed frequency-distribution or if it just reflects the situation for which optimality is desired.

When the intensity, ν, of a change tends to zero, the weights $w(s,t)$ of the partial likelihoods do not depend on t, and the limit $G(s)$ of the LR method does not depend on s. Shiryaev (1963) and Roberts (1966) suggested the Shiryaev–Roberts method (mentioned in Section 1.5.1), for which an alarm is triggered at the first time s, such that

$$\sum_{t=1}^{s} L(s,t) > G,$$

where G is a constant. The method can be seen as the limit of the LR method when ν tends to zero. The Shiryaev–Roberts method can also be derived as the LR method with a noninformative prior for the distribution of τ. Both the LR method and the Shiryaev–Roberts method can be expressed recursively. One valuable property of these methods is an approximately constant predictive value Frisén and Wessman (1999), which allows the same interpretation of early and late alarms.

The LR method is optimized for the values of the change size and for the change intensity. In the case of a normal distribution, the LR method gives an alarm at

$$t_{\mathrm{A}} = \min\left\{s;\ \sum_{t=1}^{s} \mathrm{P}(\tau = t)\exp\{t\mu^2/2\}\exp\{\mu\sum_{u=t}^{s} Y(u)\}\right.$$
$$\left. > \exp\{(s+1)\mu^2/2\}\mathrm{P}(\tau > s)\frac{K}{1-K}\right\}$$

where the constant K determines the false alarm probability.

As mentioned before, several methods can be described by approximations or combinations of likelihood ratios (Frisén 2003). Linear approximations of the LR method are of interest for two reasons – first, for obtaining a method which is easier to use and analyse but whose properties are as good as those of the LR method, and second, for getting a tool for the analysis of the approximate optimality of other methods as in Frisén (2003).

1.6 Special aspects of surveillance for financial decisions

1.6.1 General approaches which can be used in complex situations

Situations in finance are often complex. Thus, some general approaches for surveillance in more complicated situations than those of the earlier sections are of interest. When the models are completely specified both before and after the change, the likelihood components $L(s, t)$ can usually be derived or approximated. Then, these components can be combined by any of the general information aggregation methods mentioned in Section 1.5. Lai (1995, 1998) and Lai and Shan (1999) argue that the good minimax properties of generalisations of the CUSUM method make the CUSUM suitable for complicated problems. The likelihood ratio method, LR, with its good optimality properties can also be used. Pollak and Siegmund (1985) argue that the martingale property (for continuous time) of the Shiryaev–Roberts method makes this more suitable for complicated problems than the CUSUM method. The LR method also has this property, but the CUSUM method does not.

1.6.2 Evaluation by return

The return from buying an asset at $t = 0$ and selling at time t is $r(t) = x(t) - x(0)$, where X is a monotonic function of the price.

The expected return $\mathrm{E}[r(t_{\mathrm{A}})]$ of selling at the alarm time t_{A} is maximal when $\mathrm{E}[X(t_{\mathrm{A}})]$ is maximal. Thus, a sell signal should ideally come at time τ which corresponds to a peak of the price. In Chapter 3 on technical analysis this surveillance problem is analysed.

1.6.3 Surveillance of dependent data

Financial time series often have complicated time dependencies. The theory for surveillance of dependent data is not simple. The general approaches in Section 1.6.1 can be applied to obtain methods with known optimality properties. This was made for example by Petzold *et al.* (2004). The most common approach to surveillance in the case of models with dependencies is to monitor the process of residuals. Pettersson (1998) demonstrated that for an autoregressive process, this is an approximation of the LR method. Another common approach (also used by Pettersson 1998) is to adjust the alarm limit in order to adjust the false alarm risk resulting from ignoring the dependency. Chapters 5–7 and the references in these chapters contain important contributions to this very sparsely discussed area.

1.6.4 Surveillance of discrete distributions

Most of the theory of surveillance is derived for normal distributions, but a bibliography of surveillance of attribute data is given by Woodall (1997).

1.6.5 Gradual changes

Most of the literature on surveillance treats the case of an abrupt change. In many cases in finance, however, the change is gradual. The change is thus more complicated than the standard situation of a sudden shift in a parameter from one value to another. Important characteristics should be captured by the statistic under surveillance. In the presence of a nuisance parameter, a general approach is to use a pivot statistic. Krieger, Pollak and Yakir (2003) suggest the CUSUM and Shiryaev–Roberts methods based on a statistic which does not depend on the unknown parameters in a case of an unknown prechange regression. Arteaga and Ledolter (1997) compare several procedures with respect to ARL properties. One of the suggestions in the paper is a window method based on the likelihood ratio and isotonic regression techniques. In general, window methods (see Section 1.5.4) are inefficient for detecting gradual changes (Järpe

2000). Yashchin (1993) discusses generalizations of the CUSUM and EWMA methods to detect both sudden and gradual changes.

It may be hard to model the shape of a gradual change exactly or even to estimate the baseline accurately. Then, the timely detection of a change in monotonicity is of interest. The start of an increase is of course of special interest, but also the decline may be of interest to get timely sell and buy signals.

When the knowledge on the shape of the curve is uncertain, nonparametric methods are of interest. Frisén (2000) suggested surveillance that is not based on any parametric model but only on monotonicity restrictions. This surveillance method was described and evaluated by Andersson (2002) and is further described in Chapter 3.

1.6.6 Changes between unknown levels

After a change, the level of the statistic under surveillance (for example the variance) is seldom known. However, this is not a serious problem. The false alarm properties will remain the same even if the level after the change is not known. The method could be designed to be optimal for a change of a specific size, but this is not required. The unknown parameters can be handled within different frameworks corresponding to different restrictions on possible optimality.

To control false alarms is usually more important than to optimize the detection ability. Knowledge of the prechange conditions is important. The baseline is often estimated and used as a plug-in value in the method. The estimated baseline value will affect the performance of the method. In the situation where we want to detect an increase, we will get more false alarms if the baseline is underestimated than if the true value had been used. The opposite is true if the baseline is overestimated.

One way to avoid the problem of unknown parameters is to transform the data to invariant statistics. Frisén (1992) and Sullivan and Jones (2002) use the deviation of each observation from the average of all previous ones. Gordon and Pollak (1997) use invariant statistics combined by the Shiryaev–Roberts method to handle the case of an unknown prechange mean of a normal distribution. Krieger *et al.* (2003) use invariant statistics combined by the CUSUM and Shiryaev–Roberts methods for surveillance of a change in regression.

When both the baseline and the change are unknown, the aim of the surveillance could be to detect a change in a stochastically larger distribution. Bell, Gordon and Pollak (1994) suggested a nonparametric method geared to the exponential distribution. The nonparametric method of Chapter 3, designed for the detection of a change in monotonicity, also avoids the problem of unknown values of the baseline and the change.

The use of the maximum difference (measured for example by the likelihood ratio) between the baseline and the changed level is a useful approach. The GLR method (Lai 1995, 1998) uses the maximum likelihood estimator of the value after the change. Kulldorff (2001) used the same technique for the detection of clustering in spatial patterns.

Another general approach for unknown levels is the Bayesian one. The MLR method suggested by Pollak and Siegmund (1975) uses priors for the unknown parameters in the CUSUM method. Lawson (2004) used priors for the unknown parameters to calculate the posterior means in a Bayesian space-time interaction model.

1.6.7 Multivariate surveillance

We may have several data streams containing information. This is the case for example in portfolio optimization. We may also have several statistics, such as both the mean and the variance, to monitor. In Chapters 5–7 multivariate techniques for financial problems are extensively discussed.

If the model can be completely specified both before and after the change, then it is possible to derive the likelihood components $L(s, t)$ and aggregate them by a method which guarantees optimality. In complicated problems, however, this is seldom realistic. Instead, a reduction of the multivariate surveillance problem is common (Sonesson and Frisén 2005).

A reduction of the dimensionality of the problem is a natural first approach. Principal components could be used to reduce the dimensionality, but Lowry and Montgomery (1995) argued that unless the principal components can be interpreted, a surveillance method based on them may be difficult to interpret. In Rosolowski and Schmid (2003) and Chapters 5–7 the Mahalanobis distance is used to reduce the dimensionality of the statistic, thus expressing the distance from the target of the mean and the autocorrelation in a multivariate time series.

The most common way to handle multivariate surveillance is to reduce the information to one statistic and then monitor this statistic in time. Wessman (1998) proved that this is a sufficient reduction when changes occur simultaneously in all variables.

Another commonly used approach is to make parallel surveillance for each variable and make a general alarm when there is an alarm for any of the components, see Stoumbos, Reynolds Jr, Ryan and Woodall (2000). Any univariate surveillance method could be used. Parallel CUSUM methods were used by Marshall *et al.* (2004). The false alarms were controlled by using the False Discover Rate (FDR) from Benjamini and Hochberg (1995). For evaluating the detection ability, the probability of successful detection (see Section 1.4.3) was used.

The more advanced approach of vector accumulation is an in-between of the reduction by time and the reduction by variable. Here the accumulated information on each component is used to transform the vector of component-wise alarm statistics into a scalar alarm statistic and make an alarm if this statistic exceeds a limit, see for example Rogerson and Yamada (2004). It is also possible to construct the multivariate method while aiming at satisfying some global optimality criterion. Järpe (1999) suggested an ED optimal surveillance method of clustering in a spatial log-linear model. Lowry, Woodall, Champ and Rigdon (1992) proposed a multivariate extension of the univariate EWMA method, which is referred to as MEWMA. This can be described as the Hotelling T^2 control chart applied to univariate EWMA statistics instead of the original data from only the current time point and is thus a vector accumulation method. Crosier (1988) suggested the MCUSUM method, where a statistic consisting of univariate CUSUMs for each component is used. This is similar to the MEWMA statistic, which corresponds to a vector accumulation method. However, the way in which the components are used is not the same. An alternative way to construct a vector accumulating multivariate CUSUM is given by Pignatiello and Runger (1990). The methods use different weighting of the variables. One important feature of these two methods is that the characteristic zero-return of the CUSUM technique is constructed in a way that is suitable when all the components change at the same time point.

Different aspects on approaches for multivariate surveillance were given by Frisén (2003) and Sonesson and Frisén (2005). The multivariate methods can be evaluated by the measures and criteria described above or by generalized measures. Wessman (1999) suggested a generalization of the ARL measure to allow for the possibility of different change times for different variables. Controlling the false discovery rate is of interest when making conclusions about several variables and is used for example by Wong, Moore, Cooper and Wagner (2003). However, the question of optimality is always complex in multidimensional cases.

1.7 Content and the relation between chapters

Each chapter is self-contained. It is thus not necessary to read the chapters in the order they appear in the book. In this section, the relation between the chapters will be described in order to facilitate the choice of reading order. This section will also demonstrate the wide spectrum of problems and methods in financial surveillance. The short description of each chapter given here may also give information on which chapter to choose for a specific purpose.

Different financial series are analysed in the different chapters. In Chapter 3 the Hang Seng index is analysed with respect to optimal times for trading. This is a market-value weighted index of the 33 largest companies on the Hong Kong stock market. In Chapter 4 a period of Standard and Poor's 500 stock market index is analysed to investigate the ability of some suggested methods to detect a documented change in volatility. In Chapters 5 and 6 there are several analyses of the Morgan Stanley Capital International index (MSCI) for Germany. In Chapter 5 also the MSCI indices for the US and Japan are analysed. In Chapter 7 the optimal allocations of a portfolio of the US market, a non-US index, oil and gold are determined. In addition, most chapters contain simulation studies to further describe the properties of the suggested methods.

In finance much effort has been made to find models which are suitable for financial time series. Special demands have to be met. Chapter 2 gives an overview of problems in finance and models which have been found useful to describe financial data for various purposes. However, Chapter 2 is somewhat more mathematically demanding than the other chapters. Thus, it might be a good idea to start by just taking a quick look at Chapter 2, as an introduction to the kind of models which are considered to capture the characteristics of financial data. A more thorough reading of Chapter 2 can be made after the other chapters. It will then serve as an overview of a great number of models. This gives the possibility to compare methods. Although this book does not discuss surveillance with reference to all these models, it might inspire new research on the development of methods for surveillance of financial models of great interest. This will be discussed in Chapter 9. Some of the models described in Chapter 2 will be used later. However, chapters where a specific model is used will contain a description of that model. Chapters 3–8 each give a description of surveillance with reference to some specific models. Each chapter on surveillance contains all the necessary information about the model used in that chapter. However, it might be a good idea to look also at the part of Chapter 2 where the specific model and similar ones are treated.

Chapter 3 links traditional technical analysis to the new methods for surveillance. Even though the chapter contains new research material it is not very demanding mathematically. Methods for traditional technical analysis could be ad hoc without derivation from financial or statistical science. Thus, methods for technical analysis are not always considered scientific. New mathematical methods have a better reputation even though they are sometimes based on assumptions which might be doubted. Technical analysis has a background in common sense about how to find the right time to make transactions in order to increase profit. In Chapter 3 these methods are justified by a demonstration of how some common methods can be derived from simple stochastic models, natural expression of a good return and good methods for surveillance. Thus, common methods for technical analysis are linked to surveillance theory. In this way one can also find optimal ways to maximize the return for the simple models as well as ways to improve the methods of technical analysis by using a bit more complicated methods. This will be further discussed in Chapter 9.

In technical analysis the history of the price of a stock is assumed to contain information, and technical analyses are methods for extracting this information. The natural aim of technical analysis is to maximize the return. The possibility of achieving this has been much doubted during the last decades. The theory of stochastic finance has been based on an assumption of an efficient market where the financial markets are arbitrage-free and there is no point in trying to determine the optimal transaction time. Even though this view is generally accepted nowadays there are also some doubts that it is generally applicable. When the information about the process is incomplete, as for example when a change point could occur, there may be an arbitrage opportunity, as demonstrated by Shiryaev (2002). Also, there are agents who reject the efficient market hypothesis, and some studies give support for the profitability of the prospective framework. Technical analysis is used in practice, and an analysis of its relation to other models and methods may thus be of interest.

When the efficient market hypothesis is accepted, the interest is drawn to the variability of the price. A low variability is desired. Methods for trading strategies which minimize the variance in different situations are described in Chapters 4–7. In Chapter 4 an iid process (except for a step change in the variance) is assumed, and a sufficient statistic for the ordinary variance is under surveillance. When time dependence is allowed, the measure of variance becomes more complicated. In Chapter 5 the variance in a linear time series is under surveillance. In Chapter 6 nonlinear time series (GARCH, generalized autoregressive conditional heteroscedasticity) models are studied. In Chapter 7 surveillance of the minimum variance portfolio is discussed.

In the theory of surveillance, detection of a parametric step change in the model is the most developed part. While the other chapters deal with this problem, Chapter 3 deals with a nonparametric approach to detect a gradual change.

Most methods for surveillance are based on independent observation (given τ). In Chapters 5–7 several general techniques to obtain good methods for dependent time series are used. In Chapter 5 the impact of neglecting the time series structure is examined within an extensive simulation study.

In Chapters 3 and 4, the methods are expressed by likelihood ratios and the time of change, τ, is regarded as a stochastic variable. This facilitates the construction of methods which are optimal with respect to the expected delay (ED), as described in Section 1.2 of this chapter. This gives a summarizing measure of timeliness. In the other chapters the change point is regarded as a fix but unknown constant. In these chapters evaluations are made for different values of τ and especially for $\tau = 1$ which gives the commonly used measure ARL[1]. This approach avoids assumptions on the distribution of τ. The delay for the least favourable value of τ corresponds to the minimax criteria for which the CUSUM method has optimality properties.

Even though the statistics studied may be quite different, all methods use some way of aggregating the information over time in order to enable a timely decision. The methods differ as to the way in which the different partial likelihood ratios are weighted, as described in Section 1.2 of this chapter. In Chapter 3 a new maximum likelihood method is compared to the hidden Markov method, HMM, and the CUSUM method. In Chapter 4 the methods studied are the full likelihood ratio, Shiryaev–Roberts, Shewhart and CUSUM methods. Chapters 5–7 are dominated by the EWMA approach, but also the CUSUM method is evaluated in Chapter 5. The EWMA statistic cannot be expressed directly, but Frisén (2003) demonstrated that it can be seen as an approximation of the likelihood ratio method.

In Chapters 5–7 the problem of multivariate surveillance is treated while the other chapters deal with univariate surveillance. Several general techniques for multivariate surveillance are adopted in order to derive appropriate methods. The trading strategy may also involve the optimal times for updating the portfolio. The portfolio problem is treated in Chapter 7, where an optimal method and a case study of the multivariate problem of portfolio optimization are given.

Chapter 8 deals with surveillance of a continuous time model in contrast to the other chapters, which analyse discrete time models. Continuous time models are popular for describing some financial processes. However, since the models tend to be complicated, there is still a lack of methods for surveillance of continuous financial series. The estimation of parameters which can serve as indicators of important changes may be hard to achieve. Here, computer programs are provided for this task. Even though the model is specified in continuous time, the decisions are, in practice, made in discrete time. The influence of the time between observations and other problems of continuous time surveillance are discussed in Chapter 8. Decisions in continuous time are in fact easier than decisions in discrete time, since the latter type causes an overshoot. Many results are first derived for continous time and only later for the discrete time which applies in practice (see for example Shiryaev 1963 and Srivastava and Wu 1993).

In Chapter 9 some conclusions are given and the prospect of future progress is discussed. Some open problems for further research are suggested. The level as regards mathematics and citations to scientific literature is designed to be a reasonable starting point for statisticians and quantitatively minded people in finance. Knowledge of statistical concepts corresponding to about one year of study is required for understanding the text.

A companion website to this book is available at: http://www.wiley.com/go/frisen

References

Andersson, E. (2002). Monitoring cyclical processes – a nonparametric Approach. *Journal of Applied Statistics*, **29**, 973–990.

Arteaga, C. and Ledolter, J. (1997). Control charts based on order-restricted tests. *Statistics & Probability Letters*, **32**, 1–10.

Beibel, M. (2000). A note on sequential detection with exponential penalty for the delay. *The Annals of Statistics*, **28**, 1696–1701.

Bell, C., Gordon, L. and Pollak, M. (1994). An efficient nonparametric detection scheme and its application to surveillance of a Bernoulli process with unknown baseline. In *Change-Point Problems*, eds. E. Carlstein, H.-G. Muller and D. Siegmund, Hayward, California: IMS Lecture Notes, Monograph Series, pp. 7–27.

Benjamini, Y. and Hochberg, Y. (1995). Controlling the false discovery rate: A practical and powerful approach to multiple testing. *Journal of the Royal Statistical Society B*, **57**, 289–300.

Bock, D. (2008). Aspects on the control of false alarms in statistical surveillance and the impact on the return of financial decision systems. *Journal of Applied Statistics,* **35**.

Chu, C.-S. J., Stinchcombe, M. and White, H. (1996). Monitoring structural change. *Econometrica*, **64**, 1045–1065.

Cizek, P., Härdle, W. and Weron, R. (eds) (2005). *Statistical Tool for Finance and Insurance.* Springer-Verlag, Berlin.

Crosier, R. B. (1988). Multivariate generalizations of cumulative sum quality-control schemes. *Technometrics*, **30**, 291–303.

Crowder, S. V. (1989). Design of exponentially weighted moving average schemes. *Journal of Quality Technology*, **21**, 155–162.

Domangue, R. and Patch, S. C. (1991). Some omnibus exponentially weighted moving average statistical process monitoring schemes. *Technometrics*, **33**, 299–313.

Franke, J., Härdle, W. and Hafner, C. (2004). *Statistics of Financial Markets. An Introduction.* Springer-Verlag, Berlin.

Frisén, M. (1992). Evaluations of methods for statistical surveillance. *Statistics in Medicine*, **11**, 1489–1502.

Frisén, M. (1994), Statistical surveillance of business cycles, Technical Report 1994:1 (Revised 2000).

Frisén, M. (2003). Statistical surveillance. Optimality and methods. *International Statistical Review*, **71**, 403–434.

Frisén, M. and de Maré, J. (1991). Optimal Surveillance. *Biometrika*, **78**, 271–280.

Frisén, M. and Gottlow, M. (2003). Graphical evaluation of statistical surveillance. Technical Report Research Report 2003:10, Statistical Research Unit, Göteborg University.

Frisén, M. and Sonesson, C. (2006). Optimal surveillance based on exponentially weighted moving averages. *Sequential Analysis*, **25**, 379–403.

Frisén, M. and Wessman, P. (1999). Evaluations of likelihood ratio methods for surveillance. Differences and Robustness. *Communications in Statistics, Simulation and Computation*, **28**, 597–622.

Föllmer, H. and Schied, A. (2002). *Stochastic Finance. An Introduction in Discrete Time*. de Gruyter, Berlin.

Gan, F. F. (1993). An optimal design of EWMA control charts based on median run-length. *Journal of Statistical Computation and Simulation*, **45**, 169–184.

Gordon, L. and Pollak, M. (1997). Average run length to false alarm for surveillance schemes designed with partially specified pre-change distribution. *The Annals of Statistics*, **25**, 1284–1310.

Gourieroux, C. and Jasiak, J. (2002). *Financial Econometrics: Problems, Models and Methods*. University Presses Of California, Columbia and Princeton, New Jersey.

Hamilton, J. D. (1989). A new approach to the economic analysis of nonstationary time series and the business cycle. *Econometrica*, **57**, 357–384.

Harrison, P. J. and Stevens, C. F. (1976). Bayesian Forecasting, with discussion. *Journal of the Royal Statistical Society B*, **38**, 205–247.

Hawkins, D. M. and Olwell, D. H. (1998). *Cumulative Sum Charts and Charting for Quality Improvement*. Springer-Verlag, New York.

Härdle, W., Kleinow, T. and Stahl, G. (eds) (2002). *Applied Quantitative Finance. Theory and Computational Tools*. Springer-Verlag, New York.

Järpe, E. (1999). Surveillance of the interaction parameter in the Ising model. *Communications in Statistics. Theory and Methods*, **28**, 3009–3025.

Järpe, E. (2000). On univariate and spatial surveillance. PhD thesis. Göteborg University, Department of Statistics.

Knoth, S. (ed.) (2006). *The Art of Evaluating Monitoring Schemes – How to Measure the Performance of Control Charts?* (Vol. 8), eds. H.-J. Lenz and P.-T. Wilrich, Physica Verlag, Heidelberg.

Knoth, S. and Schmid, W. (2002). Monitoring the mean and the variance of a stationary process. *Statistica Neerlandica*, **56**, 77–100.

Krieger, A. M., Pollak, M. and Yakir, B. (2003). Surveillance of a simple linear regression. *Journal of the American Statistical Association*, **98**, 456–469.

Kulldorff, M. (2001). Prospective time periodic geographical disease surveillance using a scan statistic. *Journal of the Royal Statistical Society A*, **164**, 61–72.

Lai, T. L. (1995). Sequential changepoint detection in quality-control and dynamical systems. *Journal of the Royal Statistical Society B*, **57**, 613–658.

Lai, T. L. (1998). Information bounds and quick detection of parameters in stochastic systems. *IEEE Transactions on Information Theory*, **44**, 2917–2929.

Lai, T. L. and Shan, Z. (1999). Efficient recursive algorithms for detection of abrupt changes in signals and control systems. *IEEE Transactions on Automatic Control*, **44**, 952–966.

Lawson, A. (2004). Some considerations in spatial-temporal analysis of public health surveillance data. In *Monitoring the Health of Populations: Statistical Principles & Methods for Public Health Surveillance*, eds R. Brookmeyer and D. F. Stroup. Oxford University Press, Oxford, pp. 289–314.

Lowry, C. A. and Montgomery, D. C. (1995). A review of multivariate control charts. *IIE Transactions*, **27**, 800–810.

Lowry, C. A. Woodall, W. H., Champ, C. W. and Rigdon, S. E. (1992). A multivariate exponentially weighted moving average control chart. *Technometrics*, **34**, 46–53.

Lucas, J. M. and Saccucci, M. S. (1990). Exponentially weighted moving average control schemes: Properties and enhancements. *Technometrics*, **32**, 1–12.

Marshall, C., Best, N., Bottle, A. and Aylin, P. (2004). Statistical issues in the prospective monitoring of health outcomes across multiple units. *Journal of the Royal Statistical Society A*, **167**, 541–559.

Page, E. S. (1954). Continuous inspection schemes. *Biometrika*, **41**, 100–114.

Pettersson, M. (1998). Monitoring a freshwater fish population: Statistical surveillance of biodiversity. *Environmetrics*, **9**, 139–150.

Petzold, M., Sonesson, C., Bergman, E. and Kieler, H. (2004). Surveillance in longitudinal models. Detection of intrauterine growth retardation. *Biometrics*, **60**, 1025–1033.

Pignatiello, J. J. and Runger, G. C. (1990). Comparisons of multivariate CUSUM charts. *Journal of Quality Technology*, **22**, 173–186.

Pollak, M. and Siegmund, D. (1975). Approximations to the expected sample size of certain sequential tests. *The Annals of Statistics*, **3**, 1267–1282.

Pollak, M. and Siegmund, D. (1985). A diffusion process and its applications to detecting a change in the drift of Brownian motion. *Biometrika*, **72**, 267–280.

Poor, V. H. (1998). Quickest detection with exponential penalty for delay. *The Annals of Statistics*, **26**, 2179–2205.

Roberts, S. W. (1966). A comparison of some control chart procedures. *Technometrics*, **8**, 411–430.

Rogerson, P. A. and Yamada, I. (2004). Monitoring change in spatial patterns of disease: Comparing univariate and multivariate cumulative sum approaches. *Statistics in Medicine*, **23**, 2195–2214.

Rosolowski, M. and Schmid, W. (2003). EWMA charts for monitoring the mean and the autocovariances of stationary Gaussian processes. *Sequential Analysis*, **22**, 257–285.

Ryan, T. P. (2000). *Statistical Methods for Quality Improvement* (2nd edn). John Wiley & Sons, Ltd, New York.

Scherer, B. and Martin, R. D. (2005). *Introduction to Modern Portfolio Optimization with Nuopt^{tm}, S-Plus and S^{+} Bayes^{tm}*. Springer-Verlag, New York.

Schmid, W. and Schöne, A. (1997). Some properties of the EWMA control chart in the presence of autocorrelation. *The Annals of Statistics*, **25**, 1277–1283.

Shewhart, W. A. (1931). *Economic Control of Quality of Manufactured Product*. MacMillan and Co., London.

Shiryaev, A. N. (1963). On optimum methods in quickest detection problems. *Theory of Probability and Its Applications*, **8**, 22–46.

Shiryaev, A. N. (1999). *Essentials of Stochastic Finance: Facts, Models, Theory.* World Scientific, Singapore.

Shiryaev, A. N. (2002). Quickest detection problems in the technical analysis of financial data. In *Mathematical Finance – Bachelier Congress 2000*, eds. H. Geman, D. Madan, S. Pliska and T. Vorst. Springer-Verlag, Berlin.

Siegmund, D. and Venkatraman, E. S. (1995). Using the generalized likelihood ratio statistic for sequential detection of a change-point. *The Annals of Statistics*, **23**, 255–271.

Smith, A. F. and West, M. (1983). Monitoring renal transplants: An application of the multiprocess Kalman filter. *Biometrics*, **39**, 867–878.

Sonesson, C. (2003). Evaluations of some exponentially weighted moving average methods. *Journal of Applied Statistics*, **30**, 1115–1133.

Sonesson, C. and Frisén, M. (2005). Multivariate surveillance. In *Spatial Surveillance for Public Health*, eds. A. Lawson and K. Kleinman. John Wiley & Sons, Ltd, New York. pp. 169–186.

Srivastava, M. S. (1994). Comparison of CUSUM and EWMA procedures for detecting a shift in the mean or an increase in the variance. *Journal of Applied Statistical Science*, **1**, 445–468.

Srivastava, M. S. and Wu, Y. (1993). Comparison of EWMA, CUSUM and Shiryaev–Roberts procedures for detecting a shift in the mean. *The Annals of Statistics*, **21**, 645–670.

Stoumbos, Z. G., Reynolds Jr, M. R., Ryan, T. P. and Woodall, W. H. (2000). The state of statistical process control as we proceed into the 21st century. *Journal of the American Statistical Association*, **95**, 992–998.

Sullivan, J. H. and Jones, L. A. (2002). A self-starting control chart for multivariate individual observations. *Technometrics*, **44**, 24–33.

Wessman, P. (1998). Some principles for surveillance adopted for multivariate processes with a common change point. *Communications in Statistics. Theory and Methods*, **27**, 1143–1161.

Wessman, P. (1999). Studies on the surveillance of univariate and multivariate processes. PhD thesis, Göteborg University, Department of Statistics.

Wong, W. K., Moore, A., Cooper, G. and Wagner, M. (2003). WSARE: Whats Strange About Recent Events? *Journal of Urban Health*, **80**, I66–I75.

Woodall, W. H. (1997). Control charts based on attribute data: bibliography and review. *Journal of Quality Technology*, **29**, 172–183.

Yashchin, E. (1993). Statistical control schemes – methods, applications and generalizations. *International Statistical Review*, **61**, 41–66.

2

Statistical models in finance

Helgi Tómasson

University of Iceland, Faculty of Economics and Business Administrations, Oddi v/Sturlugotu, IS-101 Reykjavik, Iceland

2.1 Introduction

The aim of this review is to give a brief review of the statistical tools, models and fundamental concepts that are available for financial data analysis. The approach is set up as an index of basic concepts for the quantitatively minded. This review is inevitably very brief as both finance and statistics are large subjects.

Finance is: 'the science that describes the management of money, banking, credit, investments, and assets, basically, finance looks at anything that has to do with money and the market' (http://financial-dictionary.thefreedictionary.com/finance). Still a slightly different definition is: 'A discipline concerned with determining value and making decisions. The finance function allocates resources, including the acquiring, investing, and managing of resources' (Harvey and Morgenson 2002).

One definition of statistics is: 'The mathematics of the collection, organization, and interpretation of numerical data, especially the analysis of population characteristics by inference from sampling' (http://www.thefreedictionary.com/statistics). At http://dictionary.laborlawtalk.com/Statistics the definition given is: 'Statistics is the science and practice of developing human knowledge through the use of empirical data. It is based on statistical theory which is a branch of applied mathematics'. Within statistical theory, randomness and uncertainty are modelled by probability theory. Because one aim of statistics

Financial Surveillance Edited by Marianne Frisén

is to produce the 'best' information from available data, for eventual policy making, some authors consider statistics a branch of decision theory. Statistical practice includes the planning, summarizing, and interpreting of observations, allowing for variability and uncertainty. Statistics can be seen as a probability application where the main task is data analysis and giving general statements about an unknown reality. Because finance is a branch of economics, the practice of deriving estimators and tests in this context, is frequently referred to as financial econometrics. Analysing properties of estimators and tests are important subjects in the statistical discipline.

The discipline of finance has been expanding in the direction of mathematics in the second half of the twentieth century. Probably the first stochastic modelling approach in finance took place by Bachelier (1900) in an attempt of formalizing unpredictability of price movements. In physics, Einstein (1905) used similar ideas for describing unpredictability of movements of particles. It was, though, not until Wiener (1923) gave a proper mathematical background to the ideas of Bachelier and Einstein and proved that the Wiener process was a well defined mathematical concept. The ability to forecast in the stock market would be very valuable if it was possible. The efficient market makes the scope for profitability of forecasting very limited. Academically the problem of forecasting prices and poor performance of experts in financial markets is treated at least as early as Cowles (1933). The notion of unpredictability of prices is present in the statistical literature as early as Kendall (1953). The mathematical approach initiated by Bachelier, Einstein and Wiener in the early twentieth century, was rediscovered in a famous article by Black and Scholes (1973). The advance culminated in the 1997 Nobel Prize of economics, when Merton and Scholes were awarded the prize 'for a new method tool determine the value of derivatives'.

This development has shifted the focus of finance theory and increased its level of mathematical sophistication. A great number of books has appeared on use of stochastic models in finance, such as Merton (1990), Karatzas and Shreve (1991), Hull (1993), Wilmott, Howison and Dewynne (1995), Neftci (1996), Duffie (1996), Shiryaev (1999), Revuz and Yor (1999), Cvitanic and Zapatero (2004), Björk (2004), etc., the list is very long.

Progress in computer technology and telecommunications has made data much more accessible and transferable than before. The combination of mathematics, data and computer power has opened the field of applied statistics to financial applications. Methods that before were considered utopian both for reasons of lack of data and mathematical complexity are now feasible.

The statistical discipline of linking data and models has responded to the availability of new data and new theory. Collections of articles relating statistical methods to finance are, e.g. Maddala and Rao (1996), Hand and Jacka (1998), and Chan, Keung and Tong (2000). Examples of recent textbooks on applied data analysis in finance are Tsay (2002), and Zivot and Wang (2003).

At http://en.wikipedia.org/wiki/Finance the following definition is given: 'Finance studies and addresses the ways in which individuals, businesses and organizations raise, allocate, and use monetary resources **over time**, taking into account the risks entailed in their projects'. Therefore, the statistical methods for analysis of financial data are highly focused on the importance of time. This is also the case in surveillance where the emphasis is on making timely decisions. As the time factor is extremely important in finance, statistical methods involving time dependency are crucial for financial data analysis. Statistical models involving time dependency rely on the probability theory of stochastic processes. Modern finance also relies on the theory of stochastic processes. Therefore basic knowledge of stochastic processes is essential for the quantitative financial analyst, as well as for the theoretical one.

Stochastic processes can be classified by the nature of the state space and the nature of the time index. The state can be continuous or discrete and the time index can as well be continuous or discrete. For statistical analysis the most common, and best known tool, is time series theory, which often refers to the case where the state is continuous, but time is discrete.

The organization of this review is as follows. Section 2.2 gives a brief background of financial theory for financial markets. In Section 2.3 a brief review of the principal concepts of classical linear equispaced time series is given. For the theory of finance, volatility (standard deviation) of price is important, both for pricing and for risk management. Therefore models for second moments are of interest. The popular discrete-time approach in modelling is the ARCH class of models. Since the appearance of the ARCH model in Engle (1982), many models based on the same idea, focusing on second-moments (variances), have been derived. In Section 2.4 a brief review of some of the ARCH-family models is given. Discrete time series models for equispaced time series, ARMA, GARCH, etc., are now easily applicable by use of widespread computer software. The ARCH models are a special kind of nonlinear/semilinear models. The general class of nonlinear models is very large so, therefore, it is necessary to limit the functional form. The general class of nonlinear models is simply too large. In Section 2.5 some simple nonlinear or semilinear models that have been suggested in the literature are reviewed.

The modern finance literature is dominated by continuous-time models. The continuous-time mathematics approach offers a powerful tool for logical reasoning about a dynamic environment. The benefit of an empirical approach to continuous-time statistical models is that the interpretation will correspond to a theoretical model. In practice, however, the approach is problematic. In practice a time-continuous pattern is never observed, there is always some discretization involved. Even if one could obtain a continuous observation, one would have to integrate a continuous pattern to obtain a value of the likelihood function. The continuous-time models can be classified as continuous-path models or models with jumps.

Bergstrom (1988, 1990) gives an historic overview on the use of continuous time-models in econometrics. Bergstrom (1990) mentions that the beginning of statistical analysis of continuous-time models might be traced back to Bartlett (1946), but the mainstream econometric literature might have missed this result, perhaps due to the then recent discovery of the treatment of simultaneous models by Haavelmo (1943), which dominated econometric methodology for 30 years. In recent years impressive progress has been made in statistical treatment of discretely observed diffusion processes. Recently, statistical approaches to time-continuous diffusion models have also become feasible. Some properties of these models are briefly reviewed as well as approaches for estimating unknown parameters based on real, discretely observed data. An outline of some ideas is given in Section 2.6.

Traditional duration-data models/transition-data/survival models analyse jump-only processes, i.e. the process moves between a finite number of alternative states. Transition data models and extreme value models arise naturally when statistical aspects of financial data are analysed. In Section 2.7 transition data models are briefly reviewed.

The interest of second moments, or equivalently, volatility (standard deviation) is frequently connected with the fashionable term 'risk'. The ARCH-type models deal with dynamics of second moments. A related type of risk is the 'extreme value' analysis, the probabilistic nature of extremes. Just as in engineering, catastrophes in finance take place, firms go bankrupt, etc. In Section 2.8 some references to financial applications of extreme-value theory are reviewed.

The finance literature has recently taken steps towards memoryless processes that are partly continuous, i.e. the jump-diffusions or Lévy processes. It is in the nature of statistical inference that it will always be very hard to distinguish between a jump, a steep climb and a heavy tailed distribution. Some aspects of this and some references are given in Section 2.9.

This summary is written from an econometrics/statistics point of view. It does not contain any review of the literature on technical analysis, artificial intelligence or machine learning. Quantitative analysis of financial data is performed in these fields. The author suspects that the calculations in those disciplines are to a degree similar, but the concern about the probabilistic nature of 'the model' is less apparent, and the interpretation of results will differ from the statistical way of thinking. Chapter 3 in this book gives a formal approach of technical analysis and surveillance.

All practical analysis of financial data is evidently computer-dependent. Data are obtained and treated electronically and a vast arsenal of optimization methods is now implemented in the research centers of financial data analysts. The numerical methods are of great variety. There exist deterministic, iterative, methods, such as Newton-type methods for maximization and solving equations, and some stepwise methods like the EM algorithm, auxiliary regression, double-length regression, etc. Examples of simulation-based methods are MCMC (Markov Chain Monte Carlo) and particle filtering. The emphasis in this review is on the categories of practically feasible models and only to a small degree on the technical implementation, such as, how to get output, estimates, tests, etc. for the model/data.

2.2 A brief background of financial markets problems

In theoretical finance several assumptions of a 'perfect' market are necessary (Merton 1990, p. 477). Among these assumptions several are quite unrealistic, like no transactions costs, the investor can always buy or sell as much as he wants of any asset (including borrowing or lending money) at any point in time, the market is always in equilibrium. A key assumption is the efficient market hypothesis. Roughly expressed, the efficient market hypothesis states that 'all information' is already included in the prices, and therefore that prices should not be predictable. This nonpredictability feature is formalized in the mathematical modelling by the inclusion of a nonpredictable stochastic process, the Wiener process (the Brownian motion). How this efficiency is implemented in practice is somewhat debated. Blackwell, Griffiths and Winters (2006) give the following classification. First, the strong-form efficiency, which states that all information (private and public) is embedded in the security price. Second, the semistrong efficiency, that all public information is available in the security price. And third, the weak-form efficiency, that all past information is included in the security price. In academic finance at least the weak form of efficiency is required for arbitrage-free pricing. The logic is that if prices were predictable, the agents would quickly discover that and give higher bids on securities that are likely to increase in price. That kind of behaviour would eliminate predictability.

It turns out that enforcing arbitrage-free conditions in financial modelling results in pricing functions where the variance function is a fundamental part. Therefore the variance function is of principal interest for investment-strategy, risk-management, etc. Surveillance of the mean function in finance is, though, not uninteresting. For example financial inspection authorities have the role of monitoring the possibility of insider trading. For surveillance, therefore both surveillance of mean and variance are of interest for surveillance in finance.

2.3 Linear time series analysis

A data-set, $(x_1, \ldots, x_T)$, is a realization of set of random variables, $(X_1, \ldots, X_T)$. The simplest case is when the random variables are iid. The theory of time series deals with the situation where there is a dependency structure based on the time sequence of the sampling. In order to obtain consistent estimates it is necessary to assume that dependency fades away in the sense that observations far apart in time are almost independent. The mathematical term for this property is ergodicity. As only one realization is available, some stability property of X_t is also necessary. The theoretical concept for that is stationary, which means that the dependency structure is invariant over time. In practice weak stationarity is assumed for reasons of convenience. Weak stationarity is characterized by covariance stationarity, i.e. the mean and autocovariance functions are assumed to be constant over time.

$$\begin{aligned} &\mathrm{E}(X_t) = \mu, \\ &\mathrm{E}(X_t - \mu)(X_s - \mu) = \gamma(|t-s|) = \gamma(k), \quad k = |t-s|. \end{aligned}$$

The ergodicity assumption states that the dependence between X_t and X_s should decrease as $|t-s|$ increases, i.e. that random variables very far apart in time should be virtually independent. Several definitions of ergodicity exist, but for estimation of mean and autocovariance it is essential to require mean ergodicity:

$$\frac{1}{T}\sum_{t=1}^{T} X_t \xrightarrow[T\to\infty]{\text{a.s}} \mu,$$

where $\xrightarrow[T\to\infty]{\text{a.s}}$ denotes almost sure convergence, and autocovariance ergodicity (see, Wei 1990):

$$\frac{1}{T}\sum_{t=k+1}^{T} (X_t - \mu)(X_{t-k} - \mu) \xrightarrow[T\to\infty]{\text{a.s}} \gamma(k).$$

If the data are assumed to be a sample from a stationary normal process all information about the process is expressed by the mean, $\boldsymbol{\mu}$, and the autocovariance matrix, $\boldsymbol{\Gamma}_T$. Optimal prediction is in theory easily derived from the properties of the conditional normal distribution. If the vector $\boldsymbol{X}$ partitioned, $\boldsymbol{X}_1$ and $\boldsymbol{X}_2$ (future and past), as:

$$\boldsymbol{X} = \begin{bmatrix} \boldsymbol{X}_1 \\ \boldsymbol{X}_2 \end{bmatrix} \sim N\left(\begin{bmatrix} \boldsymbol{\mu}_1 \\ \boldsymbol{\mu}_2 \end{bmatrix}, \begin{bmatrix} \boldsymbol{\Gamma}_1 & \boldsymbol{\Gamma}_{12} \\ \boldsymbol{\Gamma}'_{12} & \boldsymbol{\Gamma}_2 \end{bmatrix} \right) \tag{2.1}$$

then the conditional distribution of the future, $\boldsymbol{X}_1$, given the past, $\boldsymbol{x}_2$, is:

$$\boldsymbol{X}_1 | \boldsymbol{X}_2 = \boldsymbol{x}_2 \sim N(\boldsymbol{\mu}_1 + \boldsymbol{\Gamma}'_{12}\boldsymbol{\Gamma}_2^{-1}(\boldsymbol{x}_2 - \boldsymbol{\mu}_2), \boldsymbol{\Gamma}_1 - \boldsymbol{\Gamma}'_{12}\boldsymbol{\Gamma}_2^{-1}\boldsymbol{\Gamma}_{12}). \tag{2.2}$$

Equation (2.2) is easily interpretable but there are technical difficulties concerning representation and computations for a stationary process; even if the autocovariance function, $\gamma(k)$, was known, as the formula involves inverting a large matrix. A popular representation of a stationary process is by approximating it with an ARMA(p,q) process:

$$\begin{aligned}(X_t - \mu) = \phi_1(X_{t-1} - \mu) + \cdots + \phi_p(X_{t-p} - \mu) \\ + \varepsilon_t - \theta_1\varepsilon_{t-1} - \cdots - \theta_q\varepsilon_{t-q}.\end{aligned} \tag{2.3}$$

A more compact notation by use of the backward operator, B, $BX_t = X_{t-1}$, is frequently used. Sometimes the backward operator is called the lag-operator, L. Equation (2.3) is written more compactly in polynomials of the B operator.

$$\begin{aligned}\Phi(B)(X_t - \mu) &= \Theta(B)\varepsilon_t, \\ \Phi(z) &= 1 - \phi_1 z - \phi_2 z^2 - \cdots - \phi_p z^p, \\ \Theta(z) &= 1 - \theta_1 z - \theta_2 z^2 - \cdots - \theta_q z^q.\end{aligned}$$

This parameterization is not unique, in the sense that another set of ϕ's and θ could generate the same autocovariance function. Requiring *invertibility*, i.e. that the roots of the polynomial $\Theta(z)$ lie outside the unit circle, solves that issue. Requiring that the roots of $\Phi(z)$ lie outside the unit circle guarantees stationarity. It is also required that the polynomials $\Phi(z)$ and $\Theta(z)$ do not have any common factors. The autocovariance function, $\gamma(k)$, for a particular ARMA(p,q) process, is a complicated function of the parameters $(\phi_1, \ldots, \phi_p, \theta_1, \ldots, \theta_q, \sigma)$. It can be calculated for example by the Durbin–Levinson algorithm, see, e.g. Brockwell and Davis (1991). The process X_t is a *filtered* version of the ε_t process. Sometimes the properties of an ARMA process are better visualized by means of spectral methods. The spectrum is defined as the Fourier transform of the autocovariance function,

$$f(\lambda) = \frac{1}{2\pi} \sum_{k=-\infty}^{\infty} e^{-ik\lambda} \gamma(k).$$

The spectral density function $f(\lambda)$ is an excellent tool for describing the cyclical properties of X_t. A peak in $f(\lambda)$ at λ_0 indicates a cycle of length $2\pi/\lambda_0$. A feature that is much harder to visualize in terms of the autocovariance function. The ARMA(p,q) representation is a flexible way of parameterizing a stationary process. This flexibility has made ARMA models very popular in applied work, based on the idea of approximating stationary processes by ARMA(p,q). The seminal book by Box and Jenkins (1970) is essentially a cookbook, based on statistical principles, on how to proceed from data to forecast. Box and Jenkins (1970) extended the idea to be applicable to processes that could be transformed into ARMA(p,q). Their recommendations were to transform the process by a variance stabilizing transform, e.g. taking logarithm, and the date differences. The idea is to approximate a transformed version of a process by an ARIMA(p,d,q) process:

$$(1 - B)^d \Phi(B)(X_t - \mu) = \Theta(B)\varepsilon_t.$$

Their approach was pragmatic rather than theoretical. It consists of steps; the first step which they called *identification*, consisted of choosing a proper variance stabilizing transform, the number of differences, d, to take, and values of p, and q. The second step they named *estimation* and consisted of obtaining estimates of the ϕ's and θ's. The third step was called *diagnostics* and consisted of analysing $\hat{\varepsilon}_t = \hat{\varepsilon}_t(\hat{\boldsymbol{\phi}}, \hat{\boldsymbol{\theta}})$. The modelling process was considered a success if the properties of empirical residuals seemed similar to what was assumed about the theoretical residuals, i.e. white-noise. If the diagnostic step was passed one proceeded to the fourth and final step, the *forecasting* step. The forecasting step consisted of calculating point forecast and corresponding interval forecasts.

Both the minimum mean-square error prediction, X_{T+h} and its conditional variance, the variance of the prediction error, e_{T+h}, are in theory easily derived from the fundamental results on normal models given by Equation (2.1). Numerically it is however more practical to use some recursive methods, the Durbin–Levinson algorithm, the innovation algorithm, or use the Kalman filter recursions or use the Cholesky decomposition of the variance–covariance matrix (Brockwell and Davis 1991).

Since the early BJ practice progress has taken place in the numerical procedures. For the normal model an algorithm for calculating the likelihood of an ARMA process was given by Gailbraith and Gailbraith (1974) and shortly after computationally efficient algorithms were given by, e.g. Ansley (1979) and Melard (1983). The Kalman filter algorithm also offers an easy way of calculating the likelihood value.

The approach of approximating nonstationary series with an ARIMA process suggested the idea of a fractionally integrated process:

$$(1 - B)^d \Phi(B) X_t = \Theta(B) \varepsilon_t$$

where d need not be an integer, see e.g. Granger and Joyeux (1980). If d is in the interval $(-0.5, 0.5)$ the process is stationary. The behaviour of the autocorrelation function differs from the usual, $d = 0$, case where the autocorrelation function decays exponentially:

$$|\rho(k)| < r^{-k},$$

and is instead the autocorrelation function decays polynomially:

$$|\rho(k)| < k^{2d-1}.$$

One may distinguish between the case $d < 0$ where $\sum_{k=-\infty}^{\infty} |\rho(k)| < \infty$ and the case $0 < d < 0.5$ where $\sum_{k=-\infty}^{\infty} |\rho(k)| = \infty$. Brockwell and Davis (1991) label the former case as an intermediate-memory process and the latter as a long-memory process. The maximum-likelihood estimation is treated in Sowell (1992) and Beran (1995). Beran (1994) has published a detailed book on long-memory modelling. A recent empirical investigation of the usefulness of ARFIMA, the AR fractionally IMA, for macroeconomics and finance is given by Bhardwaj and Swanson (2006).

The availability of cheap computing power and efficient algorithms for calculating the likelihood function has made exact maximum-likelihood estimation feasible and to a degree made the BJ scheme less formal, i.e. the identification step, the estimation step and the principle of parsimony have essentially merged into one.

One of the virtues of the BJ scheme was simplicity. When the BJ scheme is generalized to the multivariate case the simplicity disappears. One representation of a multivariate ARMA model is:

$$\boldsymbol{X}_t = \boldsymbol{\Phi}_1 \boldsymbol{X}_{t-1} + \cdots + \boldsymbol{\Phi}_p \boldsymbol{X}_{t-p} + \boldsymbol{\varepsilon}_t - \boldsymbol{\varepsilon}_{t-1} - \cdots - \boldsymbol{\varepsilon}_{t-q}, \tag{2.4}$$

$$\mathrm{E}(\boldsymbol{\varepsilon}_t) = \mathbf{0}, \quad \mathrm{E}(\boldsymbol{\varepsilon}_t \boldsymbol{\varepsilon}_t') = \boldsymbol{\Sigma}, \quad \mathrm{E}(\boldsymbol{\varepsilon}_t \boldsymbol{\varepsilon}_s') = \mathbf{0} \quad t \neq s. \tag{2.5}$$

The autocorrelation function $\boldsymbol{\Gamma}(k) = E(\boldsymbol{X}_t \boldsymbol{X}_{t-k}')$ is now a sequence of matrices so the plotting of autocorrelation function is a nontrivial matter. Equation (2.4) in polynomials of the backward operator B is:

$$\boldsymbol{\Phi}(B) \boldsymbol{X}_t = \boldsymbol{\Theta}(B) \boldsymbol{\varepsilon}_t.$$

Requiring that the roots of the polynomials $|\mathbf{\Phi}(z)|$ and $|\mathbf{\Theta}(z)|$ lie outside the unit circle and that the polynomials $\mathbf{\Phi}(z)$ and $\mathbf{\Theta}(z)$ have no common left factors are conditions that are inherited from the univariate case. A way to get a minimal multivariate ARMA representation is to minimize the McMillan degree (Hannan and Deistler 1988).

The computational complexity of multivariate ARMA models is evidently much higher than that of the univariate case, but calculation of likelihood, predictors, etc. is just a technicality. Multivariate extension of the Durbin–Levinson algorithm and the innovation algorithms exist and can be programmed in a modern programming language. State-space representation and the Kalman filter provide an approach that is perhaps the easiest to implement (Brockwell and Davis 1991; Harvey 1989, 1993; Lütkepol 1991). Another possible approach is to consider the multivariate system as a periodic ARMA, PARMA process (Lund and Basawa 2000; McLeod 1993; Pagano 1978).

The nonstationary case for multivariate is more complicated than in the univariate case. The degree of nonstationarity can vary across coordinates. In the univariate case the class of ARIMA(p, d, q) has proven to be class of nonstationary models that is large enough to be interesting. BJ suggested a very crude method of estimating an integer value of d. Since the article by Dickey and Fuller (1979) there has been a huge development of unit-root tests. Granger was awarded the Nobel Prize in economics in 2003, 'for methods of analyzing economic time series with common trends (cointegration)', which is a way of analysing the relations between nonstationary time series. The literature on cointegration is now huge.

Classical time series analysis in the ARMA spirit is essentially about linear filters,

$$X_t = \sum_{k=-\infty}^{\infty} \psi_k \varepsilon_{t-k}. \tag{2.6}$$

The case were the input, ε_t, is iid-normal is completely treated in textbooks. The literature on deviations from iid-normal is basically in two directions: (a) ε_t uncorrelated, but somehow dependent and (b) ε_t independent, but the distribution not normal. One of the stylized facts on financial time series is that their tails are heavier than would be allowed by a normal model. A simple way to incorporate that into (2.6) is to assume that ε_t is heavy-tailed, e.g. some t-distribution. Then X_t will be distributed as a weighted sum of independent t-distributions, which is certainly not a t-distribution, but a rather messy compound. Inference in such model is though relatively straightforward, because the likelihood can be calculated recursively. If the input series, ε_t, has finite variance, then the output series X_t will consist of a weighted sum of independent finite variance components and might therefore look 'more normal' than the input due to central limit theorem arguments.

In the case of normal input, the filtered process, X_t, is also normal because the normal family is closed under addition. For continuous random variables this property defines the family of stable distributions, i.e., the sum of iid variables from the family belongs also to the family. The Cauchy distribution has also this property. The normal distribution is the only continuous stable distribution which has finite variance. Therefore if the input in (2.6) is iid stable, but not normal, X_t does not have finite variance. Obviously, a criterion like MMSEP (minimum mean-square error of prediction) will not make sense in such cases. The density of the stable distributions is not available in closed form, but the logarithm of the characteristic function is of the form:

$$\log[\mathrm{E}(e^{iuX})] = \begin{cases} iu\beta - d|u|^{\alpha}(1 - i\theta\,\mathrm{sgn}(u)\tan(\pi\alpha/2)) & \alpha \neq 1, \\ iu\beta - d|u|(1 + 2i\theta/\pi)\,\mathrm{sgn}(u)\log(|u|) & \alpha = 1. \end{cases}$$

The interpretation of the parameters is, β indicates location, $d^{1/\alpha}$ indicates scale, θ indicates symmetry and α indicates tail behaviour. Recently algorithms for numerically calculating the density, generating random numbers have become available. (Lambert and Lindsey 1999). The method of Lambert and Lindsey (1999) has been implemented in the R-package (R Development Core Team 2005) stable. A textbook treatment of linear-filters with infinite variance input is given in Brockwell and Davis (1991), section 13.3. For more advanced treatment, see, e.g. Hall, Peng and Yao (2002),

2.4 Conditional heteroskedacity

In 2003, R.F. Engle was awarded the Nobel Prize in economics 'for methods of analyzing economic time series with time-varying volatility (ARCH)'. It became apparent that the dependency structure of a process was more than just the autocovariance function. Granger (1983) enlightened that some properties of a white-noise process could be predicted. Engle (1982) analysed a particular case and introduced the concept of Auto-Regressive Conditional Heteroskedacity (ARCH) models. A simple version of ARCH is:

$$\varepsilon_t = v_t\sqrt{\alpha_0 + \alpha_1\varepsilon_{t-1}^2}, \tag{2.7}$$

with

$$\mathrm{E}(v_t) = 0, \quad \mathrm{E}(v_t^2) = 1, \quad v_t \sim \mathrm{N}(0, 1).$$

By some rearranging,

$$\varepsilon_t|\varepsilon_{t-1},\ldots \sim \mathrm{N}(0,h_t), \quad h_t = \alpha_0 + \alpha_1\varepsilon_{t-1}^2,$$
$$\varepsilon_t^2 = \alpha_0 + \alpha_1\varepsilon_{t-1}^2 + \underbrace{(\varepsilon_t^2 - h_t)}_{\text{residual}}. \tag{2.8}$$

Equation (2.8) shows the similarity between ARCH models an AR models. The distribution of the residual in (2.8) is restricted by the dynamic structure and the fact that $\varepsilon_t^2 > 0$. It is therefore, necessarily nonnormal. The second moments of ε_t have a nonzero autocorrelation structure but the first moments do not. A straightforward extension of (2.7) is the ARCH(p):

$$\varepsilon_t = v_t\sqrt{\alpha_0 + \alpha_1\varepsilon_{t-1}^2 + \cdots + \alpha_p\varepsilon_{t-p}^2}.$$

The ARCH(p) process is easily interpreted as an AR(p) process for second moments an ARMA version of that is GARCH(p,q), generalized ARCH:

$$h_t = \alpha_0 + \alpha_1\varepsilon_{t-1}^2 + \cdots + \alpha_p\varepsilon_{t-p}^2 + \beta_1 h_{t-1} + \cdots + \beta_q h_{t-q}. \tag{2.9}$$

Equation (2.8) can be generalized and the GARCH(p,q) process, (2.9), can be written as:

$$\Phi^*(B)(\varepsilon_t^2 - \alpha_0) = \Theta^*(B)\ \text{residual}_t$$

where Φ^* and Θ^* are polynomials. So the development of the GARCH models can easily be interpreted as a spin of from ARMA modelling for squared processes. An AR(p) process with p large can be well approximated by an ARMA(p,q) process with $p+q$ small. The BJ principle of parsimony also applies to GARCH processes and practitioners have therefore often preferred GARCH(1,1) because a reasonable fit to real data by ARCH(p) usually requires a large p. Figure 2.1 illustrates the features of models from the ARCH/GARCH family. The process generates volatility clusters, the autocorrelation structure in X_t is weak whereas it is clear in X_t^2. Other ideas like the persistence property, i.e. unit root, a basic idea of ARIMA modelling can be incorporated in the ARCH framework such as the integrated GARCH, IGARCH (Engle and Bollerslev 1986). The simplest form is the IGARCH(1,1), which is a constrained version of (2.9):

$$h_t = \alpha_0 + \alpha_1\varepsilon_{t-1}^2 + (1-\alpha_1)h_{t-1}.$$

The ARIMA contains a unit root and is nonstationary. The IGARCH contains a unit root but can be strictly stationary (Nelson 1990). The long-memory idea of ARIMA, the ARFIMA, has also found its way into the volatility modelling and

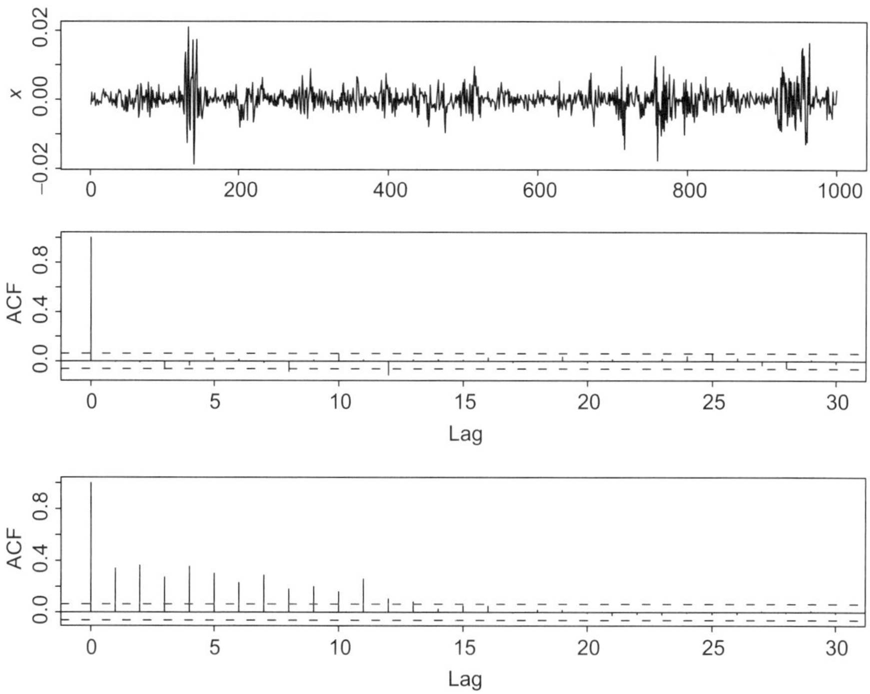

Figure 2.1 A simulated GARCH(1,1) process (top), its autocorrelation (middle) and the autocorrelation of the series squared (bottom).

has been termed FIGARCH (fractionally integrated GARCH) (Baille, Bollerslev and Mikkelsen 1996).

The step from a univariate model to multivariate model for GARCH is badly hit by the curse of dimensionality. The multivariate generalization ARCH refers to an n-dimensional vector, $\boldsymbol{\varepsilon}_t$, which is conditionally heteroskedastic:

$$H_t = V(\boldsymbol{\varepsilon}_t | \boldsymbol{\varepsilon}_{t-1}, \boldsymbol{\varepsilon}_{t-2}, \ldots).$$

Defining the dynamics of the sequence of matrices H_t, analogously to the one-dimensional case, results in the following recursions for the (k, l) element of H_t:

$$h_{k,l,t} = c_{k,l} + \sum_{i=1,q} \left[\sum_{m=1}^{n} \sum_{s=1}^{n} \alpha_{k,l,m,s,i} \varepsilon_{m,t-i} \varepsilon_{s,t-i} \right] \qquad (2.10)$$

$$+ \sum_{i=1}^{p} \left[\sum_{m=1}^{n} \sum_{s=1}^{n} \beta_{k,l,m,s,i} h_{m,s,t-i} \right].$$

There are some natural constraints on the parameters in Equation (2.10) due to the fact that H_t is a covariance matrix and therefore symmetric positive-definite. But nevertheless, the number of parameters grows dramatically with the number of dimensions. For a multivariate ARCH(1) model the number of parameters is $1 + n(n+1)/2$ (Gourieroux 1997). Therefore more parsimonious versions have been used. A variant suggested by Bollerslev (1990) is the constant conditional correlation GARCH, CCC-GARCH:

$$h_{i,i,t} = c_{i,i} + \alpha_{ii}\varepsilon_{i,t-1}^2 + \beta_{i,i}h_{i,i,t-1},$$

$$h_{i,j,t} = \rho_{i,j}h_{i,i,t}^{1/2}h_{j,j,t}^{1/2}, \qquad \text{for } i \neq j.$$

More constrained version exist, e.g. DVEC-GARCH (Yang and Allen 2005). The multivariate GARCH models are treated in another chapter of the present book.

2.5 Nonlinear time series models

The ARMA model is a linear filter, i.e. the current value, X_t, is a linear function, a weighted sum, of past values of the series, $X_{t-1}, X_{t-2}, \ldots$ and current and past noise values, $\varepsilon_t, \varepsilon_{t-1}, \ldots$. A generalization of this would be a nonlinear filter, so that a nonlinear univariate time series, X_t can be represented as, e.g.

$$X_t = f(X_{t-1}, X_{t-2}, \ldots, \varepsilon_t, \ldots), \qquad \text{where } \varepsilon_t \text{ is some noise or,} \tag{2.11}$$

$$\varepsilon_t = g(X_t, X_{t-1}, \ldots), \qquad \text{or} \tag{2.12}$$

$$X_t = h(\varepsilon_t, \varepsilon_{t-1}, \ldots). \tag{2.13}$$

The function in (2.11) represents a mixed nonlinear ARMA representation, a kind of nonlinear AR is represented by (2.12) and a nonlinear MA in Equation (2.13). All three are in a nonanticipative form, i.e. they represent X_t only as a function of its past values and past and current ε_t's. It is evidently quite hopeless to estimate a very general function of the above form from a single realization of a time series. Some intelligent starting point is needed as well as sensible bounds on how complicated the function can be. A starting point is to use h in (2.13) and assume that a Taylor expansion is allowed. Following a classic monograph on the fundamentals of nonlinear time series, Priestley (1991), h in (2.13) is Taylor expanded:

$$X_t = \mu + \sum_{i_1=0}^{\infty} g_i\varepsilon_{t-i} + \sum_{i_1=0}^{\infty}\sum_{i_2=0}^{\infty} g_{i_1,i_2}\varepsilon_{t-i_1}\varepsilon_{t-i_2} \tag{2.14}$$

$$+\sum_{i_1=0}^{\infty}\sum_{i_2=0}^{\infty} g_{i_1,i_2,i_3}\varepsilon_{t-i_1}\varepsilon_{t-i_2}\varepsilon_{t-i_3}+\cdots$$

The Volterra expansion (2.14) suggests that a reasonable starting point could be a kind of bilinear model:

$$X_t = \mu + \varepsilon_t + \alpha_1\varepsilon_{t-1} + \alpha_{12}\varepsilon_{t-1}\varepsilon_{t-2} \tag{2.15}$$

or as Granger and Andersen (1978) suggest:

$$X_t = \varepsilon_t + \alpha\varepsilon_{t-1}X_{t-2},$$

which have zero autocorrelation and are therefore not linearly predictable, but might be nonlinearly predictable. The optimal predictor for (2.15) is $\alpha_{12}\hat{\varepsilon}_{t-1}\hat{\varepsilon}_{t-2}$ if $\hat{\varepsilon}_t$ is constructed recursively by:

$$\hat{\varepsilon}_t = X_t - \alpha_1\hat{\varepsilon}_{t-1} - \alpha_{12}\hat{\varepsilon}_{t-1}\hat{\varepsilon}_{t-2}.$$

From the look of (2.14) it is clear that a search for a parsimonious nonlinear representation of a process in the spirit of BJ for linear processes will be complicated, as well as problematic, in terms of identification. Therefore the design of nonlinear time series modelling has developed towards models that have the ability of capturing particular stylized features of data. One such example is the threshold-auto-regressive (TAR) model. The idea is that the dynamics of a process X_t is different when the level of X_t is high than when it is low. A simple first-order TAR(1) model is:

$$X_t = \begin{cases} \phi_{\mathrm{H},1}X_{t-1} + \varepsilon_{\mathrm{H},t} & \text{if } X_{t-1} \geq d, \\ \phi_{\mathrm{L},1}X_{t-1} + \varepsilon_{\mathrm{L},t} & \text{if } X_{t-1} < d, \end{cases}$$

where $\phi_{\mathrm{H},1}$, $\phi_{\mathrm{L},1}$, $\varepsilon_{\mathrm{H},t}$, $\varepsilon_{\mathrm{L},t}$ refer to the high-level and the low-level case, respectively. Obviously generalizations to more complicated TAR are theoretically straightforward. The practical implementation gets quickly difficult, choosing the number of thresholds, lag-length, etc. Tong (1983) gives strategies for practical modelling, such as limiting the decision to jump to a single specific time-lag, common for all thresholds. Tong (1983) calls this SETAR (self-excited threshold autoregressive) (Priestley 1991). Properties such as stationary distributions, autocovariance function, and properties of estimators are nontrivial (Jones 1978; Klimko and Nelson 1978; Tong 1983). Many modern researchers would use computer intensive methods, such as bootstrap or simulation. The TAR model is varying-parameter model, i.e., the parameter jumps if the series passes a certain value. Some may find that feature unwanted and therefore some alternatives have been developed. One version is the EAR (exponential autoregressive) model, a simple version of EAR(2) is:

$$X_t = \phi_1(X_{t-1}) + \phi_2(X_{t-2}) + \varepsilon_t,$$
$$\phi_1(X_{t-1}) = \alpha_{1,1} + \alpha_{2,1}\exp(-\alpha_{3,1}X_{t-1}^2),$$
$$\phi_2(X_{t-2}) = \alpha_{2,1} + \alpha_{2,2}\exp(-\alpha_{3,2}X_{t-2}^2).$$

The EAR(2) can behave very similar to an AR(2). When the characteristic polynomial of the AR(2) has complex roots, the process tends to show cyclical behavior. The EAR is to a degree similar to TAR but the coefficients, $\phi_1, \phi_2, \ldots$ evolve smoothly between the minimum ($\alpha_{1,i}$) and the maximum value ($\alpha_{1,i} + \alpha_{2,i}$). Therefore, the EAR can generate amplitude-dependent cycles, jump-like behaviour, and limit cycles. For more details see, e.g. Ozaki (1982, 1985). Identifiability is obviously an issue, e.g. it will be difficult to estimate $\alpha_{2,i}$ and $\alpha_{3,i}$ when $\alpha_{2,i}$ is small and $\alpha_{3,i}$ is large. The EAR is likely to capture similar features of a series, such as the TAR. Many practitioners use AIC/BIC criteria, or to a degree, common sense, in choosing between models. Still a slightly modified version is the STAR (smooth transition autoregressive) model. A simple form is:

$$X_t = \alpha_1 X_{t-1} + X_{t-1}F(X_{t-1}) + \varepsilon_t \tag{2.16}$$

where F is a suitable function. A possible choice of F is the logistic function and then the model (2.16) is labelled LSTAR (logistic STAR).

The above models, TAR/EAR/STAR, etc. have in common that they switch regimes depending on the observed time series X_t. A related idea is to let another process rule the regime switching. An idea is the Markov switching regime (Hamilton 1989):

$$X_t = \begin{cases} \text{model 1} & S_t = 1 \\ \text{model 2} & S_t = 2, \end{cases}$$

where the state S_t is ruled by a Markov chain with transition probabilities:

$$\boldsymbol{P} = \begin{bmatrix} p_{11} & 1 - p_{22} \\ 1 - p_{11} & p_{22} \end{bmatrix}.$$

The modelling consists of estimating the number of states, the transition matrix as well as the dynamic model in each state. Hamilton (1994) reviews some aspects of a practical approach, estimating parameters, singularities in the likelihood, etc.

Table 2.1 A subset of the ARCH family.

PGARCH	AGARCH	TGARCH	IGARCH
EGARCH	AR-ARCH	ARCH-M	BEKK
CCC	FIGARCH	PSD-VECH	DVEC
FIEGARCH	NGARCH	VGARCH	QGARCH

The usual GARCH model can be thought of as a kind of linear filter of second moments. The nonlinear ideas above have been brought over to volatility modelling. Nelson (1991) suggests that:

$$\log(h_t) = \alpha_t + \sum_{k=1}^{\infty} \beta_k g(\varepsilon_{t-k}),$$

in order to accommodate for the asymmetric relation between volatility, $\sigma_t = \sqrt{h_t}$, and prices, i.e. markets react differently to negative shocks than to positive shocks. Nelson (1991) suggest that by choosing:

$$g(\varepsilon_t) = \theta\varepsilon_t + \gamma(|\varepsilon_t| - \mathrm{E}(|\varepsilon_t|)),$$

the series σ_t can be a well-behaved process, depending on the choice of g. This turns out to be very similar to, e.g. the TAR process (Tsay 2002). A related version is the PGARCH (power GARCH) mixing the long-memory property with nonlinearity:

$$\sigma_t^d = \alpha_0 + \sum_{i=1}^{p} \alpha_i(|\varepsilon_{t-i}| + \gamma_i\varepsilon_{t-i})^d + \sum_{j=1}^{q} \beta_j \sigma_{t-j}^d.$$

The PGARCH form includes many members of the GARCH family as special cases.

The tremendous creativity in the models in this category has been driven by the wish to represent the stylized facts of a financial market in simple ARMA-type formulas. First and second moments of series are modelled and of course the two are mixed, e.g. AR-GARCH, ARCH-M, etc. A brief list of the ARCH family is shown in Table 2.1

2.6 Continuous time diffusions

The field of finance has evolved strongly in the past decades. A cornerstone element of the theory is the Wiener process, or Brownian motion. The Wiener process is a continuous-path, memoryless process. If the time-horizon is $t \in [0, 1]$:

$$W(0) = 0,$$

$$W(t_4) - W(t_3) \text{ indpendent of } W(t_2) - W(t_1), \quad t_1 < t_2 < t_3 < t_4,$$

$$V(W(t)|W(0)) = t,$$

$$W(t) \text{ continous with probability 1.}$$

The process can be defined similarly for any time interval. According to the functional central limit theorem $W(t)$ is normally distributed. The Wiener process is in the literature often called Brownian motion after the biologist Brown (1827), who was describing movements he observed in his microscope. The term Wiener process is due to the mathematician Wiener (1923) who proved that the process was mathematically well defined. The mathematical literature on Wiener/Brownian process is huge.

The theory of differential equations is aimed at describing dynamics, i.e. movement of a particle in time. Time series models in discrete time can be thought of as difference equations containing a stochastic term, i.e. the input noise. Following the same spirit in the continuous-time case gives rise to the need to define the concept of stochastic differential equation (SDE). This is done by defining SDE through the concept of the stochastic integral. The SDE is written as:

$$\mathrm{d}X(t) = \mu[X(t), t]\mathrm{d}t + \sigma[X(t), t]\mathrm{d}W(t). \tag{2.17}$$

The interpretation of (2.17) is that it has a solution on the form:

$$X(t) = X(t_0) + \underbrace{\int_{t_0}^{t} \mu[X(s), s]\mathrm{d}s}_{\text{term 1}} + \underbrace{\int_{t_0}^{t} \sigma[x(s), s]\mathrm{d}W(s)}_{\text{term 2}}. \tag{2.18}$$

The functions μ and σ are called trend and diffusion, respectively. The first term in equation (2.18) is an ordinary Riemann integral. The second term is a stochastic integral. The most commonly used concept for a stochastic integral is the Ito integral. If the function σ was a step function which only jumped in $t_1, \dots, t_n = t$, the Ito-integral is defined as:

$$\int_{t_0}^{t} \sigma(s)dW(s) = \sum_{k=1}^{n} \sigma(t_{k-1})[W(t_k) - W(t_{k-1})].$$

The key issue here is to define the integral based on the value of σ the left end of the interval. That way the independent increment property of W makes formulas simpler, such as the variance:

$$V\left[\int_{t_0}^{t} \sigma(s)dW(s)\right] = \sum_{k=1}^{n}[\sigma(t_{k-1})]^2(t_k - t_{k-1}),$$

all covariance terms disappear due to the independent increments of $W(t)$ and the forward increment feature of the definition of the integral. The definition is then extended to functions that can be approximated by a step function so that the Ito-integral is defined for a class of 'well-behaved' functions. This mathematical background means that (2.17) is just another way of writing a stochastic integral. Another definition of a stochastic integral is the Stratonovic integral. Having a working definition of the stochastic integral activates a vast mathematical machinery. In many ways the continuous-time approach is more tractable than the discrete-time one. The $\mathrm{d}W(t)$ term in (2.18) plays the role of ε_t, i.e. the white noise in the discrete-time models. The term $\mathrm{d}W(t)$ is often called the continuous-time white noise; even though it does not exist mathematically, it is still a useful form which refers directly to the integral. A useful property of the Ito integral is that it is a martingale. Another virtue of using the Ito integral as a definition of the stochastic integral is the practically of the Ito lemma; if the dynamics of $X(t)$ is given by (2.18) then the dynamics of $Y(t) = g[X(t), t]$ is given by:

$$\mathrm{d}Y(t) = \mu^*[X(t), t]\mathrm{d}t + \sigma^*[X(t), t]\mathrm{d}W(t),$$

$$\mu^*[X(t), t] = \left\{\frac{\partial g}{\partial t}[X(t), t] + \mu[X(t), t]\frac{\partial g}{\partial x}[X(t), t] + \frac{1}{2}\sigma^2[X(t), t]\frac{\partial^2 g}{\partial x^2}[X(t), t]\right\}$$

$$\sigma^*[X(t), t] = \sigma[X(t), t]\frac{\partial g}{\partial x}[X(t), t].$$

If g is a well-behaved process the dynamics of transformed version of the process is also driven by a 'normal white noise', $\mathrm{d}W(t)$. This is very different from the discrete-time case. If a discrete-time model is driven by white noise, a function of it is in general not driven by white noise. Having defined the white-noise term, $\mathrm{d}W(t)$, in terms of the Ito integral it is possible to define a continuous-time linear filter, i.e. a continuous-time ARMA.

As a contrast to the discrete-time case, the step from linear dynamics to nonlinear dynamics is more manageable in the continuous-time case. Many one-dimensional nonlinear SDE are tractable. The mathematical literature on SDE and its application to finance is huge. The mathematical conditions on the existence of a solution of a SDE depend on the nature of the functions μ and σ. One distinguishes between strong and weak solutions. Good mathematical references are Karatzas and Shreve (1991), Øksendal (1998) and Revuz and Yor (1999).

Visualization of a diffusion process can be done by simulation. The simulated process is essentially, a step function with frequent jumps, i.e. the diffusion process is approximated by a process that jumps at discrete time points, $t_1, \ldots, t_n$. A well known simulation scheme is the Euler scheme, which is simply based on substituting independent standard pseudonormal random variables, Z_i, into Equation (2.18),

$$X(t_i) = X(t_{i-1}) + \mu[X(t_{i-1}), t_{i-1}]\Delta_i + \sigma[X(t_{i-1}), t_{i-1}]\underbrace{Z_i\sqrt{\Delta_i}}_{\mathrm{d}\widehat{W(t_i)}}, \quad (2.19)$$

$$\Delta_i = t_i - t_{i-1}, \quad V[\mathrm{d}\widehat{W(t_i)}] = \Delta_i. \quad (2.20)$$

The feature $V(\mathrm{d}W(t)) = \mathrm{d}t$ is reflected in the simulation by equation (2.20). The quality of the simulation depends on how fine the mesh, Δ_i is, as well the complexity of the process, i.e. μ and σ. An improved version, which is based on Taylor approximations, is the Milstein scheme:

$$\begin{aligned} X(t_i) = X(t_{i-1}) &+ \mu[X(t_{i-1}), t_{i-1}]\Delta_i + \sigma[X(t_{i-1}), t_{i-1}]Z_i\sqrt{\Delta_i} \\ &+ \frac{1}{2}\sigma[X(t_{i-1}), t_{i-1}]\frac{\partial\sigma}{\partial x}[X(t_{i-1}), t_{i-1}][(Z_i\sqrt{\Delta_i} - Z_{i-1}\sqrt{\Delta_{i-1}})^2 - \Delta_i]. \end{aligned}$$

Kloeden and Platen (1992) give higher order approximations as well as multivariate version of the approximations. These are of course only approximations. The simulated process is constant or somehow interpolated between the time points. In the general case simulation is hindered due to the fact that the transition density is generally unknown, and therefore direct sampling from the transition density is impossible. Beskos and Roberts (2005) and Beskos, Papaspiliopoulos, Roberts and Fearnhead (2006) give an algorithm for sampling from the exact transition density.

The practical situations of estimation is more complicated than in the discrete-time case. The process is continuous, but in practical cases only discrete observations are available. A continuous-time approach based on deriving estimators for case that data consists of an entire path is shown in Kutoyants (1984). The log likelihood-function of a continuously observed process is given by (2.21):

$$\log[L(\boldsymbol{\theta}|t_0, t)] = c + \int_{t_0}^{t} \frac{\mu[X(s), s]}{\sigma^2[X(s), s]}\mathrm{d}X - \frac{1}{2}\int_{t_0}^{t} \frac{\mu^2[X(s), s]}{\sigma^2[X(s), s]}\mathrm{d}s. \quad (2.21)$$

Implementation of methods such as maximizing (2.21) would require dense observations. In the (usual) case of discrete observations:

$$x(t_1), \ldots, x(t_n), \quad t_1 < t_2 < \cdots < t_n,$$

an idea could be to simulate the pattern between observations and replace the integrals with sums. The traditional statistical approaches for seeking estimators, method of moments, least squares, maximum likelihood and Bayesian methods are all complicated. The existence of moments and which moments to match is complicated. Calculation of the likelihood, $L(\boldsymbol{\theta})$, by recursively calculating the transition density $f[x(t_i)|x(t_{i-1})]$, using:

$$L[\boldsymbol{\theta}|x(t_1), \ldots, x(t_n)] = f[x(t_1), \boldsymbol{\theta}] \prod_{i=2}^{n} f[x(t_i)|x(t_{i-1}), \boldsymbol{\theta}],$$

is possible due to the nature of the process. But as closed form for the transition density are only available for some particular processes, in general, some approximations are necessary.

Typically the parameter space is restricted. It is usually easy to decide whether a one-dimensional diffusion process has a stationary distribution. When modelling dynamic phenomena that are believed to have some stationary features, some simple stationary diffusions could be used as starting points.

In some cases calculations of moments are easy. The form (2.18) is essentially a way of writing the integral:

$$X(t) = X(t_0) + \int_{t_0}^{t} \mu[X(s), s]\mathrm{d}s + \int_{t_0}^{t} \sigma[X(s), s]\mathrm{d}W(s). \tag{2.22}$$

Using the forward nature of the Ito integral, the conditional expected value of $X(t)|X(t_0)$ can be calculated by taking expectation through (2.22). So:

$$\mathrm{E}[X(t)|X(t_0)] = X(t_0) + \int_{t_0}^{t} \mathrm{E}\{\mu[X(s), s]\}\mathrm{d}s. \tag{2.23}$$

The Ito lemma give the dynamics of $Y(t) = X(t)^2$ and taking expectation again gives:

$$\mathrm{E}[X(t)^2|X(t_0)] =$$
$$X(t_0)^2 + \int_{t_0}^{t} \left[2\mathrm{E}\{X(s)\mu[X(s), s]\} + \mathrm{E}\{\sigma^2[X(s), s]\}\right] \mathrm{d}s.$$

In the case when $\mu(x)$ is linear in x the first conditional moment is derived by using Ito lemma and solving a differential equations, e.g. if $\mu(x) = \alpha(\beta - x)$ then (2.23) becomes:

$$\begin{aligned} m(t) &= \mathrm{E}[X(t)|X(t_0)] \\ &= X(t_0) + \int_{t_0}^{t} \alpha[\beta - \mathrm{E}(X(s)|X(t_0))]\mathrm{d}s \end{aligned}$$

$$= X(t_0) + \int_{t_0}^{t} \alpha[\beta - m(s)]\mathrm{d}s,$$

$$m'(t) = \alpha[\beta - m(t)], \text{ i.e. } m(t) = X(t_0)\mathrm{e}^{-\alpha(t-t_0)} + \beta(1 - \mathrm{e}^{-\alpha(t-t_0)}).$$

For some particular $\sigma(x)$ the derivation of conditional second moments can be just as simple, i.e. a question of solving a differential equation. In the general case this approach is not feasible.

For a one-dimensional diffusion the invariant distribution is (in the case that it exists) on the form:

$$f(x) \propto \frac{1}{\sigma^2(x)} \exp\left\{\int_c^x \frac{2\mu(s)}{\sigma^2(s)}\mathrm{d}s\right\}.$$

Again this might be hard to evaluate in some cases. Mao, Yuan and Yin (2005) give some numerical methods. When using stationary diffusion processes in modelling the stationary distribution should reflect features to be matched with the scientific phenomenon of interest.

Moment-based methods have their appeal because for some models moments can be calculated in closed form, even if transition probabilities cannot. Moment-based versions are, e.g. GMM, EMM, SMM and some methods based on estimating functions. Bibby, Jacobsen and Sørensen (2004) give a review on the use of estimating functions.

Some nonparametric, semiparametric, partly parametric approaches have been tried. If the existence of an invariant distribution $f(x)$ is assumed, then the relation between the invariant distribution, the drift function and the diffusion function is:

$$\frac{\mathrm{d}}{\mathrm{d}x}[\sigma^2(x)]f(x) = 2\mu(x)f(x),$$

$$\sigma^2(x) = \frac{1}{f(x)} \int_{-\infty}^{x} 2\mu(s)f(s)\mathrm{d}s$$

Aït-Sahalia (1996) suggests estimating the drift parametrically, estimating the invariant distribution with a kernel method and then plug $\hat{\mu}(x)$ and $\hat{f}(x)$ into the invariant distribution formula to get a nonparametric estimate of the diffusion function $\sigma(x)$. Aït-Sahalia (1999, 2002) approximates the transition density by Taylor expanding the Kolmogorov forward equation, and substitutes the approximation for the true likelihood and maximizes it numerically. For affine diffusions the characteristic function is often manageable. It is possible to use numerical Fourier inversion to derive the likelihood. An example of an estimation procedure based on using the characteristic function for affine diffusions is given by Singleton (2001).

2.7 Analysis of duration and transition data

A lot of financial activity consists of waiting for a particular event to occur such as a shift of state. A principle statistical discipline of analysing waiting time is survival analysis. The survival time, T, the waiting time for a shift in the state of survival is a nonnegative random variable. A standard form for describing the risk for a change of state is the hazard function, $\lambda(t)$:

$$\lambda(t) = P(t < T < t + \mathrm{d}t | T > t) = \frac{f(t)}{1 - F(t)},$$

$$F(t) = P(T \leq t) = 1 - \mathrm{e}^{\int_0^t \lambda(s)\mathrm{d}s}, \quad f(t) = F'(t).$$

The hazard function denotes the instantaneous risk of change of state conditional on the state up to time t. There exists extensive literature on survival analysis where the focus is on studying the hazard of the one-way transition from life to death (Andersen, Borgan, Gill and Keiding 1993; Fleming and Harrington 1991).

The hazard function can in principle be any nonnegative function. It should be noted that if the integral

$$\int_0^\infty \lambda(s)ds$$

is finite, then that means that there is positive probability of eternal life. When dealing with biological data it is usually not realistic to analyse a sequence of survival times for an individual. In sociological data it is conceivable to observe, e.g. the sequences between jail sentences for criminals but, in general there are so few time spells that it is not fruitful to derive a dynamic structure.

Engle and Russell (1998) defined the ACD, an autoregressive conditional duration model for explaining dynamics of waiting times between transactions in a financial market. If transactions take place at time points $t_1, t_2, \ldots$, the ACD approach is basically:

$$x_i = t_i - t_{i-1},$$

$$E(x_i | x_{i-1}, x_{i-2}, \ldots) = \psi_i(\boldsymbol{\theta}, x_{i-1}, x_{i-2}, \ldots). \quad (2.24)$$

Equation (2.24) denotes the conditional expectation of duration number i and is a function of past durations and a parameter vector $\boldsymbol{\theta}$. The conditional probability model for X_i is supposed to be of the form:

$$X_i = \psi_i \varepsilon_i,$$

where ε_i is a sequence of iid variables with a parametric distribution and parameter $\boldsymbol{\phi}$, e.g. an exponential or a Weibull distribution. The dynamic functional form of ψ_i is given by:

$$\psi_i = \omega + \sum_{j=1}^{m} \alpha_j x_{i-j} + \sum_{j=0}^{q} \beta_j \psi_{i-j}.$$

The abbreviation for this type of model is ACD(m, q). There is a striking resemblance in the derivation of the ACD model and the derivation of the ARCH model. As in the case of ARCH models many new abbreviations have been generated for describing the various forms of autocorrelated durations. Examples are AACD of Fernandes and Grammig (2006) which is based on applying the idea of Box–Cox transformation (flexible forms), to the ACD dynamics:

$$\psi_i^{\lambda} = \omega + \alpha \psi_{i-1}^{\lambda} \left[|\varepsilon_{i-1} - b| + c(\varepsilon_{i-1} - b) \right]^{\nu} + \beta \psi_{i-1}^{\lambda}. \qquad (2.25)$$

Equation (2.25) is an example of AACD(1,1). The idea is that the AACD should contain a lot of other ACD variants as special cases. The form is decided by the parameters, (b, c, λ, ν). Still another way of relaxing the functional form of the duration dynamics is, e.g. the semiparametric autoregressive conditional proportional hazard (SACPH) model in Gerhard and Hautcsh (2002).

Lancaster (1990) gives a review of methods for duration/transition-data analysis. The standard survival model generalizes to models with competing risks or multiple hazards, i.e. the situation where exit from a state can occur due to various reasons. In the competing risk model there are many waiting-time variables, $T_1, \ldots, T_k$, but there is only opportunity to observe one:

$$T = \min(T_1, \ldots, T_n).$$

The case where the competing durations are independent is relatively easy, because then the total hazard is just the sum of the competing hazard. A financial study of lender/borrower waiting times is given by Lambrecth, Perraudin and Satchell (2003). Some effort in exploring the possibility of dependent competing risks is given in Lindeboom and Van den Berg (1994). A version of the competing risk model with possible exit to many destinations is given a set of hazard functions, λ_{ij}, describing the hazard for transition from state i to state j (Lancaster 1990). A typical multi-state phenomenon in finance is a rating system, AAA, Aaa, etc. Examples of analysis of transition rates between ratings are in Lando and Skødeberg (2002) and Bladt and Sørensen (2006). Observed multivariate durations in multiple states, e.g. bivariate a distribution (T_1, T_2), refer to the process of waiting for many events. An example application to financial data is by An, Christensen and Gupta (2003) where the pension and retirement of spouses are analysed. An application to financial market data is given by Quoreshi (2006) using the analogy of count-processes and durations.

A typical characteristic of duration/transition data is censoring. The type of censoring, functional form of the impact of regressors, the choice of probability model; all these issues affect the choice of estimation strategy.

2.8 Extreme value analysis

The term 'risk' has been frequently used in financial context in recent years. The term 'risk' can reflect many different apsects. Sometimes it seems that risk refers to lack of certainty, sometimes volatility in a price process, sometimes it seems to be risk for a particular event, e.g. bankruptcy. The formal notion of the statistic of interest is:

$$X_{\max,n} = \max_{0 \le i \le n} X(i) \quad \text{or} \quad X_{\min,n} = \min_{0 \le i \le n} X(i).$$

For the discrete-time case of iid data the finite-sample maximum is simply:

$$F_{\max}(x_{\max}) = P(X_i \le x_{\max}| \text{ for } i \le n) = F(x_{\max})^n.$$

The classical central limit theorem can be interpreted as a large sample result for a sum of iid random variables. If a sequence of iid random variables, $X_1, \ldots, X_n$ with $E(X_i^2) < \infty$, the sum converges in distribution to a normal distribution, i.e.

$$S_n = \sum_{i=1}^{n} X_i \xrightarrow{d} N(a_n, b_n), \qquad \text{with } a_n = nE(X_i), \quad b_n = nV(X_i).$$

In many statistical textbooks in chapters on order statistics there are similar results for the maximum (and minimum) of iid random variables. The question is whether we have limit result for the function maximum (or minimum), just as we have a limit result for function sum. If there exists sequences of constants, a_n, b_n, such that $(X_{\max,n} - a_n)/b_n$ converges in distribution to a nontrivial limit, then the form of limiting distribution has to one of the following (Embrechts, Klüppelberg and Mikosch 1997; Mood, Graybill and Boes 1974):

$$F_1(x) = I_{(0,\infty)}(x)\mathrm{e}^{-x^{-\gamma}}, \quad \gamma > 0, \tag{2.26}$$

$$F_2(x) = I_{(-\infty,0)}(x)\mathrm{e}^{-|x|^{\gamma}} + I_{[0,\infty)}(x), \tag{2.27}$$

$$F_3(x) = \mathrm{e}^{-\mathrm{e}^{-x}}. \tag{2.28}$$

The limiting distribution of (2.26) is obtained if (and only if)

$$\frac{1 - F(x)}{1 - F(\tau x)} \xrightarrow[x \to \infty]{} \tau^{\gamma}.$$

The limit distribution of (2.27) is obtained if and only if

$$F(x_0) = 1 \text{ for some } x_0 \text{ and } F(x_0 - \varepsilon) < 1, \text{ for all } \varepsilon > 0.$$

The limit distribution of (2.28) is obtained if and only if

$$n[1 - F(b_n x + a_n)] \xrightarrow[n \to \infty]{} \mathrm{e}^{-x}.$$

The limit distribution is called an extreme-value distribution and its properties are decided by the behaviour of the distribution function F in the tails. The result in Equations (2.26–2.28) is sometime referred to as the Fisher–Tippett theorem for limit laws of maxima and the distributions are called Frechet, Weibull and Gumbell, respectively.

If the X_i's form a dependent sequence, results similar to CLT are available, i.e. if the dependency fades away with increasing time lag, A formal way of expressing such a fadeout is the concept of m-dependence. Some examples are in Mcneil (1997), Resnick (1997) and Johansson (2003).

Multivariate generalizations are complicated and perhaps only practical in special situations. Starica (1999) treats the multivariate case of constant conditional correlations. It is in the nature of extremes that multivariate extremes are hard to deal with. Typical, practical statistical problems in extreme-value theory are quantile estimation and tail-index estimation. Extreme values are rare so getting an accurate estimate of a high quantile is not possible. A direct citation from Embrechts, Klüppelberg and Mikosch (1997) is: 'There is no free lunch when it comes to high quantile estimation'; they also give sample properties of some estimators. Mikosch (2004) gives some guidelines. For more advanced treatment of extreme-value theory see, e.g. Embrechts, Klüppelberg and Mikosch (1997) and Resnick (1987). In the statistical package R (R Development Core Team 2005) it is possible to do some univariate and bivariate extreme-value calculations using the package `evd`.

In recent years the approach of using copulas has become a popular tool for representing multivariate distributions. The idea is to get a formal tool of defining dependence between two variables. A copula is a multivariate density function with uniform margins. Any random variable can in theory be transformed to a univariate random variable by applying the univariate distribution function to it. If a bivariate normal random variable (X_1, X_2) was transformed to have support on $[0, 1] \times [0, 1]$ by applying the univariate inverse distribution to X_1 and X_2, respectively, the result consists of two uniform $U(0, 1)$ random variables with a particular dependency structure. Conversely, if we have two dependent $U(0, 1)$ random variables, U and V, with cumulative distribution function (cdf) $F(u, v)$, and apply the inverse of a univariate normal cdf to U and V, respectively, then we have a bivariate random variable with normal margins, but in general this bivariate random variable is not bivariate normal. The motive is that in finance, variables may seem weakly correlated most of the time, but when something serious happens, the catastrophe happens to both variables. A textbook on copula methods in finance is Cherubini, Luciano and Vecchiato (2004). Statistically speaking, when it comes to estimating copulas from data, it is a question of estimating a multivariate distribution. Estimation takes either place through parsimonious parameterization or through (kernel) smoothing where accuracy is decided by the choice of bandwidth. In the case of multivariate extremes there is *no free lunch*, data will be thin and estimates will be inaccurate. A sober view of the importance of copulas in statistical analysis of multivariate extremes is given by Mikosch (2005).

2.9 Jump processes, and further topics

The popularity of continuous-time finance has generated an understanding of the nature of diffusion processes among data analysts. They have realized that some of the movements that are seen in the financial markets are not likely to be the outcome of a continuous-state process Therefore the diffusion models have been modified to allow jumps. This calls for a definition of the nature of jumps. In the words of Merton (1990) this is *continuous-path with rare events*. The idea is to add a weighted Poisson-type process, $N(t)$, to the diffusion.

$$\mathrm{d}X(t) = \mu[X(t), t]\mathrm{d}t + \sigma[X(t), t]\mathrm{d}W(t) + J(t)\mathrm{d}N(t).$$

The function $J(t)$ denotes the size of the jump. To be operational, it is necessary, in addition to the Wiener process (normal distribution), to define the probability distribution of $N(t)$, the event of jump and that of the jump size, $J(t)$. For maximum-likelihood estimation there will be identification problems that will require restrictions on the parameter space (Honoré 1998). Evidently the urge for tractability will influence the model choice. Kou (2002) gives a model that should be able to capture stylized facts, being analytically well-behaved at the same time: ‘A jump diffusion model for option pricing with three properties: leptokurtic feature, volatility smile, and analytical tractability’. The model is defined in equations (2.1) and (2.2) in his paper:

$$\frac{\mathrm{d}S(t)}{S(t)} = \mu \mathrm{d}t + \sigma \mathrm{d}W(t) + \mathrm{d}\left(\sum_{i=1}^{N(t)} (V_i - 1)\right), \tag{2.29}$$

$N(t)$ Poisson $E[N(t)] = \lambda t$,

V_i iid double exponential.

The solution of (2.29), conditioned on $S(0)$:

$$S(t) = S(0)\mathrm{e}^{(\mu - \frac{1}{2}\sigma^2)t + \sigma W(t)} \prod_{i=1}^{N(t)} V_i,$$

contains three different probability distributions, the normal, the Poisson and the double exponential. The memoryless property of the Wiener process, the Poisson process, and the double exponential are helpful in deriving closed form solutions of some option pricing problems. The spirit of this model is similar to that in Merton (1976, 1990). There are many more examples of applications of jump diffusions in finance (Wong and Li 2006).

Another modern approach of incorporating discontinuity in continuous-time process is the use of Lévy processes. Sato (2001) defines $L(t)$, the Lévy process, as:

$L(t_2) - L(t_1)$ and $L(t_4) - L(t_3)$ are independent if $t_1 < t_2 < t_3 < t_4$,

$L(0) = 0$,

The distribution of $L_{t+s} - L_t$ is not a function of t,

$$P(|L_{s+t} - L_t| > \varepsilon) \underset{s \to 0}{\longrightarrow} 0, \quad \forall \varepsilon > 0,$$

$L(t)$ is right continuous with left limits.

The class of Lévy processes is very large. The Poisson process and Wiener process are both Lévy processes. In the Poisson process all movements occur in jumps of size 1, in the Wiener process all movement is along a continuous pattern. Eberlein (2001) gives a simple formula for representing a Lévy process which is generated with a distribution that has finite first moments:

$$X(t) = \sigma W(t) + Z(t) + \alpha t,$$

where $W(t)$ is the standard Wiener process and $Z(t)$ is a purely discontinuous martingale independent of $W(t)$. The formal notion of a Lévy process allows for a formal definition of a stochastic differential equation that is driven by a Lévy process instead of the more traditional Wiener process, e.g. an Ornstein–Uhlenbeck type process:

$$dX(t) = \alpha(\beta - X(t))dt + \sigma dL(t). \tag{2.30}$$

Equation (2.30) represents a dynamic system that in finance is called the Vasicek model. It is conceivable to have $L(t)$ as purely discontinuous, such that $X(t)$ will remain positive. That way it can be a plausible model for a dynamic variance (or volatility). Barndorff-Nielsen and Shephard (2001) give a thorough treatment of ideas in that spirit. There is substantial interest in creating a positive process for modelling the volatility process; Tsai and Chan (2005) give a comment on how to ensure non-negativity. Marquardt and Stelzer (2007) discuss a multivariate CARMA driven with Lévy Process. Many of the other models can be nested within the formal framework of working with Lévy processes. Brockwell and Marquardt (2005) treat the fractionally integrated CARMA. Brockwell, Chadraa and Lindner (2006) derive a continuous-time GARCH process.

2.10 A comment on model building

A key issue in statistical analysis is the 'model'. Data is interpreted through a model. The model is a kind of mathematical idealization of some real-world phenomena, and the general conclusion to be drawn from data refers to that particular model. There are several principles in designing an interesting model. Draper and Smith (1966) give three principal categories; (a) the functional model, which explains the nature of the underlying process; (b) the control model, i.e. a model where control does not rely on full understanding of the underlying process; and (c) the predictive model, which aims at giving a method for statements about future observations. A famous phrase is 'all models are wrong but some are useful' (Box 1979), describes the situation facing the applied data analyst. Practical choice of statistical model is mainly based on a combination of the following; (a) mathematical tractability; (b) some theoretical basis; and (c) some functions that are likely to fit stylized features of data. A model should be plausible, e.g. we do not like to get negative values from something that should be positive. It should also be possible to interpret the model. It is preferable that the model can explain something, and last but not least it should be possible to reject a model if data look very incompatible to possible output of the model.

A frequent approach in practical analysis is to employ some kind of pretesting. The estimation procedure is then essentially a two-step procedure, consisting of a test of a particular hypothesis and if the hypothesis is rejected, then a particular estimation procedure is performed. For example:

$$\boldsymbol{\theta}_{\text{Pre-test}} = I_{H_0}(\boldsymbol{X})\boldsymbol{\theta}_0 + I_{H_1}(\boldsymbol{X})\boldsymbol{\theta}_{\text{ML}}. \tag{2.31}$$

In Equation (2.31), $I_{H_0}(\boldsymbol{X})$ and $I_{H_1}(\boldsymbol{X})$ are indicator-functions of data taking the value 1 or 0, depending on which hypothesis is supported. An early literature review is given by Judge and Bock (1978). A recent PhD thesis on the subject is Danilov (2003). Extensive pretesting schemes have been developed. One scheme is based on starting with a big model and try to test away model components of minor importance. Sometimes this is labelled stepwise-backward or general-to-specific. The system PCGETS described in Hendry and Krolzig (2001) is an example of this. Another approach is sometimes called stepwise-forward or specific-to-general is based on starting with a simple model and try to include 'significant' variables. The system RETINA (Perez-Amaral, Gallo and White 2003) is an example of this approach.

The pretest procedure is biased, as there is a bias towards H_0. It is to be expected that this bias is beneficial, relative to say least-squares estimation, when the truth is close to H_0. It is also to be expected that biased estimation is harmful if the truth is far from H_0. It turns out that for some loss functions it is possible to dominate the ordinary least-squares estimators in linear models. Examples of such estimators are the Stein rule estimators. An early review is given by Judge and Bock (1978). The idea is to shrink the ML estimates towards an a priori defined subspace of the parameter space. The pretest strategy is a kind of jump-shrinking strategy, whereas the Stein-family estimators and ridge-regression type estimators are examples of continuous shrinkage. Tómasson (1986) applies these ideas to ARMA models. It is clear that how to shrink depends on the characteristics of the application. Stein estimators can be derived as a kind of empirical Bayes procedure. An empirical Bayes procedure is based on a Bayes estimator where parts of the prior are estimated from data. A later review is given by Saleh (2006). Many practitioners now do this by minimizing the AIC or the BIC.

$$\text{AIC} = -Tl(\hat{\boldsymbol{\theta}}) + k,$$

$$\text{BIC} = -Tl(\hat{\boldsymbol{\theta}}) + \frac{k}{2}\log(T),$$

where T is the number of observations, $l(\hat{\boldsymbol{\theta}})$ is the log likelihood evaluated at the ML estimate and k is the number of estimated parameters. For linear models the AIC behaves similar to a pretest estimator with a fixed rejection level, whereas the BIC behaves similar to pretest estimator where the rejection level depends on the sample size.

2.11 Summary and discussion

The aim of financial data analysis is to make inference based on data. Typically data consist of a single realization of a time series. Stationarity and ergodicity are necessary for obtaining consistent estimates. The linear-normal discrete-time model is the most widely used and best understood model. Univariate ARMA, or multivariate, VARMA, regression based-variants, ARMAX, where it is allowed to condition on some explanatory variables, all belong to that class. Common to those is that, like prediction, missing data, systematic sampling, calculation of likelihood, etc., most treatments of interest are possible by using recursive algorithms like the Kalman filter. The ARMA is a linear filter. An important feature of the linear filter is that if the input is normal, then the output is also normal. If the input to a MA process is a sequence of iid stable distributions then the output is also stable. The normal distribution is the only stable distribution with finite variance. In general the closed form of the density of a stable distribution is unknown, but in some cases the characteristic function can be written down. There exist computer programs for simulating a sample from stable distributions. Numerical methods of the inverting the Fourier transform can be used to obtain the likelihood function, that then can be numerically maximized. Even quantile-regression and simulation based methods could be used. In many dimensions the situation is more complicated. There exist methods for multivariate stable distributions, but the computational aspect is difficult. When the input to a linear filter is a finite variance nonnormal noise, then the output is not normal, but due to central limit theorem type arguments, the output is 'more normal' than the input. In many applications, it is natural to assume that the level of a series is much more normal than the series of innovation. Assuming some parametric form of the distribution of innovations makes it possible to calculate the likelihood function recursively, estimate parameters, perform tests, etc. The usual ARMA model has an exponential decaying autocorrelation function. The ARFIMA is a way of allowing slower decay of the autocorrelation without abandoning the stationarity assumption.

Doing constructive nonlinear modelling requires a firm idea of the form of nonlinearity, e.g. a process jumping in first or second moments. The ARCH family aims at modelling second moments of a process by applying the auto regressive concept to the second moments of a measured process. The distributional properties are somewhat complicated. The general multivariate ARCH is extremely complicated. Many ideas of ordinary time series, threshold models, long memory, etc. have found their way to the ARCH literature. As an example, in ARCH-M and AR-ARCH models both the dynamics of first and second moments are modelled.

The continuous-time models based on the Wiener process have a mathematical appeal, and many models can be motivated by theory, e.g. pricing methods based on no arbitrage. The univariate diffusions may also have a stationary invariant distribution that can be interpreted in a real world context. A feature of the univariate diffusion model is that it can combine short-term dynamics and long-term equilibrium into one formula. High dimensional diffusion process have in common with the multivariate nonlinear discrete-time models that they are hard to visualize and in practical cases it is necessary to have a good idea about the nature of the functional relationship between variables. For a high-dimensional diffusion system the issue of an invariant stationary distribution is no longer simple, both in its existence and in how to find it in case of existence.

In the continuous-time discrete state space, any positive random variable can play the role of a waiting time (duration) for a shift between states. The interpretable form of the distribution of the waiting-time is the hazard function. Monitoring the hazard function over time mounts to monitoring the risk for change of state. The hazard rate itself can be thought of as a positive stochastic process, with a dependency structure over time and possibly a function of external regressors.

The extreme value theory is evolving, but even if multivariate models become computationally tractable, their usability will be limited by lack of data.

Some data analysts have found that the elegant mathematical idealization of the diffusion process does not give a realistic picture of what is happening in financial markets. Stylized facts such as the volatility smile, etc. have led researches to work with modifications such as jump diffusions and Lévy-driven stochastic differential equations.

When time series are analysed it is of importance to understand the source of the data, e.g. whether we have stock data or flow data. A lot of the models in the literature focus on the return data in a financial market. Trading data consists of time, price and volume.

In the author's mind, the following are most important building blocks for surveillance of time-dependent data. First the linear-filter with normal input. Then an understanding of the importance of the distribution. The diffusion models are an elegant approach of linking theory and data. The theory of diffusion models can give good motivation for a choice of a linear model.

The univariate extreme value theory is reasonably simple and will give good approximation. The multivariate extreme value model will require judgment and parsimonious parameterization. The continuous-time models driven with Lévy process have not yet reached practitioners, so for a while they will be doing ad hoc jump diffusion models. The ARCH models have been around for some time. Academics and practitioners have built up experience on their benefits and drawbacks. Their future is dependent on the emerge of realistic alternatives. Starica (2004) has raised some questions about their ability to cope with short-term properties and long-term properties of financial time series. His criticism is roughly, that even though the ARCH-type model can capture some short-term volatility dynamics, it comes at the price of overestimating long-term variance. The practitioner has always to choose a model that serves a certain aim. The choice of model is a compromise between its simplicity and its ability to capture important features of real life.

References

Aït-Sahalia, Y. (1996). Non-parametric pricing of interest rate derivative securities. *Econometrica*, **64**, 527–560.

Aït-Sahalia, Y. (1999). Transition densities for interest rate and other nonlinear diffusion. *Journal of Finance*, **54**(4), 1361–1395.

Aït-Sahalia, Y. (2002). Maximum likelihood estimation of discretely sampled diffusions: A closed-form approximation approach. *Econometrica*, **70**(1), 223–262.

An, M. Y., Christensen, B. J. and Gupta, N. D. (2003). On pensions and retirement: Bivariate mixes proportional hazard modelling of joint retirement. Working paper no 163. Center of analytical finance, University of Aarhus, Aarhus Business School.

Andersen, P. K., Borgan, Ø., Gill, R. D. and Keiding, N. (1993). *Statistical Models Based on Counting Processes*. Springer-Verlag, Berlin.

Ansley, C. F. (1979). An algorithm for the exact likelihood of a mixed autoregressive moving average process. *Biometrika*, **66**, 59–65.

Bachelier, L. (1900). Theorie de la speculation. *Annales de l'Ecole Normale Superiore*, pages 21–86.

Baille, R. T., Bollerslev, T. and Mikkelsen, H. O. (1996). Fractionally integrated generalized autoregressive conditional heteroskedacity. *Journal of Econometrics*, **74**, 3–30.

Barndorff-Nielsen, O. E. and Shephard, N. (2001). Non-Gaussian Ornstein-Uhlenbeck-based models and some of their uses in financial economics. *Journal of the Royal Statistical Society: Series B (Statistical Methodology)*, **64**(2), 167–241.

Bartlett, M. S. (1946). On the theoretical specification and sampling properties of autocorrelated time series. *Journal of the Statistical Society Supplement*, **8**, 27–41.

Beran, J. (1994). *Statistics for Long-Memory Processes*. Chapman & Hall, London.

Beran, J. (1995). Maximum-likelihood estimation of the differencing parameter for invertible short and long ARIMA models. *Journal of the Royal Statistical Society Series B*, **57**(4), 659–672.

Bergstrom, A. R. (1988). The history of continuous-time econometric models. *Econometric theory*, **4**, 365–383.

Bergstrom, A. R. (1990). *Continuous-Time Econometric Modelling*. Oxford University Press, Oxford.

Beskos, A., Papaspiliopoulos, O., Roberts, G. O. and Fearnhead, P. (2006). Exact and efficient likelihood-based inference for discretely observed diffusion processes with discussion. *Journal of the Royal Statistical Society, series B*, **68**, 333–382.

Beskos, A. and Roberts, G. O. (2005). Exact simulation of diffusions. *Annals of Applied Probability*, **15**, 2422–2444.

Bhardwaj, G. and Swanson, N. R. (2006). An empirical investigation of the usefulness of ARFIMA models for predicting macroecnomic and financial time series. *Journal of Econometrics*, **131**, 539–578.

Bibby, B. M., Jacobsen, M. and Sørensen, M. (2004). Estimating functions for discretely sampled diffusion-type models. Preprint 2004-4, Department of Applied Mathematics and Statistics, University of Copenhagen.

Björk, T. (2004). *Arbitrage Theory in Continuous Time*. Oxford University Press, Oxford.

Black, F. and Scholes, M. (1973). The pricing of options and corporate liabilities. *Journal of Political Economy*, pages 635–654.

Blackwell, D., Griffiths, M. and Winters, D. (2006). *Modern Financial Markets*. John Wiley & Sons, Ltd, New York.

Bladt, M. and Sørensen, M. (2006). Efficient estimation of transition rates between credit ratings from observations at discrete time points. Preprint No. 2006-2, Department of Applied Mathematics and Statistics, University of Copenhagen.

Bollerslev, T. (1990). Modelling the coherence in short-run nominal exchange rates: A multivariate generalized ARCH approach. *Review of Economics and Statistics*, **72**, 498–505.

Box, G. E. P. (1979). Robustness in the strategy of scientific model building. In Launer, R. L. and G N. Wilkinson, E. (Eds.), *Robustness in Statistics*. Academic Press, New York.

Box, G. E. P. and Jenkins, G. M. (1970). *Time Series Analysis, Forecasting and Control*. Holden Day, San Fransisco.

Brockwell, P., Chadraa, E. and Lindner, A. (2006). Continuous-time GARCH processes. *The Annals of Probability*, **16**(2), 790–826.

Brockwell, P. J. and Davis, R. A. (1991). *Time Series: Theory and Methods*. Springer-Verlag, Berlin.

Brockwell, P. J. and Marquardt, T. (2005). Levy-driven and fractionally integrated ARMA processes with continuous time parameter. *Statistica Sinica*, **15**, 477–494.

Brown, R. (1827). *A brief account of microscopical observations*. Unpublished, London.

Chan, W.-S., Keung, W. and Tong, H. (Eds.). (2000). *Statistics and Finance: An Interface*. Imperial College Press, London.

Cherubini, U., Luciano, E. and Vecchiato, W. (2004). *Copula Methods in Finance*. John Wiley & Sons, Ltd, New York.

Cowles, A. (1933). Can stock market forecasters forecast? *Econometrica*, **1**(3), 309–324.

Cvitanic, J. and Zapatero, F. (2004). *Introduction to the Economics and Mathematics of Financial Markets*. MIT Press, Cambridge.

Danilov, D. (2003). The effects of pretesting in econometrics with applications in finance. PhD thesis, Tilburg University.

Dickey, D. A. and Fuller, W. A. (1979). Distribution of the estimators for autoregressive time series with unit root. *Journal of the American Statistical Association*, **74**, 427–431.

Draper, N. and Smith, H. (1966). *Applied Regression Analysis*. John Wiley & Sons, Ltd, New York.

Duffie, D. (1996). *Dynamic Asset Pricing Theory*. Princeton University Press, Princeton.

Eberlein, E. (2001). Application of generalized hyperbolic levy motions to finance. In Barndorff-Nielsen, O. E., Mikosch, T. and Resnick, S. I. (Eds.), *Lévy Processes, Theory and Applications*. Birkhäuser, Boston.

Einstein, A. (1905). On the movement of small particles suspended in a stationary liquid by the molecular-kinetic theory of heat. *Annalen der Physik*, pages 549–560.

Embrechts, P., Klüppelberg, C. and Mikosch, T. (1997). *Modelling Extremal Events for Insurance and Finance*. Springer-Verlag, Heidelberg.

Engle, R. F. (1982). Autoregressive conditional heteroskedacity with estimates of the variance of United Kingdom inflation. *Econometrica*, **50**, 987–1007.

Engle, R. F. and Bollerslev, T. (1986). Modeling the persistence of conditional variances. *Econometric Reviews*, **5**, 1–50.

Engle, R. F. and Russell, J. R. (1998). Autoregressive conditional duration a new model for irregularly-spaced transaction data. *Econometrica*, **66**(5), 1127–1162.

Fernandes, M. and Grammig, J. (2006). A family of autoregressive conditional duration models. *Journal of Econometrics*, **130**(1), 1–23.

Fleming, T. R. and Harrington, D. P. (1991). *Counting Processes and Survival Analysis*. John Wiley & Sons, Ltd, New York.

Gailbraith, R. F. and Gailbraith, J. I. (1974). On the inverse of some patterned matrices arising in the theory of stationary time series. *Journal of Applied Probability*, **11**, 63–71.

Gerhard, F. and Hautcsh, N. (2002). Semiparametric autoregressive conditional proportional hazard models. No 2002-W2, Economics Papers from Economics Group, Nuffield College, University of Oxford.

Gourieroux, C. (1997). *ARCH models and Financial Applications*. Springer-Verlag, New York.

Granger, C. (1983). *Forecasting White Noise, in, Applied Time Series Analysis of Economic Data, Proceedings of the Conference on Applied Time Series Analysis of Economic Data (October 1981), Editor. A. Zellner*. US Government Printing Office.

Granger, C. W. J. and Andersen, A. P. (1978). *An Introduction to Bilinear Time Series Models*. Vandenchoeck & Ruprect, Göttingen.

Granger, C. W. J. and Joyeux, R. (1980). An introduction to long memory time series models and fractional differencing. *Journal of Time Series*, **1**, 15–29.

Haavelmo, T. (1943). The implications of a system of simultaneous equations. *Econometrica*, **11**, 1–12.

Hall, P., Peng, L. and Yao, Q. (2002). Prediction and nonparametric estimation for time series with heavy tails. *Journal of Time Series*, **23**(3), 313–331.

Hamilton, J. D. (1989). A new approach to the economic analysis of nonstationary time series and the business cycle. *Econometrica*, **57**, 357–384.

Hamilton, J. D. (1994). *Time Series Analysis*. Princeton University Press, Princeton.

Hand, D. J. and Jacka, S. D. (Eds.). (1998). *Statistics in Finance*. Arnold, London.

Hannan, E. J. and Deistler, M. (1988). *The Statistical Theory of Linear Systems*. John Wiley & Sons, Ltd, New York.

Harvey, A. C. (1989). *Forecasting, Structural Time Series Models and the Kalman Filter*. Cambridge University Press, Cambridge.

Harvey, A. C. (1993). *Time Series Models*. Harverster Wheatsheaf, London.

Harvey, C. R. and Morgenson, G. (2002). *The New York Times Dictionary of Money and Investing: The Essential A-to-Z Guide to the Language of the New Market*. Times Books, New York.

Hendry, D. F. and Krolzig, H.-M. (2001). *Automatic Econometric Model Selection Using PcGets*. Timberlake Consultants Press.

Honoré, P. (1998). Pitfalls in estimating jump-diffusion models. Working paper, University of Aarhus.

Hull, J. (1993). *Options, Futures and other Derivatives* 2nd edn. Prentice-Hall, Englewood Cliffs.

Johansson, N. C. J. (2003). Moment estimation using extreme value techniques. PhD thesis, Chalmers University of Technology and Göteborg University.

Jones, D. A. (1978). Non-linear autoregressive processes, series A. *Proc. Roy. Soc. London*, **360**, 71–95.

Judge, G. G. and Bock, M. E. (1978). *The Statistical Implications of Pre-Test and Stein-rule Estimators in Econometrics*. North-Holland, Amsterdam.

Karatzas, I. and Shreve, S. E. (1991). *Brownian Motion and Stochastic Calculus,* 2nd edn. Springer-Verlag, Berlin.

Kendall, M. G. (1953). The analysis of economic time-series part I: Prices. *Journal of the Royal Statistical Society. Series A (General)*, pages 11–25.

Klimko, L. A. and Nelson, P. I. (1978). On conditional least squares estimation for stochastic processes. *Annals of Statistics*, **6**, 629–642.

Kloeden, P. E. and Platen, E. (1992). *Numerical Solution of Stochastic Differential Equations*. Springer-Verlag, Berlin.

Kou, S. (2002). A jump diffusion model for option pricing. *Management Science*, **48**, 1086–1101

Kutoyants, Y. A. (1984). *Parameter Estimation for Stochasic Processes*. Helderman, Berlin.

Lambert, P. and Lindsey, J. K. (1999). Analysing financial returns using regression models based on non-symmetric stable distributions. *Applied Statistics*, **48**, 409–424.

Lambrecth, B. M., Perraudin, W. R. M. and Satchell, S. (2003). Mortgage default and possession under recourse: A competing hazard approach. *Journal of Money, Credit and Banking*, **35**(3), 425–442.

Lancaster, T. (1990). *The Econometric Analysis of Transition Data*. Cambridge University Press, Cambridge.

Lando, D. and Skødeberg, T. M. (2002). Analyzing rating transitions and rating drift with continuous observations. *Journal of Banking and Finance,* **26**(2-3), 423–444.

Lindeboom, M. and Van den Berg, G. J. (1994). Heterogeneity in bivariate duration models: The importance of mixing distribution. *Journal of the Royal Statistical Society; B*, **56**, 49–60.

Lund, R. and Basawa, I. (2000). Recursive prediction and likelihood evaluation for periodic ARMA models. *Journal of Time Series Analysis*, **21**, 75–93.

Lütkepol, H. (1991). *Introduction to Multiple Time Series Analysis*. Springer-Verlag, Berlin.

Maddala, G. S. and Rao, C. R. (Eds.) (1996). *Handbook of Statistics, Volume 14: Statistical Methods in Finance*. Elsevier, Amsterdam.

Mao, X., Yuan, C. and Yin, G. (2005). Numerical method for stationary distribution of stochastic differential equations with Markovian switching. *J. Comput. Appl. Math.*, **174**(1), 1–27.

Marquardt, T. and Stelzer, R. (2007). Multivariate CARMA processes. *Stochastic Processes and their Applications*, **117**, 96–120.

McLeod, A. (1993). Parsimony, model adequacy and periodic correlation in forecasting time series. *International Statistical Review*, **61**, 387–393.

Mcneil, A. J. (1997). Estimating the tails of loss severity distributions using extreme value theory. *Astin Bulletin*, **27**(1), 117–137.

Melard, G. (1983). A fast algorithm for the exact likelihood of moving average models. *Applied Statistics*, **33**, 104–114.

Merton, R. C. (1976). Option pricing when underlying stock returns are discontinuous. *Journal of Financial Economics*, **3**, 125–144.

Merton, R. C. (1990). *Continuous-Time Finance*. Blackwell, Oxford.

Mikosch, T. (2004). How to model multivariate extremes if one must? Maphysto, Research Report no. 21.

Mikosch, T. (2005). Copulas tales and facts. Discussion paper, International Conference on Extreme Value Analysis in Gothenburg.

Mood, A. M., Graybill, F. A. and Boes, D. C. (1974). *Introduction to the Theory of Statistics*, 3rd edn. McGraw-Hill, New York.

Neftci, S. N. (1996). *An Introduction to the Mathematics of Financial Derivatives*. Academic Press, San Diego.

Nelson, D. (1991). Conditional heteroskedacity in asset returns. *Econometrica*, **59**, 347–370.

Nelson, D. B. (1990). Stationarity and persistence in the GARCH(1,1) model. *Econometric theory*, **6**, 318–344,

Øksendal, B. (1998). *Stochastic Differential Equations: An Introduction with Applications*, 5th edn. Springer-Verlag, Berlin.

Ozaki, T. (1982). The statistical analysis of perturbed limit cycle processes using nonlinear time series models. *Journal of Time Series*, **3**, 29–41.

Ozaki, T. (1985). Nonlinear time series models and dynamical systems. In Hannan, E. and Krishnaiah, P. (Eds.), *Handbook of Statistics*, volume 5. North-Holland, Amsterdam.

Pagano, M. (1978). On the periodic and multiple autoregressions. *Annals of Statistics*, **6**, 1310–1317.

Perez-Amaral, T., Gallo, G. and White, H. (2003). A flexible tool for model building: The relevant transformation of the inputs network approach (RETINA). *Oxford Bulletin of Economics and Statistics*, pages 821–838.

Priestley, M. B. (1991). *Non-Linear and Non-Stationary Time Series Analysis*. Academic Press, New York.

Quoreshi, A. M. M. S. (2006). Bivariate time series modeling of financial count data. *Communications in Statistics: Theory and Methods*, **35**(7), 1343–1358.

R Development Core Team (2005). *R: A language and environment for statistical computing*. R Foundation for Statistical Computing, Vienna, Austria.

Resnick, S. I. (1987). *Extreme Values Regular Variations, and Point Processes*. Springer-Verlag, New York.

Resnick, S. I. (1997). Discussion on the data on large fire insurance losses. *Astin Bulletin*, **27**(1), 139–151.

Revuz, D. and Yor, M. (1999). *Continuous Martingales and Brownian Motion* 3rd edn. Springer-Verlag, Berlin.

Saleh, A. K. M. E. (2006). *Theory of Preliminary Test and Stein-Type Estimation with Applications*. John Wiley & Sons, Ltd, New York.

Sato, K. (2001). Basic results on Lévy processes. In Barndorff-Nielsen, O. E., Mikosch, T. and Resnick, S. I. (Eds.), *Lévy processes: Theory and Application*. Birkhäuser, Boston.

Shiryaev, A. N. (1999). *Essentials of Stochastic Finance*. World Scientific, Singapore.

Singleton, K. J. (2001). Estimation of affine asset pricing models using the empirical characteristic function. *Journal of Econometrics*, **102**(2), 111–141.

Sowell, F. (1992). Maximum-likelihood estimation of stationary univariate fractionally integrated time series models. *Journal of Econometrics*, **53**, 165–188.

Starica, C. (1999). Multivariate extremes for models with constant conditional correlations. *Journal of Empirical Finance*, **6**, 515–553.

Starica, C. (2004). Is GARCH(1,1) as good a model as the Nobel Prize accolades would imply? Econometrics 0411015, EconWPA. available at http://ideas.repec.org/p/wpa/wuwpem/0411015.html.

Tómasson, H. (1986). Prediction and estimation in ARMA models. PhD thesis, University of Gothenburg.

Tong, H. (1983). *Threshold Models in Non-Linear Time-Series Analysis*. Springer-Verlag, New York.

Tsai, H. and Chan, K. S. (2005). A note on non-negative continuous time processes. *Journal of the Royal Statistical Society: Series B (Statistical Methodology)*, **67**(4), 589–597.

Tsay, R. S. (2002). *Analysis of Financial Time Series*. John Wiley & Sons, Ltd, New York.

Wei, W. W. S. (1990). *Time Series Analysis–Univariate and Multivariate Methods*. Addison-Wesley, Reading.

Wiener, N. (1923). Differential space. *Journal of Mathematical Physics*, **2**, 131–174.

Wilmott, P., Howison, S. and Dewynne, J. (1995). *The Mathematics of Financial Derivatives*. Cambridge University Press, Cambridge.

Wong, H. Y. and Li, C. P. (2006). Estimating jump diffusion structural credit risk models. Working Paper, Chinese University, Hong Kong.

Yang, W. and Allen, D. E. (2005). Multivariate GARCH hedge ratios and hedging effectiveness in Australian futures markets. *Accounting and Finance*, **45**(2), 301–321.

Zivot, E. and Wang, J. (2003). *Modeling Financial Time Series with S-plus*. Springer-Verlag, New York.

3

The relation between statistical surveillance and technical analysis in finance

David Bock[a], Eva Andersson[a,b] and Marianne Frisén[a]

[a] *Statistical Research Unit, School of Business, Economics and Law, Göteborg University, PO Box 660, SE 405 30 Göteborg, Sweden*

[b] *Department of Occupational and Environmental Medicine, Sahlgrenska University Hospital, Box 414, SE 405 30 Göteborg, Sweden.*

3.1 Introduction

The purpose of this chapter is to investigate the inferential differences and similarities between some methods of statistical surveillance and some prospective decision rules used in finance, and to give a brief review of these financial decision rules from a statistical viewpoint. Furthermore, evaluation measures and utility functions used in statistical surveillance are compared with those used in financial settings.

In the financial market, a natural aim is to try to maximize profit. This requires optimal sequential decisions. An indicator is monitored with the aim of detecting the optimal time to trade. The indicator could be the price itself or related to the price. Optimal times to trade are related to regime shifts in the stochastic properties of the indicator. Thus, finding the optimal time to trade is equivalent to the timely detection of a regime shift.

Financial Surveillance Edited by Marianne Frisén

According to the efficient market hypothesis, the financial markets are arbitrage-free and there is no point in trying to determine the optimal transaction time. But when the information about the process is incomplete, as for example when a change point could occur, there may be an arbitrage opportunity, as demonstrated by Shiryaev (2002). Many agents reject the efficient market hypothesis and many studies give support for the profitability of the prospective framework, see for example Sweeney (1986) and Lo (2000). In technical analysis the history of the price of a stock is assumed to contain information and technical analyses are methods for extracting this information. In this chapter we do not claim to verify or reject the hypothesis of an arbitrage-free market, we merely notice that technical analysis is used in practice and that the efficient market hypothesis is not universally accepted. The aim in this chapter is to compare some of the suggested methods for technical analysis to methods for statistical surveillance.

Since timeliness is crucial in a trading setting, the incoming data should be analysed online and sequential trading decisions made. In this chapter we investigate methods which aim to identify regime shifts, especially turning points. The inference situation is one of surveillance where the aim is to detect a change quickly and safely. It is characterized by repeated decisions and by never accepting the null hypothesis. For general reviews on statistical surveillance, see Frisén and de Maré (1991), Srivastava and Wu (1993), Lai (1995), and Frisén (2003)) Other names are statistical process control and monitoring.

Neftci (1991), Dewachter (2001), and others have pointed out that many trading rules are ad hoc and that their statistical properties are often unknown. Since the properties of methods of statistical surveillance have been investigated extensively, an integration of surveillance theory and financial decision rules could prove fruitful. Schmid and Tzotchev (2004) state that applications of surveillance methods in finance have been scarce. Lam and Yam (1997) claimed to be the first to link methods of surveillance to technical trading rules. Recent studies using surveillance for financial applications are discussed in Section 3.3.

'Optimal stopping rules' are based on the assumption that the model is completely known. Then probability theory can be used to find optimal trading times, see e.g. Shiryaev, Kabanov, Kramkov and Melnikov (1994), Shiryaev (1999), Jönsson, Kukush and Silvestrov (2004), and Lai and Lim (2005). In this chapter, however, the process includes unknown statistical parameters, and we consider an inferential approach where we want to infer from data whether a regime shift has occurred or not.

In Section 3.2, regime shifts and model specifications are exemplified. In Section 3.3 we briefly describe the methodology of statistical surveillance and discuss similarities and differences between methods proposed in the literature on finance and methods of statistical surveillance. In Section 3.4 some of the methods are applied to the Hang Seng Index. Concluding remarks are given in Section 3.5.

3.2 Indicators and regime shifts

The indicator to be monitored can be constructed from one or several processes. The price level of an asset may itself be the indicator. To detect an increased risk level, the estimated variance can be used. The Leverage effect, see Black (1976), motivates simultaneous monitoring of the mean and variance, as in Schipper and Schmid (2001a). The arrival time of transactions can be monitored to detect a change in intensity, see for example Zhang, Russell and Tsay (2001) and Sonesson and Bock (2003). Often so-called marks (for example the volume traded) are available (see, for example Dufour and Engle 2000), which again motivates multivariate surveillance. For reviews on multivariate surveillance, see Wessman (1999) and Frisén (2003). A general type of indicator is the residuals of a time series model where the change is in the stochastic properties of the residuals. In order to focus on the decision system we use an anonymous indicator X. In finance, turning points are of special interest, since it is profitable to sell at highs and buy at lows. The time of the turn is denoted τ, which is a discrete-valued random variable with probability function $\pi_i = \mathrm{P}(\tau = i)$, for $i = 1, 2, \ldots$. When a distribution of τ is needed, a geometric distribution on $\{1, 2, \ldots\}$ is used. When necessary a simple standard model conditioned on $\tau = i$ is used

$$X(t) = \mu^i(t) + \varepsilon(t), \tag{3.1}$$

where $\mu^i(t) = E[X(t)|\tau = i]$ is the trend cycle and $\varepsilon(t) \sim$ iid N[0, σ^2], $t = 1, 2, \ldots$ conditionally on τ. The index i in $\mu^i(t)$ is suppressed when obvious. Model (3.1) may be too simple for some financial data. However, the model is often used, and here it is used to emphasize the inferential issues of on-line turning point detection. The optimal methods are derived for the simple model, but they are evaluated under realistic conditions in Section 3.4. The vector μ is determined by τ. At a peak we have

$$\begin{cases} \mu(1) \leq \cdots \leq \mu(t), & t \leq \tau \\ \mu(1) \leq \cdots \leq \mu(\tau) \text{ and } \mu(\tau+1) \geq \cdots \geq \mu(t), & t > \tau. \end{cases} \tag{3.2}$$

In the second row of (3.2), at least one inequality should be strict in the second part. The left-hand side set or the right-hand side set can be empty for some values of τ.

One parametric specification of μ is a piecewise linear regression,

$$\mu(t) = \begin{cases} \beta_0 + \beta_1 \cdot t, & t \leq \tau \\ \beta_0 + \beta_1 \cdot \tau + \beta_2 \cdot (t - \tau), & t > \tau \end{cases} \tag{3.3}$$

where $\beta_1 \geq 0$ and $\beta_2 \leq 0$.

For a differentiated process $(Y(t) = X(t) - X(t-1))$, a specification of $\mathrm{E}[Y(t)] = \mu_Y(t)$ is

$$\mu_Y(t) = \begin{cases} \beta_1, & t \leq \tau \\ \beta_2, & t > \tau. \end{cases} \tag{3.4}$$

The assumption in (3.4) is used by e.g. Layton (1996), Ivanova, Lahiri and Seitz (2000), and Layton and Katsuura (2001). For a differentiated process with expected values as in (3.4), the undifferentiated process can be either independent over time with an expected value as in (3.3) or a random walk with drift.

At each decision time s, a system is used to decide whether data indicate a turn in μ (at an unknown time τ) in the time period $\{1, 2, \ldots, s\}$ or not. Thus, the aim is to discriminate between $\{\tau \leq s\}$ and $\{\tau > s\}$. Sometimes this is expressed as discriminating between one process before the shift (when $\tau > s$) and another process after the shift (when $\tau \leq s$). The following notation can be used for the situation in (3.3):

$$X(t) = \begin{cases} Z(t) = \beta_0 + \beta_1 \cdot t + \varepsilon(t), & t \leq \tau \\ Z(t) + (\beta_2 - \beta_1) \cdot (t - \tau), & t > \tau \end{cases} \tag{3.5}$$

where $\beta_1 \geq 0$ and $\beta_2 \leq 0$. In the expression (3.5), $Z(t)$ might be named the in-control process. In this chapter, however, we will explicitly state both processes. Since the change of interest here is a turning point in μ, we will be expressing the change by $\mu(t)$, as is done in (3.2) and (3.3).

3.3 Statistical surveillance and strategies suggested for technical analysis

3.3.1 Measures for evaluations

When evaluating systems for on-line detection of changes it is important to consider the timeliness of the alarms. In finance, the expected return from investments is often used as a measure of performance. The return is often defined as

$$r(t) = x(t) - x(0), \tag{3.6}$$

where X is the (logarithm of the) price, and measures the return of buying at $t = 0$ and selling at time t.

The expected return $E[r(t_A)]$ is maximized when $E[X(t_A)]$ is maximized (t_A denotes the time of alarm). This occurs when we call a sell alarm at the peak, in this case at $t_A = \tau$. Maximizing $E[r(t_A)]$ is equivalent to minimizing $|t_A - \tau|$. This is also the aim in statistical surveillance.

The probability of successful detection used for example by Frisén (1992) and Frisén and Wessman (1999) measures the ability to detect a change within m time units from τ,

$$\mathrm{PSD}(m, i) = \mathrm{P}(t_A - \tau \leq m | t_A \geq \tau, \tau = i). \tag{3.7}$$

The return in (3.6) is measured along the log-price scale, whereas $\mathrm{PSD}(m, i)$ in (3.7) is measured along the time scale. Consider the expected difference $\delta(t_A, i) = \mathrm{E}[r(t_A) - r(\tau)|\tau = i]$. When $\mathrm{E}[X(t)]$ is piecewise linear, then $\mathrm{P}(\delta(t_A, \tau) \geq m \cdot \beta_2 | t_A \geq \tau, \tau = i)$ is equivalent to $\mathrm{PSD}(m, i)$, where β_2 is the post-turn slope.

When we consider transaction costs, we have a penalty for each alarm, which favours infrequent trading. The influence of transaction costs was discussed by Lam and Wei (2004).

In the specification of utility by Shiryaev (1963),

$$u(t_A, \tau) = \begin{cases} h(t_A - \tau), & t_A < \tau \\ a_1 \cdot (t_A - \tau) + a_2, & t_A \geq \tau, \end{cases} \tag{3.8}$$

the gain of an alarm is a linear function of the expected delay. The loss of a false alarm is an arbitrary function of the same difference. A specification of h which is of interest in finance gives

$$u(t_A, \tau) = \begin{cases} b_1 \cdot (t_A - \tau) + b_2, & t_A < \tau \\ a_1 \cdot (t_A - \tau) + a_2, & t_A \geq \tau, \end{cases}$$

where $b_1 > 0$, $a_1 < 0$ and where b_2 and a_2 would depend, for example, on the transaction cost. When $b_2 = a_2 = 0$ and $\mathrm{E}[X(t)]$ is piecewise linear it follows that $\mathrm{E}[u] = \mathrm{E}[r(t_A)] - \mathrm{E}[r(\tau)]$, where the expectation is taken with respect to the disturbance. Thus, maximizing the expected utility is the same as maximizing the expected return.

Also other special cases of (3.8) are of interest. If the function h in (3.8) is set as a constant and $a_2 = 0$, when the utility is linear combination of $P(t_A > \tau)$

and $E(t_A - \tau)^+$. Shiryaev (2004) demonstrated that maximizing this utility is equivalent to minimizing the expected value of $|t_A - \tau|$ as considered by Karatzas (2003) for the case of a Wiener process and an exponential distribution of τ.

In statistical surveillance the type I error is usually characterized by the average run length until the time of alarm, t_A, at no change, $ARL^0 = E[t_A|\tau = \infty]$. A widely used optimality criteria in the literature on quality control is the minimal $ARL^1 = E[t_A|\tau = 1]$ for a fixed ARL^0. Drawbacks with this criterion are discussed by Frisén (2003). ARL^1 only considers immediate changes ($\tau = 1$), whereas the conditional expected delay, $CED(i) = E(t_A - \tau|t_A \geq \tau = i)$, considers different change points. A minimax criterion is the minimum of the maximal $CED(i)$, with respect to $\tau = i$ and $X_{\tau-1}$, where $X_{\tau-1} = \{X(1), X(2), \ldots, X(\tau - 1)\}$.

3.3.2 Methods

Recall that the aim in statistical surveillance is to detect a turn in μ as soon as possible after it has occurred. Here we introduce some notation: the event $\{\tau \leq s\}$ is denoted by $C(s) = \{\bigcup_{i=1}^{s} C_i\}$, where $C_i = \{\tau = i\}$, and the event $\{\tau > s\}$ is denoted by $D(s)$. We want to discriminate between $C(s)$ and $D(s)$ and this discrimination is done by means of a surveillance system, consisting of an alarm statistic and an alarm limit.

3.3.2.1 Bench mark: The full likelihood ratio method

The full likelihood ratio method (LR) is optimal in terms of (3.8) and fulfils several other optimality criteria, see for example Frisén (2003), and serves here as a benchmark. The alarm criterion of the LR method is

$$\frac{f_{X_s}[x_s|C(s)]}{f_{X_s}[x_s|D(s)]} = \sum_{i=1}^{s} w(i) \cdot L(s, i) > g_{LR}(s), \tag{3.9}$$

where $L(s, i) = f_{X_s}(x_s|\mu = \mu^{C_i})/f_{X_s}(x_s|\mu = \mu^{D})$ is the partial likelihood ratio when $\tau = i$, and $w(i) = P(\tau = i)/P(\tau \leq s)$. The vector μ is on the form μ^{C_i} when $\tau = i$ (for state C_i) and on the form μ^D when $\tau > s$ (for state $D(s)$). Thus, the vector μ is known given the state, but the state is random.

A likelihood ratio method based on a small change intensity ($\nu = P(\tau = i|\tau \geq i) \to 0$) is the Shiryaev–Roberts (SR) method (Shiryaev 1963; Roberts 1966). In a Bayesian framework the SR method can be seen as based on a noninformative generalized prior for the change time since equal weights are used for all components in the likelihood.

3.3.2.2 Turn detection and the SRnp method

When discussing 'sign prediction', Dewachter (1997) argued that it is the direction of the evolution that is important in finance. Frisén (1994) suggested a surveillance approach which is nonparametric in the sense that it is not based on a parametric curve shape but based only on the monotonicity and unimodality restrictions in (3.2). Combined with the weights of the SR method, this is the SRnp method (with g_{SRnp} as a constant),

$$\frac{\sum_{i=1}^{s} f_{X_s}(x_s|\mu = \hat{\mu}^{C_i})}{f_{X_s}(x_s|\mu = \hat{\mu}^{D})} > g_{\text{SRnp}}. \tag{3.10}$$

The vector $\hat{\mu}^D$ is the estimate of μ under monotonicity restriction D (no turn), and $\hat{\mu}^{C_i}$ is the estimate under restriction C_i (turn at i). The estimators give maximum likelihood when the disturbance has a Gaussian distribution (Frisén 1986). SRnp was described and evaluated in Andersson (2002, 2006), and Andersson, Bock and Frisén (2005, 2006). So far it has not been used as a financial trading rule, but its possible application for this purpose will be examined in Section 3.4.

3.3.2.3 Forecasts and the Shewhart method

Modelling of the financial process as a base for the trading strategy is important. The modelling can be used to forecast the next value, and the difference between the forecast and the last value can then be used in a trading rule (see for example Franke 1999 and Neely and Weller 2003).

The Shewhart method signals an alarm as soon as the last partial likelihood ratio $L(s, s)$ exceeds a constant g, i.e. for the model in (3.1) we have

$$x(s) - \mu^D(s) < g_{\text{Shewhart}}. \tag{3.11}$$

The method is optimal in terms of (3.8) when $C(s) = \{\tau = s\}$.

3.3.2.4 CUSUM and Filter rules

The CUSUM method of Page (1954) gives a signal as soon as the maximum of the partial likelihood ratios $L(s, i)$, exceeds a limit. For an independent Gaussian process Y, with constant μ_Y^D and μ_Y^C as in (3.4), the alarm criterion for a downward shift is

$$\sum_{j=s-i+1}^{s} (y(j) - \mu_Y^D) < -(g_{\text{CUSUM}} + k \cdot i), \tag{3.12}$$

for some $i = 1, 2, \ldots, s$ where g_{CUSUM} is a constant. The optimal value of k can be expressed as $(\mu_Y^D - \mu_Y^C)/2$. The CUSUM method satisfies the minimax criterion in Section 3.1.

Lam and Yam (1997) propose a generalized filter rule (GFR). At decision time s, a peak signal for the process X is given when

$$\frac{x(s) - x(s-i)}{x(s-i)} < -(g_{\mathrm{GFR}} + k_{\mathrm{GFR}} \cdot i) \tag{3.13}$$

for some $i = 1, 2, \ldots, s$, where g_{GFR} and k_{GFR} are chosen constants. We will now show that GFR is approximately equivalent to the CUSUM method. If we let $Y(t) = \ln X(t) - \ln X(t-1)$, the alarm criterion of the CUSUM method can be expressed as

$$\frac{x(s) - x(s-i)}{x(s-i)} < \exp\{-g_{\mathrm{CUSUM}}\} \cdot \exp\{-(k - \mu_Y^D \cdot i)\} - 1. \tag{3.14}$$

Let $\exp\{-g_{\mathrm{CUSUM}}\} = (1 + g_{\mathrm{GFR}})^{-1}$ and $\exp\{-(k - \mu_Y^D) \cdot i\} = (1 + k_{\mathrm{GFR}})^{-i}$. We approximate $(1 + k_{\mathrm{GFR}})^{-i}$ by $(1 - k_{\mathrm{GFR}} \cdot i)$ and $(1 + g_{\mathrm{GFR}})^{-1}$ by $(1 - g_{\mathrm{GFR}})$. If we make these substitutions in (3.14), then

$$\frac{x(s) - x(s-i)}{x(s-i)} < (1 - g_{\mathrm{GFR}}) \cdot (1 - k_{\mathrm{GFR}} \cdot i) - 1.$$

If $g_{\mathrm{GFR}} \cdot k_{\mathrm{GFR}} \cdot i \approx 0$, then (3.14) approximately equals (3.13). Thus, it follows that GFR is approximately minimax optimal under certain conditions.

A special case of GFR is obtained when $k_{\mathrm{GFR}} = 0$. This is the widely used filter rule, FR (Alexander 1961; see also Taylor 1986 and Sweeney 1986). According to Lam and Yam (1997), the FR calls a peak alarm when

$$\frac{\max_{t \le s}\{x(t)\} - x(s)}{\max_{t \le s}\{x(t)\}} > g_{\mathrm{FR}}$$

where g_{FR} is a constant. Lam and Yam showed that FR, used on $X(t)$, is equivalent to a special case of CUSUM used on $Y(t) = \ln X(t) - \ln X(t-1)$. Hence, when $Y(t)$ is independent with a Gaussian distribution and $\mu_Y^D = (\mu_Y^D - \mu_Y^C)/2$, FR has the same properties as the CUSUM method. The $\mu_Y^D = (\mu_Y^D - \mu_Y^C)/2$ implies that $\mu_Y^C = -\mu_Y^D$, i.e. symmetry. The performance of the CUSUM depends on which shift size measured by $(\mu_Y^D - \mu_Y^C)$ the method is designed to detect, and thus the FR method is minimax optimal for a symmetric turn. The FR is the same as the trading range break.

The performances of GFR and FR were evaluated by Lam and Yam (1997) for different combinations of g_{CUSUM} and k, using 24 years of daily data on the Hang Seng Index. For some combinations, GFR had a better return than FR. In the case of GFR, however, no discussion was made regarding the relation between k and the size of the shift.

3.3.2.5 Moving averages in surveillance and finance

The moving average method of surveillance (see for example Schmid and Okhrin 2003) gives an alarm when

$$\sum_{i=s-m+1}^{s} (x(i) - \mu^D(s)) < g_{\mathrm{MAsur}}, \tag{3.15}$$

where m is the window width and g_{MAsur} is a constant. It can also be defined as giving an alarm when the partial likelihood ratio $L(s, s-m)$ exceeds a constant. When $C(s) = \{\tau = s - m + 1\}$, the moving average method with window width m is optimal (Frisén 2003). For an independent process with a Gaussian distribution, the alarm criterion by this definition coincides with (3.15).

Moving average rules (several variants have been suggested) may be the most commonly discussed trading rule. The rule used by e.g. Neftci (1991), Brock *et al.* (1992), and Neely (1997), calls a peak alarm (sell signal) as soon as the difference between two overlapping moving averages is below a limit:

$$\frac{1}{m} \cdot \sum_{i=s-m+1}^{s} x(i) - \frac{1}{n} \cdot \sum_{i=s-n+1}^{s} x(i) < g(s)_{\mathrm{MAR}}, \tag{3.16}$$

where the narrow window has width m and the wide window has width n. The alarm limit is usually set to zero. Dewachter (1997, 2001) referred to this rule as the 'oscillator rule'. By expressing (3.14) as $\sum_{i=s-m+1}^{s}[x(i) - \hat{\mu}^D(s)] < g'(s)_{\mathrm{MAR}}$, where

$$\hat{\mu}^D(s) = \frac{\sum_{i=s-n+1}^{s-m} x(i)}{n-m}$$

and $g'(s)_{\mathrm{MAR}} = m \cdot n \cdot g(s)_{\mathrm{MAR}}/(n-m)$, we see that (3.16) is the surveillance method in (3.15) with $\mu^D(s)$ replaced by the moving average of $(n-m)$ past observations. A special case of (3.16), which is often considered, is when $m = 1$,

$$x(s) - \frac{1}{n} \cdot \sum_{i=s-n+1}^{s} x(i) < g(s)_{\mathrm{MAR}}. \tag{3.17}$$

By expressing (3.17) as $x(s) - \hat{\mu}^D(s) < g'(s)_{\mathrm{MAR}}$, where

$$\hat{\mu}^D(s) = \frac{\left(\sum_{i=s-n+1}^{s-1} x(i)\right)}{(n-1)}$$

and

$$g'(s)_{\mathrm{MAR}} = \frac{n}{n-1} \cdot g(s)_{\mathrm{MAR}},$$

we see that (3.17) is the Shewhart situation in (3.11).

The optimality of (3.16) and (3.17) is not so clear-cut, as $\mu^D(s)$ is estimated. Andersson and Bock (2001) demonstrated that in the case of cyclical processes, a moving average does not always preserve the true time of the turning point. This causes a delay of the signal.

Another method based on moving averages is the EWMA (exponentially weighted moving averages) method. The optimality of this method is analysed by Frisén and Sonesson (2006).

3.3.2.6 Rules based on hidden Markov models

In a hidden Markov model (HMM) the process has different properties for different states and a first-order time-homogenous Markov process governs the switching between the states. Examples of the use of HMM in finance are given by Marsh (2000) and Dewachter (1997, 2001). In financial applications the hidden Markov chain often have two states, $J(t) = \{1, 2\}$, the expansion and recession phase. The process depends on the state, so that for (3.4) we have $\mathrm{E}[Y(t)|J(t) = i] = \beta_i$. Sometimes a HMM is referred to as a Markov-switching or regime switching model.

A natural alarm statistic is based on the one-step-ahead predicted expected value of the differentiated process Y conditional on past values, with alarm rule

$$\begin{aligned} \mathrm{E}[Y(s+1)|y_s] = \mathrm{P}(J(s+1) = 1|y_s) \cdot \\ \beta_1 + \mathrm{P}(J(s+1) = 2|y_s) \cdot \beta_2 < c, \end{aligned} \tag{3.18}$$

where $\mathrm{P}(J(s+1) = 1|y_s) = p_{11} \cdot \mathrm{P}(J(s) = 1|y_s) + p_{21} \cdot \mathrm{P}(J(s) = 2|y_s)$. This is hereafter denoted the HMR method (hidden Markov rule). The alarm limit c in (3.18) is usually set to zero. A related statistic is the posterior probability, $\mathrm{P}(J(s) = 2|x_s)$ (see Hamilton 1989). The posterior probability was also used by Rukhin (2002) in a retrospective setting for estimating the change point time.

Frisén and de Maré (1991) showed that when the two states are complements to each other, rules based on the posterior probability are equivalent to the LR method in (3.9). Thus, the HMR method is ED optimal. However, when HMR and LR have different aims (classification and change detection respectively), the methods imply different properties (see Andersson *et al.* 2005).

3.3.2.7 The Zarnowitz–Moore method and multivariate surveillance

The methods by Zarnowitz and Moore (1982) are explicitly stated as sequential signal systems for business cycles but have also been used as trading rules for financial series by Boehm and Moore (1991) and Moore, Boehm and Banerji (1994). Their methods are multivariate (with $X(s)$ as the process of interest and $L(s)$ as a one-dimensional leading index) and only utilize information from before the decision time s by a rule of 'natural ordering' of statements. This method can be regarded as a multivariate Shewhart method, since only the last observation is used for the alarm. For reviews on multivariate surveillance, see e.g. Wessman (1999), Ryan (2000), Andersson (2007), and Sonesson and Frisén (2005).

3.3.2.8 Methods for statistical surveillance used as trading rules in finance

Lam and Yam (1997) claimed to be the first to discuss using methods for statistical surveillance as financial trading rules. After that, Yashchin, Philips and Stein (1997) and Philips, Yashchin and Stein (2003) advocated the use of the CUSUM method for this purpose Beibel and Lerche (1997) and Shiryaev (2002) used the theory of optimal surveillance to derive trading rules for continuous time processes, and Schmid and Tzotchev (2004) used different types of EWMA methods to detect changes in a discrete time interest rate model.

Severin and Schmid (1998, 1999), Schipper and Schmid (2001a, 2001b) compared the performance of different versions of the CUSUM, EWMA, and Shewhart methods with respect to detecting changes in GARCH processes (generalized autoregressive conditional heteroscedasticity), which are used to describe volatility in financial markets. Whereas Schipper and Schmid (2001b) aimed at detecting a change in the variance of a GARCH process, the aim in Schipper and Schmid (2001a) was to simultaneously detect an additive outlier and a changed variance. In Severin and Schmid (1998, 1999), the CUSUM, EWMA, and Shewhart methods were compared with respect to a change in the mean of an ARCH(1) process Schipper and Schmid (2001b) used the CUSUM and EWMA methods on the following indicators: the squared observations, the logarithm of the squared observations, the conditional variance, and the residuals of a GARCH model. Schipper and Schmid (2001a) used the EWMA method on each of these indicators to monitor the level of the process. Here the ARL^1 for a fixed ARL^0 was used as evaluation measure.

Śliwa and Schmid (2005) suggested methods for monitoring of the cross-covariances of a multivariate ARMA process to monitor data on the Eastern European stock markets by different EWMA methods.

Steland (2002, 2003) monitor both a GARCH and an independent process with a Gaussian distribution for a change in the drift. The indicator under surveillance is a nonparametric kernel estimator of μ, and only the latest estimated value is used in the surveillance (a Shewhart-type method). The performance is measured by ARL^1 for a fixed ARL^0. In Steland (2003) results are given for the tail behaviour of the false alarm distribution and an upper bound for the delay of motivated alarms. A similar approach was used in Steland (2005) to monitor a smooth but nonlinear change.

Blondell, Hoang, Powell and Shi (2002) suggested a CUSUM method with re-estimation of parameters for the detection of the turns in a cyclical mean level with volatility regime shifts of financial time series.

3.4 Illustration by data on the Hang Seng Index

A common way to evaluate decision rules in financial literature is using one or several case studies. Here the case studies will not be used to decide which method is the best, but instead to illustrate several difficulties with evaluation by case studies. Some of the methods described above are applied to data on the Hang Seng index (HSI), which was also used by Lam and Yam (1997). HSI is a market-value weighted index of the stock prices of the 33 largest companies on the Hong Kong stock market. The weight each stock is assigned in the index is related to the price of the stock. HSI can thus be seen as the price of a stock portfolio. Usually, as here, the values reported are the logarithms of the prices. The values of HSI for the period from February 10, 1999 to June 26, 2002 are shown in Figure 3.1. The series is divided into two periods so that several aspects of methods and evaluations can be illustrated. All days are not trading days and the trading days are numbered consecutively. Period I goes from February 10, 1999 to May 28, 1999 (day 0–71). Period II goes from May 31, 1999 to June 26, 2002 (day 72–828).

3.4.1 Specification of the trading rules

The statistical surveillance methods that will be evaluated are SRnp (alarm rule (3.10)) and CUSUM (alarm rule (3.12)). The CUSUM method is very similar to the financial trading rule GFR in (3.13), as was discussed in Section 3.3.2.4. The HMM-based method to be evaluated is HMR (alarm rule (3.18)).

3.4.1.1 Parameter values

The different methods need different parameters in the alarm statistic, such as the trend cycle ($\mu(t)$), the variance (σ), and the transition probabilities (p_{11}

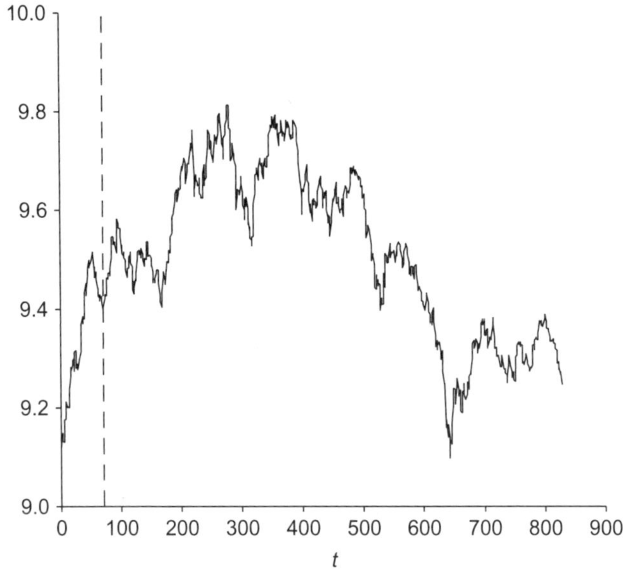

Figure 3.1 Daily observations of HSI from February 10, 1999 to June 26, 2002. The limit between the periods (I and II) is marked with a dashed vertical line.

and p_{22}). The data from Period I (the first peak) will be used to estimate the parameters in question.

For SRnp, the estimation of μ is nonparametric and noninformative weights are used for the intensity. Thus, it suffices to estimate the variance of the process.

For CUSUM(opt), the optimal CUSUM in (3.12), we obtain the value of k from the estimated parameters of μ_Y^C and μ_Y^D. For CUSUM, we also use another set of parameters. In Lam and Yam (1997) the value of k was determined from a long period to yield maximum profit. CUSUM(L&Y best) is the CUSUM method with their best values of the alarm limit and k.

In HMR, we need estimates of the transition probabilities p_{12} and p_{21}, in addition to the parameter estimates mentioned earlier. Here we use the maximum likelihood estimates under assumption of constant transition probabilities.

3.4.1.2 Controlling false alarms

In surveillance, the alarm limit is often determined so as to get control over false alarms, but also other approaches are possible.

In Lam and Yam (1997) no discussion was made regarding the false alarm rate. Instead, the alarm limit is determined to yield maximum profit in a long period. This alarm limit is used in the CUSUM(L&Y best) method. For the SRnp and CUSUM(opt) methods, the limits are determined to give a maximal total return for Period II.

The limit zero in the HMR alarm rule (3.18) is often described as a 'natural' limit and used in the evaluations below. However, other limits have been used with reference to transaction costs.

In a practical online monitoring situation, where the process is continually observed, the parameters must be estimated by observations that are available at the current time point. We illustrate this aspect by not re-estimating the parameters when evaluating the methods with respect to Period II. Thus, the parameters estimated by observations from Period I are also used in Period II.

3.4.2 Results

After having bought the stock at the start, we sell it at the first sell-signal, i.e. at the alarm for a peak. After a peak alarm is given, the aim is to detect a forthcoming trough in order to buy one unit, and so on. We assume that it is possible to buy or sell at the price at the alarm time t_A. For all methods except HMR, only observations past the previous alarm are used. For HMR all past observations within the evaluation period are used for the updating of the posterior probability. There are several aspects to consider when evaluating a monitoring system, among them timeliness. Direct measures such as those described in Section 3.1 require a precise definition of what is a turning point. For this series of observations we have not used that approach. Instead, timeliness can be measured indirectly, as the amount gained by detecting the change at the 'right' time. Such a measure is the return (3.6) which is used here.

Transaction costs and interest rate, earned by having money in the bank, are not reported here. Their exact values depend on several circumstances, and for this series they would have negligible impact compared with the stock return. Transaction costs are proportional to the number of transactions and, as seen in Table 3.2 below, the SRnp method and the HMR method would have the largest transaction costs while these would be less for the CUSUM method. Also the earned interest rate would be disadvantageous for the HMR method, as the periods when the asset is not held are shorter than for the other methods. However, both effects are too small to have any substantial influence on the comparison of returns.

Table 3.1 Summary of the returns of Period I, February 10, 1999 to May 28, 1999.

Method	Time of sell-signal	Return
SRnp	23, 27, 56	0.297
CUSUM (opt)	56	0.359
CUSUM (L&Y best)	58	0.350
HMR	62	0.327

3.4.2.1 Period I

Period I (day 0–71) is used for estimation, and therefore the illustration of the methods as regards this period can be seen as an in-sample performance. In Table 3.1 the returns for the alarm systems (CUSUM(opt), CUSUM(L&Y), HMR and SRnp) are reported. SRnp which is the only method which does not have the advantage of using in-sample estimation has less return than the others. The CUSUM(opt) has a slightly better return than the CUSUM(L&Y best) due to parameter values which are exactly adopted to that period. The HMR method yields a smaller return than the CUSUM methods because the alarm is given late.

3.4.2.2 Period II

In the comparison of the methods for Period II (day 72–828), the parameter estimates obtained in Period I are used again.

In Period II, when the parametric methods use parameter estimates which do not agree so well with the data in this period, the return is smaller than for SRnp, which is nonparametric and thus does not have the problem of being mis-specified. The CUSUM(opt) which use estimated for Period I has less return than CUSUM(L&Y) which has parameters estimated from a longer period. The HMR method is very different from the others since it does not require strong evidence for an alarm but classifies into either state. On a plateau (where the return is almost independent of any trading strategy) the expected value in (3.18) fluctuates around zero, i.e. the HMR alarm statistic fluctuates around its alarm limit which yield frequent buy and sell signals. On long and monotone upward (or downward) stretches, however (where one would get the best return by not doing any transactions), the expected value is far from zero and no alarm is given for small fluctuations. The returns for Period II are reported in Table 3.2. Since the number of transactions differs much between

Table 3.2 Summary of the return of Period II, May 31, 1999 to June 26, 2002.

Method	# of sell signals	Return	
		without transaction cost	with 0.1% transaction cost
SRnp	49	0.319	0.221
CUSUM (opt)	27	0.150	0.096
CUSUM (L&Y best)	24	0.197	0.149
HMR	41	0.300	0.218

methods, we also report a case with transaction cost of 0.1%. The ordering between methods is kept but the differences are less. For a larger transaction cost, the methods with few transactions will have a better return than the others.

3.4.2.3 Case studies versus Monte Carlo methods

The use of a single case study for evaluation can be questioned. How well does the actual data set represent the process of interest? How extreme is the outcome of the process? If it is very extreme, then the resulting alarm time is very rare. However, it is impossible to know whether an outcome is extreme or not, unless several examples are available or the process is replicated. If we perform a study where we replicate the process by Monte Carlo methods, we can make statements about the properties of a method under the assumptions made in that study.

3.4.3 Robustness of alarm systems used in finance

The interpretation of results of case studies should be made with care, and we will now point out some sensitive issues.

3.4.3.1 Transition probabilities

The HMR method depends on the value of the probabilities of staying in the current state (p_{11} for expansion and p_{22} for recession, respectively). In the HMR method, as well as in most HMM-based methods (for example Dewachter 1997, Hamilton 1989, and Marsh 2000), constant transition probabilities are used. Another approach is to allow duration dependence, which means that the intensity changes with time spent in the regime (for example Maheu and

McCurdy 2004). When only uncertain information about the intensity is available, noninformative weights for the change point time prevents the risk of serious mis-specification. This approach is used by the SRnp method.

3.4.3.2 Alarm limits

For the HMR method, the alarm limit determines the possible lack of symmetry between the two states (expansion and recession). For the other methods it determines the amount of evidence required for any action. In this latter situation, the alarm limit determines the sensitivity of the alarm statistic. A low sensitivity can be motivated by a wish to decrease the frequency of false alarms and thus the transaction costs. In the study above, transaction costs were small compared with the return.

3.4.3.3 Slopes within the phases

Marsh (2000) addressed the problem of instability of the estimates of β_1 and β_2 over different time periods. In Figure 3.1 it is seen that the cycles vary a lot. It was shown in Andersson, Bock and Frisén (2004) that even a small mis-specification of μ results in long delays for early turns.

When the parameters are determined from the same data that are being monitored, as in Period I, the methods which use these estimates give better returns than the nonparametric SRnp method (Table 3.1). However, when these estimates are used in Period II, the situation is quite different. The nonparametric method SRnp performs best, even though very few assumptions are made. The other methods, however, which all assume known parameter values, perform badly in some periods where the parameters differ much from the true ones, and the total return is not so impressive for Period II (see Table 3.2) as for Period I (see Table 3.1). When no assumption is made regarding the parametric form of the curve, as in the SRnp method, the surveillance system is robust against changes in the characteristics of the curve over time.

The optimal value for k in CUSUM is $(\mu_Y^D - \mu_Y^C)/2$. The other parameter of the alarm limit, g_{CUSUM}, can be determined to regulate the false alarm property Lam and Yam (1997) do not discuss using the optimal criterion for k. Instead, the maximum return is considered for combinations of both parameters (k and g). For Period II we conclude that the value of k, estimated from Period I, does not fit whereas the Lam and Yam value of k fits better. This underlines once more the risks of parametric assumptions.

3.4.3.4 Autocorrelation

In statistical surveillance there are several ways of dealing with dependent data (see Pettersson 1998 and Frisén 2003 for reviews).

The methods discussed in this chapter do not take account of any dependency structure in the alarm statistics. A dependency structure of the disturbance term appears to be present, however, and a first-order autoregressive process describes the disturbance term well. Lam and Yam (1997) calculated the ARL^0 analytically, but the possibility of a dependency structure was not discussed. Many methods are constructed under the assumption of independent observations. As a result, they are useful if the dependency is slight but might be misleading if the dependency is strong. The effects of such mis-specification are studied for example by Kramer and Schmid (1997) and Schmid and Okhrin (2003).

One approach is to use a method designed for independence but to adjust the alarm limit so that the false alarm property is correct for the dependent process. Such methods are often referred to as 'modified control charts' (Kramer and Schmid 1997). This may work well but is not optimal.

Another approach is to calculate the residuals of an estimated model of the dependency structure and then monitor these. Such methods are often referred to as 'residual charts' (Kramer and Schmid 1997). It was demonstrated by Pettersson (1998) that this results in an approximation of the full likelihood ratio method.

A general approach in statistical surveillance is to eliminate an unknown parameter by a pivot statistic. In finance it is common to differentiate the observations in order to eliminate the dependency. This is done under the assumption of a random walk. If the undifferentiated process is independent, however, the first difference Y will be a MA(1) process.

As is done in Section 3.2.1 for an independent process, it is possible to derive the likelihood ratio statistic for a process with a dependency structure. The LR statistic for a change in the mean of a stationary AR(1) process with normally distributed disturbances was derived in Pettersson (1998). The partial likelihood ratio $L(s, i)$ for a change in the mean of a MA process where the disturbance term has a Gaussian distribution is derived in Petzold, Sonesson, Bergman and Kieler (2004).

Further improvements can be expected by using methods of surveillance that take the dependency structure into account.

3.5 Concluding remarks

A desirable property of any prospective monitoring method is that a change is detected quickly (timeliness) and without too many unmotivated signals of change (safety). In the general theory of statistical surveillance, the aim is to optimize the method with respect to these properties. Thus, the aims of trading rules in finance agree with those of statistical surveillance.

The common strategies suggested for financial trading decisions, which have been investigated in this chapter, are all special cases of well-known methods of statistical surveillance.

The filter rule (or trend breaking rule) FR has been shown to be a special case of the CUSUM method. GFR is a close approximation of CUSUM that does take the aspect of the relation between the slopes of the upward and downward trends into account. It is well known that the CUSUM method has minimax optimality properties. Thus, the FR and GFR methods will have approximate minimax properties under certain conditions.

Methods based on moving averages are common and relatively simple to construct. One method, the oscillator method, which is widely used for trading, consists of comparing two overlapping moving averages of different lengths. In this chapter it has been demonstrated that this approach is equivalent to a moving average method described in the literature of surveillance, where the in-control mean is estimated by a moving average. A special case often considered is when the current observation is compared with a simple moving average of observations. It has been shown that this method is similar to the Shewhart method of surveillance where, again, the in-control mean is estimated by a moving average. The optimality properties of these two surveillance methods are known. They have good properties for detecting large recent changes. This does not, however, necessarily imply that the moving average trading rule is optimal, as the result depends on the properties of the moving average estimator. In any case, the method cannot be considered suited to detect small changes which occur gradually, since only the last observations are used.

Several rules for financial trading use a hidden Markov model approach. The inferential structure of the HMR method is equivalent to that of the optimal LR method when the specifications of the states that are to be discriminated between are the same. This requires, however, that we use knowledge of the type of the next turn. Thus, the method is also optimal in the sense that the expected utility is maximized. However, the results from the monitoring also depend on the knowledge of (or the method for estimating) parameters and the distribution of τ.

Since the aim of a financial decision rule generally is to maximize the expected return (adjusted for the risk exposure and transaction costs), return measures are natural to use. Most of the results in the theory of statistical surveillance have been developed solely on the use of the timeliness scale. However, timeliness and return are, as we have demonstrated, closely related.

Improper assumptions regarding the process under surveillance may have large impact on performance. Thus, single case studies are very sensitive to how representative the chosen data are. The application to the Hang Seng index demonstrated for the HMR method a lack of robustness against errors in the estimation of the transition probabilities.

When the distribution of the time of the change τ is unknown, noninformative weights can be used in order to avoid the risk of serious mis-specification. This is what the SRlin and SRnp methods use. One way of avoiding the risk of seriously mis-specifying the regression is to use the nonparametric approach of the SRnp method. The case studies illustrate the advantage of the SRnp method when the current turn has a different shape than the previous one.

Since there are many problems to deal with in the implementation of surveillance in finance, further research is needed. Much effort has been made on the modelling aspect of processes related to finance, such as volatility, the arrival time of transactions, and smooth transitions between regimes. However, much remains to be done as regards the implementation of these models in a decision system, although the most recent research is very promising. The use of knowledge on statistical surveillance for the construction of financial trading rules will certainly be of value.

Acknowledgements

The authors are grateful for valuable comments by Mattis Gottlow and Christian Sonesson. The research was supported by the Swedish Research Council, The Bank of Sweden Tercentenary Foundation, Kungliga and Hvitdfeldtska Stiftelsen, Wilhelm and Martina Lundgrens Vetenskapsfond and the West Sweden Chamber of Commerce and Industry.

References

Alexander, S. S. (1961). Price movements in speculative markets: Trends or random walks. *Industrial Management Review*, **2**, 7–26.

Andersson, E. (2002). Monitoring cyclical processes. A non-parametric approach. *Journal of Applied Statistics*, **29**, 973–990.

Andersson, E. (2006) Robust On-line Turning Point Detection. The Influence of Turning Point Characteristics. *Frontiers in Statistical Quality Control 8*. Eds. Lenz, H.-J. and Wilrich, P.-TH., pp. 223–248.

Andersson, E. (2007). Effect of dependency in systems for multivariate surveillance. Technical Report 2007:1, Statistical Research Unit, Department of Economics, Göteborg University.

Andersson, E. and Bock, D. (2001). On seasonal filters and monotonicity. Technical Report 2001:4, Department of Statistics, Göteborg University.

Andersson, E., Bock, D. and Frisén, M. (2004). Detection of turning points in business cycles. *Journal of Business Cycle Measurement and Analysis*, **1**, 93–108.

Andersson, E., Bock, D. and Frisén, M. (2005) Statistical surveillance of cyclical processes with application to turns in business cycles. *Journal of Forecasting*, **24**, 465–490.

Andersson, E., Bock, D. and Frisén, M. (2006). Some statistical aspects on methods for detection of turning points in business cycles. *Journal of Applied Statistics*, **33**, 257–278.

Beibel, M. and Lerche, H. R. (1997). A new look at optimal stopping problems related to mathematical finance. *Statistica Sinica*, **7**, 93–108.

Black, F. (1976). The pricing of commodity contracts. *Journal of Financial Economics*, **3**, 167–179.

Blondell, D., Hoang, H., Powell, J. G. and Shi, J. (2002). Detection of financial time series turning points: A new approach CUSUM applied to IPO cycles. *Review of Quantitative Finance and Accounting*, **18**, 293–315.

Boehm, E. A. and Moore, G. H. (1991). Financial market forecasts and rates of return based on leading index signals. *International Journal of Forecasting*, **7**, 357–374.

Brock, W., Lakonishok, J. and LeBaron, B. (1992). Simple technical trading rules and the stochastic properties of stock returns. *Journal of Finance*, **47**, 1731–1764.

Dewachter, H. (1997). Sign predictions of exchange rate changes: Charts as proxies for bayesian Inferences. *Weltwirtschaftliches Archiv–Review of World Economics*, **133**, 39–55.

Dewachter, H. (2001). Can Markov switching models replicate chartist profits in the foreign exchange market? *Journal of International Money and Finance*, **20**, 25–41.

Dufour, A. and Engle, R. F. (2000). Time and the price impact of a trade. *Journal of Finance*, **55**, 2467–2498.

Franke, J. (1999). Nonlinear and nonparametric methods for analyzing financial time series. In *Operation Research Proceedings 98*, eds. P. Kall and H.-J. Luethi. Springer-Verlag, Heidelberg.

Frisén, M. (1986). Unimodal regression. *The Statistician*, **35**, 479–485.

Frisén, M. (1992). Evaluations of methods for statistical surveillance. *Statistics in Medicine*, **11**, 1489–1502.

Frisén, M. (1994). Statistical surveillance of business cycles. Technical Report 1994:1 (Revised 2000), Department of Statistics, Göteborg University.

Frisén, M. (2003). Statistical surveillance. Optimality and methods. *International Statistical Review*, **71**, 403–434.

Frisén, M. and de Maré, J. (1991). Optimal surveillance. *Biometrika*, **78**, 271–280.

Frisén, M. and Sonesson, C. (2006). Optimal surveillance based on exponentially weighted moving averages. *Sequential Analysis*, **25**, 379–403.

Frisén, M. and Wessman, P. (1999). Evaluations of likelihood ratio methods for surveillance. Differences and robustness. *Communications in Statistics. Simulations and Computations*, **28**, 597–622.

Hamilton, J. D. (1989). A New approach to the economic analysis of nonstationary time series and the business cycle. *Econometrica*, **57**, 357–384.

Ivanova, D., Lahiri, K. and Seitz, F. (2000). Interest rate spreads as predictors of German inflation and business cycles. *International Journal of Forecasting*, **16**, 39–58.

Jönsson, H., Kukush, A. and Silvestrov, D. S. (2004). Threshold structure of optimal stopping strategies for American type options. I. *Theory of Probability and Mathematical Statistics*, 82–92.

Karatzas, I. (2003). A note on Bayesian detection of change-points with an expected miss criterion. *Statistics & decisions*, **21**, 3–13.

Kramer, H. and Schmid, W. (1997). Control charts for time series. *Nonlinear Analysis*, **30**, 4007–4016.

Lai, T. L. (1995). Sequential changepoint detection in quality control and dynamic systems. *Journal of the Royal Statistical Society B*, **57**, 613–658.

Lai, T. L. and Lim, T. W. (2005). Optimal stopping for Brownian motion with applications to sequential analysis and option pricing. *Journal of Statistical Planning and Inference*, **130**, 21–47.

Lam, K. and Wei, L. (2004). Is the perfect timing strategy truly perfect? *Review of Quantitative Finance and Accounting*, **22**, 39–51.

Lam, K. and Yam, H. C. (1997). CUSUM techniques for technical trading in financial markets. *Financial Engineering and the Japanese Markets*, **4**, 257–274.

Layton, A. P. (1996). Dating and predicting phase changes in the US business cycle. *International Journal of Forecasting*, **12**, 417–428.

Layton, A. P. and Katsuura, M. (2001). A new turning point signalling system using the Markov switching model with application to Japan, the USA and Australia. *Applied Economics*, **33**, 59–70.

Lo, A. W. (2000). Foundations of technical analysis: Computational algorithms, statistical inference, and empirical implementation. *Journal of Finance*, **55**, 1705–1770.

Maheu, J. M. and McCurdy, T. H. (2004). News arrival, jump dynamics, and volatility components for individual stock returns. *Journal of Finance*, **59**, 755–793.

Marsh, I. W. (2000). High-frequency Markov switching models in the foreign exchange market. *Journal of Forecasting*, **19**, 123–134.

Moore, G. H., Boehm, E. A. and Banerji, A. (1994). Using economic indicators to reduce risk in stock-market investments. *International Journal of Forecasting*, **10**, 405–417.

Neely, C. J. (1997). Technical analysis in the foreign exchange market: A layman's guide. *The Federal Reserve Bank of St. Louis Review*, **79**, 23–38.

Neely, C. J. and Weller, P. A. (2003). Intraday technical trading in the foreign exchange market. *Journal of International Money and Finance*, **22**, 223–237.

Neftci, S. N. (1991). Naive trading rules in financial-markets and Wiener-Kolmogorov prediction-theory – A study of technical analysis. *Journal of Business*, **64**, 549–571.

Page, E. S. (1954). Continuous inspection schemes. *Biometrika*, **41**, 100–114.

Pettersson, M. (1998). Evaluation of some methods for statistical surveillance of an autoregressive Process. Technical Report 1998:4, Department of Statistics, Göteborg University.

Petzold, M., Sonesson, C., Bergman, E. and Kieler, H. (2004). Surveillance of longitudinal models. Detection of intrauterine growth retardation. *Biometrics*, **60**, 1025–1033.

Philips, T. K., Yashchin, E. and Stein, D. M. (2003). Using statistical process control to monitor active managers, *Journal of Portfolio Management*, **30**, 86–95.

Roberts, S. W. (1966). A comparison of some control chart procedures. *Technometrics*, **8**, 411–430.

Rukhin, A. L. (2002). Asymptotic behavior of posterior distribution of the change-point parameter. *Journal of Statistical Planning and Inference*, **105**, 327–345.

Ryan, T. P. (2000). *Statistical methods for quality improvement* (2 ed.). John Wiley & Sons, Ltd, New York.

Schipper, S. and Schmid, W. (2001a). Control charts for GARCH processes. *Nonlinear Analysis*, **47**, 2049–2060.

Schipper, S. and Schmid, W. (2001b). Sequential methods for detecting changes in the variance of economic time series. *Sequential Analysis*, **20**, 235–262.

Schmid, W. and Okhrin, Y. (2003). Tail behaviour of a general family of control charts. *Statistics & Decisions*, **21**, 79–92.

Schmid, W. and Tzotchev, D. (2004). Statistical surveillance of the parameters of a one-factor Cox-Ingersoll-Ross model. *Sequential Analysis*, **23**, 379–412.

Severin, T. and Schmid, W. (1998). Statistical process control and its application in finance. In Contributions to Economics: *Risk Measurement, Econometrics and Neural Networks*, eds. G. Bol, G. Nakhaeizadeh and C.-H. Vollmer. Physica Verlag, Heidelberg, pp. 83–104.

Severin, T. and Schmid, W. (1999). Monitoring changes in GARCH processes. *Allgemeines Statistisches Archiv*, **83**, 281–307.

Shiryaev, A. N. (1963). On optimum methods in quickest detection problems. *Theory of Probability and its Applications*, **8**, 22–46.

Shiryaev, A. N. (1999). *Essentials of Stochastic Finance*. World Scientific, Singapore.

Shiryaev, A. N. (2002). Quickest detection problems in the technical analysis of financial data. In *Mathematical Finance – Bachelier Congress 2000*, eds. H. Geman, D. Madan, S. Pliska and T. Vorst. Springer-Verlag, Berlin.

Shiryaev, A. N. (2004). A remark on the quickest detection problems. *Statistics & Decisions*, **22**, 79–82.

Shiryaev, A. N., Kabanov, Y. M., Kramkov, O. D. and Melnikov, A. V. (1994). Toward the theory of pricing of options of both European and American types. I. Discrete time. *Theory of Probability and its Applications*, **39**, 14–60.

Słiwa, P. and Schmid, W. (2005). Monitoring the cross-covariances of a multivariate time series. *Metrika*, **61**, 89–115.

Sonesson, C. and Bock, D. (2003). A Review and discussion of prospective statistical surveillance in public health. *Journal of the Royal Statistical Society A*, **166**, 5–21.

Sonesson, C. and Frisén, M. (2005). Multivariate surveillance. in *Spatial Surveillance for Public Health*, eds. A. Lawson and K. Kleinman. John Wiley & Sons Ltd, New York, pp. 169–186.

Srivastava, M. S. and Wu, Y. (1993). Comparison of EWMA, CUSUM and Shiryayev–Roberts procedures for detecting a shift in the mean. *Annals of Statistics*, **21**, 645–670.

Steland, A. (2002). Nonparametric monitoring of financial time series by jump-preserving control charts. *Statistical Papers, Berlin*, **43**, 401–422.

Steland, A. (2003). Jump-preserving monitoring of dependent time series using pilot estimators. *Statistics & Decisions*, **21**, 343–366.

Steland, A. (2005). Optimal sequential kernel detection for dependent processes. *Journal of Statistical Planning and Inference*, **132**, 131–147.

Sweeney, R. J. (1986). Beating the foreign exchange market. *The Journal of Finance*, **41**, 163–182.

Taylor, S. (1986). *Modelling Financial Time Series*, Vol. 1., John Wiley & Sons, Ltd, Chichester.

Wessman, P. (1999). Studies on the surveillance of univariate and multivariate processes. Doctoral Thesis, Göteborg University, Sweden, Department of Statistics.

Yashchin, E., Philips, T. K. and Stein, D. M. (1997). Monitoring active portfolios using statistical process control. In *Computational Approaches to Economic Problems. Selected Papers from the 1st Conference of the Society for Computational Economics* (Vol. 193–205), ed. H. e. a. Amman. Kluwer Academic, Dordrecht.

Zarnowitz, V. and Moore, G. H. (1982). Sequential signals of recessions and recovery. *Journal of Business*, **55**, 57–85.

Zhang, M. Y., Russell, J. R. and Tsay, R. S. (2001). A nonlinear autoregressive conditional duration model with applications to financial transaction data. *Journal of Econometrics*, **104**, 179–207.

4

Evaluations of likelihood-based surveillance of volatility

David Bock
Statistical Research Unit, School of Business, Economics and Law, Göteborg University, PO Box 660, SE 405 30 Göteborg, Sweden

4.1 Introduction

Detecting changes in the volatility of asset returns are important in, e.g. portfolio management (see Chapter 7). In Chapter 6 and in Severin and Schmid (1998, 1999) and Schipper and Schmid (2001a, 2001b) several surveillance methods were compared with respect to detecting changes in GARCH (generalized autoregressive conditional heteroscedasticity) processes, which are used to describe volatility in financial markets.

The aim of this chapter is to construct and evaluate likelihood based methods for detecting a change in volatility. Many methods for surveillance are in one way or another based on likelihood ratios. The methods have mostly been constructed and evaluated in a situation where the aim is to detect a change in the level of the process. Likelihood ratio based methods are known to possess several optimality properties and the different methods are suitable for different situations. Increasing attention has however been given to the monitoring of the variance (or the standard deviation).

Here an independent Gaussian process is studied, as in most of the literature. Simultaneously surveillance of both the mean and the volatility is treated in, e.g. Knoth and Schmid (2002). The reason for studying an independent Gaussian

Financial Surveillance Edited by Marianne Frisén

process is that no explicit expression for the univariate marginal distribution of a GARCH process is known (Schipper and Schmid 2001b). Constructing the required likelihood is therfore not possible.

The methods studied are the full likelihood ratio (LR), Shiryaev–Roberts (SR), Shewhart and the CUSUM methods, presented in Section 4.4. The methods differ in what way the different partial likelihood ratios are weighted and they depend on different number of process parameters.

These methods were studied in Frisén and Wessman (1999) and Järpe and Wessman (2000) for the same process as here but a change in the level. In case of a change in the variance, earlier studies have been made of the Shewhart (see, e.g. Reynolds and Soumbos 2001), CUSUM (e.g. Srivastava 1997 and Acosta-Mejia, Pignatello and Rao 1999) and SR (Srivastava and Chow 1992) methods.

Different variants of the EWMA (exponentially weighted moving averages) method suggested by Roberts (1959) are often suggested, see, e.g. Crowder and Hamilton (1992), MacGregor and Harris (1993), Acosta-Mejia and Pignatiello (2000) and Schipper and Schmid (2001b). The EWMA method is not exactly likelihood based and is not studied here. However Frisén and Sonesson (2006) demonstrate that it can be seen as an approximation of a LR method.

The performance has often been assessed by the average run length when a change happens either immediately or never. The sole use of these two measures has however been criticized and a single measure of performance is not always enough but evaluations of different properties might be necessary, as pointed out by several authors, e.g. Frisén (1992, 2003). In Frisén and Wessman (1999) and Järpe and Wessman (2000) the methods were made comparable by having the same average run length when there is no change. Here the median run length is used.

The different parameters can be chosen to make the methods optimal for specific situations. Since information on the parameters is rarely known in practice, there is a risk of mis-specification. The effect of mis-specifications on the performance of the methods is studied.

As an illustrative example we monitor a period of Standard and Poor's 500 stock market index to investigate whether our procedures could have detected a documented change in volatility.

The plan of this chapter is as follows. Notation and specifications are given in Section 4.2. Optimality and measures of evaluation are described in Section 4.3. Methods are described in Section 4.4. Results from a simulation study are given in Section 4.5 and in Section 4.6 the methods are applied in a case study. Concluding remarks are given in Section 4.7.

4.2 The change-point problem

The process under surveillance, denoted by X is, as in most literature on quality control, measured at discrete time points, $t = 1, 2, \ldots$ and assumed to be independent Gaussian. Dependent and multivariate processes are treated in Chapter 5. Continuous-time interest rate models are treated in Chapter 8.

Both the situation with subgroups, that is samples of more than one observation are made at each time, and without subgroups, that is a single observation is made at each time, have been treated in the literature. Often both location and dispersion are monitored simultaneously.

Here we have a single observation at each time and at an unknown time point, denoted by τ, there is an increase in the variance;

$$\mathrm{Var}\,[X(t)] = \begin{cases} \sigma^2, & t < \tau \\ \Delta \cdot \sigma^2, & t \geq \tau \end{cases} \tag{4.1}$$

where $\Delta > 1$ is the unknown size of the shift. At time $t < \tau$ and $t \geq \tau$ the process is said to be in-control and out-of-control, respectively.

Sometimes this is expressed as a change from a target process, denoted by $Y(t)$. In such situations, the following notation can be used for the situation in (4.1):

$$\mathrm{Var}\,[Y(t)] = \sigma^2$$

$$X(t) = \begin{cases} Y(t), & t < \tau \\ \sqrt{\Delta} \cdot Y(t), & t \geq \tau \end{cases} \tag{4.2}$$

where $\Delta > 1$. In this chapter, however, we will explicitly state both the target process and the process after the change as in (4.1).

The aim is to detect the change as soon as possible after it has occurred. Only one-sided procedures are considered. Here σ^2 and μ are considered as known and the change point time τ is a discrete-valued random variable with intensity parameter

$$\nu_t = \mathrm{P}(\tau = t | \tau \geq t). \tag{4.3}$$

We treat the case of a constant unknown intensity ν, that is τ has a geometric distribution with density $\mathrm{P}(\tau = t) = \nu \cdot (1 - \nu)^{t-1}$ on $t = 1, 2 \ldots$ as in e.g. Shiryaev (1963) and Frisén and Wessman (1999). Without loss of generality we take $\mu = 0$. In those methods where it is required, the unknown parameters Δ and ν are replaced by values d and v, respectively. The values are chosen to be relevant for the problem at hand and the methods are optimized for these values.

At each decision time $s = 1, 2 \ldots$, we want to discriminate between $C(s)$ and $D(s)$ where $C(s)$ is the critical event implying that the process is out-of-control and $D(s)$ implies that the process is in-control. The $C(s)$ and $D(s)$ can be specified in various ways and different methods are optimal for different specifications. Sometimes it is important to see whether there has been a change since the start of the surveillance and then $C(s) = \{\tau \leq s\} = \left\{\bigcup_{i=1}^{s} C_i\right\}$ where $C_i = \{\tau = i\}$ and $D(s) = \{\tau > s\}$. The conditional variances $\mathrm{Var}[X(t)|\tau = i]$ and $\mathrm{Var}[X(t)|\tau = \infty]$ are denoted by σ_{Ci}^2 and σ_D^2, respectively.

An alarm set $A(s)$ is constructed, with the property that as soon as the vector $\mathbf{X}_s$ belongs to $A(s)$ we infer that $C(s)$ has occurred. Usually the alarm set consists of an alarm statistic $p(\mathbf{x}_s)$ and an alarm limit $g(s)$, where the time of an alarm, t_A, is defined as

$$t_A = \min\{s : p(\mathbf{X}_s) > g(s)\}. \tag{4.4}$$

4.3 Measures of evaluation and optimality criteria

A desirable property of a method is that it detects a change quickly without having too many false alarms. We must however face a trade-off between false alarms and the ability to detect a change. Likewise traditional hypothesis testing optimality of surveillance methods is assessed by the detection ability given a controlled error rate. However, as opposed to hypothesis testing is surveillance characterized by repeated decisions. Consequently, measures such as the significance level and the power need to be generalized to consider the sequential aspect.

Chu, Stinchcombe and White (1996) advocated controlling the probability of any false alarm during an infinitely long surveillance period, $\lim_{i\to\infty} \mathrm{P}(t_A \leq i|D) < 1$. This is convenient since ordinary statements of hypothesis testing can be made. It was however pointed out by Pollak and Siegmund (1975) and Frisén (1994) that the ability to detect a change deteriorates rapidly with the time of the change. Consequences of this were illustrated in Bock (2008).

A commonly used measure to summarize and control the false alarms is by the average run length, $\mathrm{ARL}^0 = \mathrm{E}(t_A|\tau = \infty)$. Hawkins (1992) and Gan (1993) suggest that the control is made by the median run length, $\mathrm{MRL}^0 = \mathrm{Median}(t_A|D)$ as it has easier interpretations for skewed distributions and much shorter computer time for calculations. A third measure is the probability of a false alarm, $\mathrm{PFA} = \mathrm{P}(t_A < \tau) = \mathrm{E}_\tau(\mathrm{P}(t_A < \tau|\tau = t))$, which can be though of as a characteristic for surveillance corresponding to the level of significance for hypothesis testing (Järpe and Wessman 2000).

The timeliness of motivated alarms can be reflected by the average run length given an immediate change, $ARL^1 = E(t_A|\tau = 1)$. This is the most commonly used measure but it is relevant to consider other change point times as well, as will be discussed later. The ability to detect a change within m time units from τ is reflected by the probability of successful detection, $PSD(m, i) = P(t_A - \tau \leq m|t_A \geq \tau, \tau = i)$, $m = 0, 1, \ldots$. It was suggested by Frisén (1992) and is an important measure if there is limited time available for rescuing action, e.g. in the surveillance of the fetal heart rate during labour or intrauterine growth retardation. Another measure is the conditional expected delay $CED(i) = E(t_A - \tau|t_A \geq \tau, \tau = i)$. The delay is summarized with respect to the distribution of τ by $ED = E_\tau\ (ED(\tau))$ where $ED(i) = CED(i) \cdot P(t_A \geq i)$. An important aspect when evaluating a method is the trust you should have in an alarm at a specific time. The predictive value of an alarm at time t, $PV(t) = P(\tau \leq t|t_A = t)$, suggested by Frisén (1992), reflects the trust of an alarm.

The most commonly used optimality criteria is minimal ARL^1 for a fixed ARL^0. In the literature on control charts for the variance, this criterion has been used with only one exception (Hawkins and Zamba 2005). This criterion might be suitable in an industrial manufacturing process where one considers various start-up problems. An advantage is that the criterion does not require an assumption regarding the distribution of τ but Frisén (2003) and Frisén and Sonesson (2006) have questioned it as a formal criterion.

In the utility function suggested by Shiryaev (1963) the gain of an alarm is a linear function of the expected delay. The loss of a false alarm is an arbitrary function of the same difference. The criterion of maximization of the expected utility, where the expectation is taken with respect to τ, is often referred to as the ED criterion (see, e.g. Frisén 2003), since the expected delay is to be minimized. In Chapter 3 it was demonstrated that for certain assumptions regarding the price on assets, fulfilling of the ED criterion is equivalent to maximizing the expected return.

When the worst possible case is important, the minimax criteria of Moustakides (1986) can be used. The criterion is minimal CED given the worst possible value of τ and the worst possible outcome of $\mathbf{X}_{\tau-1}$, given a fixed ARL^0. As only the worst possible value of CED is used, a distribution of τ is not required.

4.4 Methods for surveillance

4.4.1 Suggested statistics under surveillance

For the situation specified in Section 4.2 $\{(X(t) - \mu)^2\}$ $t = 1, 2, \ldots, s$ is a sufficient statistic for the problem as will be seen in the next section. Often a transformation of the estimated variance at each time is used in the alarm statistic. Different transformations have different motivations. Often the transformation is made such that the variable under surveillance is (approximately) Gaussian and standard charts for Gaussian variables therefore can be used. Examples such of transformations are the logarithm of the subgroup standard deviation (Crowder and Hamilton 1992) and $|X(t)/\sigma^2|^{1/2}$ (Hawkins 1981).

In the presence of a nuisance parameter, using a pivot statistic is often advocated. The subgroup range or a moving range or consecutive differences when there are no subgroups have been suggested when μ is unknown as these statistics are robust to changes in μ, see e.g. Page (1963), Rigdon, Cruthis and Champ (1994), Acosta-Mejia (1998) and Acosta-Mejia and Pignatiello (2000). On the other hand, for simultaneous surveillance of μ and the variance by a single statistic, see Domangue and Patch (1991), Chen, Cheng and Xie (2004) and Costa and Rahim (2004).

In Ncube and Li (1999) the values of the EWMA statistic are discretized by a score that is assigned different values depending on whether the process is within different intervals. The alarm statistic is formed by the cumulative score. This could be motivated from a robustness perspective but implies a suboptimal procedure as a direct loss of information owing to the discretization of the data, as pointed out by Sonesson and Bock (2003).

4.4.2 Likelihood-based methods

The methods differ with respect to how the partial likelihood ratios

$$L(s,i) = \frac{f_{X_s}(\mathbf{X}_s|\sigma^2_{C_i})}{f_{X_s}(\mathbf{X}_s|\sigma^2_D)}, i = 1, \ldots, s, \tag{4.5}$$

for a change at $\tau = i$, are weighted. The methods depend on different parameters which can be chosen to make them optimal for specific situations such as one with an intensity v and shift size d.

The method based on the full likelihood ratio, the LR method, has the alarm statistic

$$p(\mathbf{x}_s) = \frac{f_{X_s}[\mathbf{X}_s|C(s)]}{f_{X_s}[\mathbf{X}_s|D(s)]} = \sum_{i=1}^{s} w(t) \cdot L(s,i) \tag{4.6}$$

where $w(i) = \mathrm{P}(\tau = i)/\mathrm{P}(\tau \leq s)$ is the weight for $L(s,i)$. It was shown by Frisén and de Maré (1991) that the alarm rule of the LR method can be expressed in terms of the posterior probability $\mathrm{P}(C(s)|\mathbf{X}_s)$ and a positive constant limit g_{PP}. This is equivalent to the LR method with the limit

$$g_{\mathrm{LR}}(s) = \frac{g_{\mathrm{PP}}}{1-g_{\mathrm{PP}}} \cdot \frac{\mathrm{P}(D(s))}{\mathrm{P}(C(s))} = \frac{g_{\mathrm{PP}}}{1-g_{\mathrm{PP}}} \cdot \frac{\mathrm{P}(\tau > s)}{\mathrm{P}(\tau \leq s)}.$$

The LR method depends on the specified v and d, for which the method is optimized. For a geometric distribution the method is ED optimal for a process with the parameter values used.

A likelihood ratio method based on a small intensity ($\nu \to 0$) is the SR method (Shiryaev 1963; Roberts 1966) which can be used when the distribution of τ is unknown. From a Bayesian point of view this method can be seen as based on a noninformative generalized prior for τ since the weights $w(t)$ tend to a constant. Also the alarm limit $g(s)$ tend to a constant. The SR method depends only on d and can be used as an approximation to the LR method. Frisén and Wessman (1999) showed that the approximation works well, even for as large intensities as $v = 0.20$. In Chapter 3 and in Andersson, Bock and Frisén (2004, 2005, 2006) the SR method is used to detect turning points in μ. In these studies μ is unknown except for it's monotonicity restrictions and maximum likelihood ratios are used.

The CUSUM method of Page (1954) uses $p(\mathbf{X}_s) = \max_{1 \leq i \leq s}\{L(s, i)\}$ where $g(s)$ is a constant g. It depends on d and satisfies the minimax criterion described in the previous section. The Shewhart method uses $p(\mathbf{X}_s) = L(s, s)$ and a constant limit, i.e. an alarm is given as soon as the last observations exceeds the limit. It has no dependency on v and d. The Shewhart method is ED optimal when $C(s) = \{\tau = s\}$ and $D(s) = \{\tau > s\}$ because then the alarm statistic of the LR method reduces to $L(s,\ s)$.

For the situation specified in Section 4.2 where $\mu = 0$ the partial likelihood ratios (4.5) are

$$L(s, i) = d^{-(s-(i-1))/2} \cdot \exp\left\{\delta(d, \sigma_D^2) \cdot \sum_{t=i}^{s} x^2(t)\right\}, i = 1, \ldots, s \tag{4.7}$$

where $\delta(d, \sigma_D^2) = (2 \cdot \sigma_D^2)^{-1} \cdot (1 - d^{-1})$. The alarm statistic (4.6) of the LR method can then be expressed as

$$p(\mathbf{x}_s) = (d^{s/2} \cdot \mathrm{P}(\tau \leq s))^{-1} \cdot \sum_{i=1}^{s} \mathrm{P}(\tau = i) \cdot d^{(i-1)/2}.$$

$$\exp\left\{\delta(d, \sigma_D^2) \cdot \sum_{t=i}^{s} x^2(t)\right\} \tag{4.8}$$

which can be written recursively as

$$p(\mathbf{x}_s) = \frac{(\mathrm{P}(\tau \leq s - 1)}{d^{1/2} \cdot \mathrm{P}(\tau \leq s)} \cdot \{\exp\{\delta(d, \sigma_D^2) \cdot x^2(s)\} \cdot \{p(x_{s-1})$$

$$+\frac{\mathrm{P}(\tau = s)}{\mathrm{P}(\tau \leq s-1)}\Bigg\}\Bigg\}, \tag{4.9}$$
$$s = 2, 3, \ldots, p(x(1)) = d^{-1/2} \cdot \exp\left\{\delta(d, \sigma_D^2) \cdot x^2(1)\right\}.$$

The SR method has the alarm statistic

$$p(\mathbf{x}_s) = d^{-s/2} \cdot \sum_{i=1}^{s} d^{(i-1)/2} \cdot \exp\left\{\delta(d, \sigma_D^2) \cdot \sum_{t=i}^{s} x^2(t)\right\} \tag{4.10}$$

which can be written recursively as

$$p(\mathbf{x}_s) = d^{-1/2} \cdot \{\exp\{\delta(d, \sigma_D^2) \cdot x^2(s)\} \cdot \{p(x_{s-1}) + 1\}\},$$
$$s = 2, 3, \ldots, p(x(1)) = d^{-1/2} \cdot \exp\{\delta(d, \sigma_D^2) \cdot x^2(1)\}. \tag{4.11}$$

The alarm statistic of the CUSUM method can be written recursively as

$$p(\mathbf{x}_s) = \max\{0, p(x_{s-1}) + x^2(s) - k\} \tag{4.12}$$

where $p(x(0)) = 0$ and $k = (2 \cdot \delta(d, \sigma_D^2))^{-1} \cdot \ln d$. It can be shown that $\sigma_D^2 \leq k \leq d \cdot \sigma_D^2$. The alarm rule of the Shewhart method can be written as

$$p(\mathbf{x}_s) = x^2(s)/\sigma_D^2 > g. \tag{4.13}$$

4.4.3 Limiting equalities

It was proven by Frisén and Wessman (1999) that when the size of the change in the mean for which the methods are optimized for tends to infinity the stopping rules of LR, SR and CUSUM tends to the stopping rule of the Shewhart method. Below we prove the same behaviour for a change in the variance.

Theorem 1 The stopping rule of the LR method tends to that of the Shewhart method when d tends to infinity.

Proof

$$p(\mathbf{x}_s) > g_{\mathrm{LR}}(s) \Leftrightarrow (d^{s/2} \cdot \mathrm{P}(\tau \leq s))^{-1} \cdot \sum_{i=1}^{s} \mathrm{P}(\tau = i) \cdot d^{(i-1)/2} \cdot$$
$$\exp\left\{\delta(d, \sigma_D^2) \cdot \sum_{t=i}^{s} x^2(i)\right\} > \frac{g_{\mathrm{PP}}}{1 - g_{\mathrm{PP}}} \cdot \frac{\mathrm{P}(\tau > s)}{\mathrm{P}(\tau \leq s)}$$
$$\Leftrightarrow \sum_{i=1}^{s} \mathrm{P}(\tau = i) \cdot d^{(i-1)/2} \cdot \exp\left\{\delta(d, \sigma_D^2) \cdot \sum_{t=i}^{s} x^2(t)\right\}$$

$$> \left(\frac{g_{PP}}{1-g_{PP}}\right) \cdot d^{s/2} \cdot \mathrm{P}(\tau > s)$$

$$\Leftrightarrow \exp\{\delta(d,\sigma_D^2)\cdot x^2(s)\} \cdot \left[\sum_{i=1}^{s-1} \mathrm{P}(\tau = i)\cdot d^{(i-1)/2} \cdot \exp\left\{\delta(d,\sigma_D^2)\cdot \sum_{t=i}^{s-1} x^2(t)\right\} + \mathrm{P}(\tau = s)\cdot d^{(s-1)/2}\right]$$

$$> \left(\frac{g_{PP}}{1-g_{PP}}\right) \cdot d^{s/2} \cdot \mathrm{P}(\tau > s)$$

$$\Leftrightarrow \exp\{\delta(d,\sigma_D^2)\cdot x^2(s)\} > \frac{g_{PP}\cdot d^{s/2}\cdot \mathrm{P}(\tau > s)}{(1-g_{PP})\cdot \sum_{i=1}^{s-1} \mathrm{P}(\tau = i)\cdot d^{(i-1)/2}\cdot \exp\left\{\delta(d,\sigma_D^2)\cdot \sum_{t=i}^{s-1} x^2(t)\right\} + \mathrm{P}(\tau = s)\cdot d^{(s-1)/2}}$$

$$\Leftrightarrow \exp\{\delta(d,\sigma_D^2)\cdot x^2(s)\} > \frac{g_{PP}}{(1-g_{PP})\cdot v\cdot \sum_{i=1}^{s-1}(1-v)^{-(s-i+1)}\cdot d^{-(s-i+1)/2}\cdot \exp\left\{\delta(d,\sigma_D^2)\cdot \sum_{t=i}^{s-1} x^2(t)\right\} + (1-v)^{-1}\cdot d^{-1/2}}$$

$$\Leftrightarrow x^2(s) > \frac{\ln g_{PP}}{\delta(d,\sigma_D^2)} - \frac{1}{\delta(d,\sigma_D^2)} \times \ln\left\{(1-g_{PP})\cdot v\cdot \sum_{i=1}^{s-1}\left((1-v)^{-(s-i+1)}\cdot d^{-(s-i+1)/2}\cdot \exp\left\{\delta(d,\sigma_D^2)\cdot \sum_{t=i}^{s-1} x^2(t)\right\}\right) + (1-v)^{-1}\cdot d^{-1/2}\right\}$$

$$\Leftrightarrow x^2(s) > \frac{\ln g_{PP}}{\delta(d,\sigma_D^2)} - \frac{1}{\delta(d,\sigma_D^2)} \cdot \ln\{(1-g_{PP})\cdot v\cdot O(d^{-1}) + (1-v)^{-1}\cdot d^{-1/2}\}.$$

The dependency on s of the right-hand side of the last expression disappears when d tends to infinity such that the stopping rule tends to the one of the Shewhart method.

Theorem 2 The stopping rule of the SR method tends to that of the Shewhart method when d tends to infinity.

Proof In analogy with the proof of Theorem 1.

Theorem 3 The stopping rule of the CUSUM method tends to that of the Shewhart method when d tends to infinity.

Proof

$$d \to \infty \Rightarrow k = (2 \cdot \delta(d, \sigma_D^2))^{-1} \cdot \ln d \to \infty$$

$$\Rightarrow \mathrm{P}(\max\{0, p(x_{s-1}) + x^2(s) - k\} > g) \to$$

$$\mathrm{P}(x^2(s) - k > g) = \mathrm{P}(x^2(s) > g + k)$$

since $\lim_{k \to \infty} \mathrm{P}(p(x_{s-1}) > 0) = 0$.

4.5 A Monte Carlo study

In this section, we study the properties of the methods. To make the methods comparable the alarm limits are adjusted to yield the same level of MRL^0. Which level of MRL^0 that should be chosen and what size of the scale change to be studied depends on the application.

A low value of MRL^0 can be interpreted as a situation where observations are made seldom and a high value with more frequent observations. This can be interpreted as differences in time scale. How distinct the differences between the methods are depends on the scale, as pointed out by Frisén and Wessman (1999). For example, if observations are made frequent, e.g. each day (a large value of MRL^0) then there is a larger information loss of only using the last observation (Shewhart method) compared to less frequent observations, e.g. each week (a small value of MRL^0). Comparisons with different values of MRL^0 are not made here. The alarm limits are set here such that $\mathrm{MRL}^0 = 60$ which reflects roughly three months of daily data in the financial markets.

The in-control and out-of-control variance is set to 1 and 2, respectively, i.e. $\sigma^2 = 1$ and $\Delta = 2$ in Section 4.2, as in e.g. MacGregor and Harris (1993) and Acosta-Mejia *et al.* (1999). The size of the change for which the methods are optimized for, d, is set to 1.5, 2 and 2.5. For $d = 2$ the variance is correctly specified whereas for d equals to 1.5 and 2.5 the variance is under- and over specified by 50%, respectively. The value of the intensity for which the LR method is optimized for, v, is set to 0.10 and 0.20. To distinguish between the same methods with different values of v and d, the values will be given as arguments, e.g. LR(v; d).

For the Shewhart method analytical calculations were made. For the other methods simulations of 10^7 replicates were made. The limits were set such that the largest deviation between the values of $P(t_A \leq 60|\tau = \infty)$ and the intended values of 0.50 were smaller than 0.1%.

4.5.1 In-control properties

Having equal MRL^0 does not mean that the in-control run length densities are identical but they can have different shapes. The most common way to control the false alarms is by the ARL^0. $MRL^0 = 60$ corresponds to values of ARL^0 between 60 and 87 for the methods. The great variation in ARL^0 is due to the great differences in skewness seen in the in-control run length densities shown in Figure 4.1. LR(0.2) yield the smallest values of ARL^0 for all values of d and Shewhart the largest and these two methods have densities that are most symmetric and skewed, respectively. Shewhart and CUSUM have similar ARL^0. As implied by the theorems, the larger the d the more similar are the densities to that of the Shewhart method. A method designed to detect a large change quickly should allocate nearly all weight to the single last observation, as pointed out by Frisén and Wessman (1999). When the methods are optimized for detecting a small change in the variance, many observations are required to have enough evidence for a change and the densities are consequently less skewed compared to when d is large.

It seem surprising that LR(0.2) is less skewed compared to LR(0.1) and SR since a large intensity should intuitively yield a large probability of early

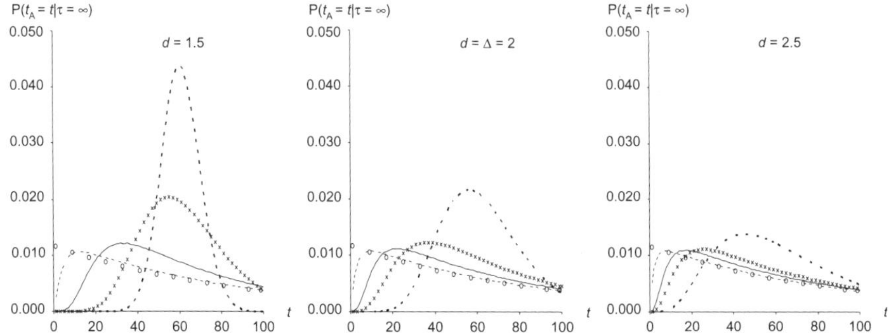

Figure 4.1 The density of the time of alarm, $P(tA = t|\tau = \infty)$. Shewhart(○), CUSUM(- - -), SR(_____), LR(0.1)(×), LR(0.2)(—-).

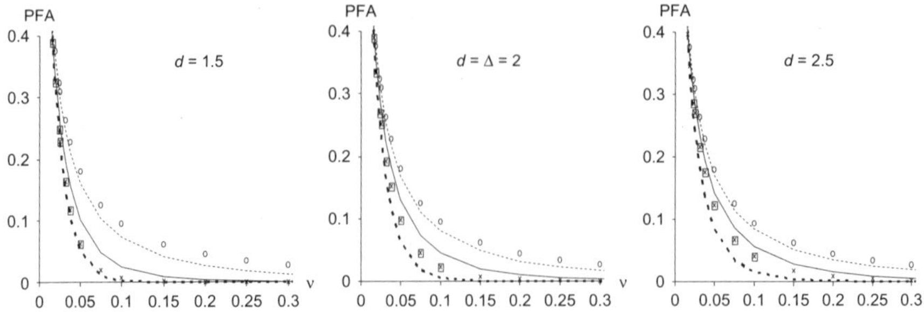

Figure 4.2 The probability of a false alarm, PFA. Shewhart(○), CUSUM(– – –), SR(_____), LR(0.1)(×), LR(0.2)(—-).

alarms. This was also noted by Frisén and Wessman (1999) who explained it by the way the false alarms are controlled. For a low intensity the right-hand tail of the run length distribution is thick. As ARL^0 was fixed, the only possibility was a high alarm probability at early times. When MRL^0 is fixed the time points of the alarms have less effect but many alarm times larger than 60 must still be compensated by high alarm probabilities early.

The probability of a false alarm, PFA, is another measure used to control the false alarms. It summarizes the false alarm distribution by weights with the distribution of τ. It is shown as a function of ν in Figure 4.2.

The differences in PFA are due to both the differences in shape of $P(t_A < \tau|D)$ and the location. As a result of the shape of the geometric distribution early alarms have a great influence on PFA. The large PFA for the Shewhart method is a result of the many early false alarms seen in Figure 4.1. Due to the opposite behaviour of the error spending of LR(0.2) it has the smallest PFA. In the same way that an equal MRL^0 apparently does not imply equal PFA and vice versa, Frisén and Wessman (1999) and Frisén and Sonesson (2006) demonstrated the same difference between ARL^0 and PFA. Consequently, comparisons between methods depend on which measure that is controlled.

4.5.2 Out-of-control properties

As was mentioned in Section 4.3 the out-of-control behaviour is often summarized by the ARL^1. In Figure 4.3 the ARL^1 is shown as a function of d. The convergence to the ARL^1 of the Shewhart method is evident.

The conditional expected delay, CED, is shown in Figure 4.4 for different values of i. For $\tau = 1$, $CED = ARL^1 - 1$. The CED clearly depends on τ for several methods, which is not revealed by the ARL^1. The worst value of CED is at $i = 1$ for the CUSUM method and CUSUM has the smallest CED(1)

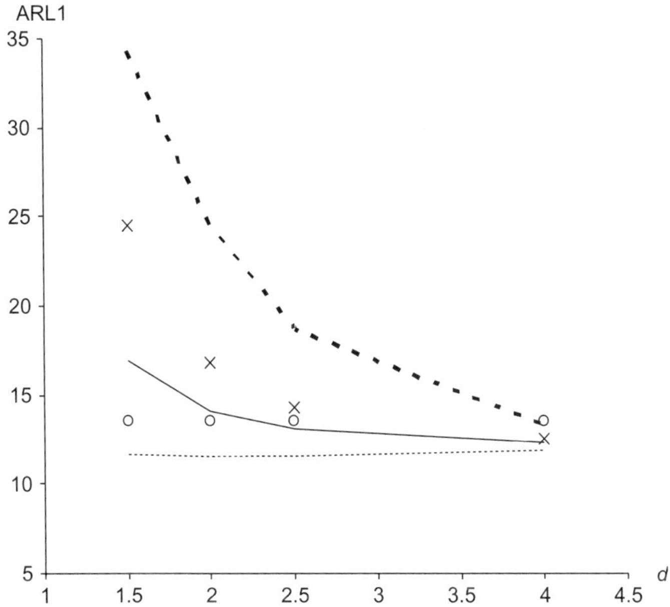

Figure 4.3 ARL1 as function of d. $\Delta = 2$. Shewhart(◯), CUSUM(− − −), SR(_____), LR(0.1)(×), LR(0.2)(—-).

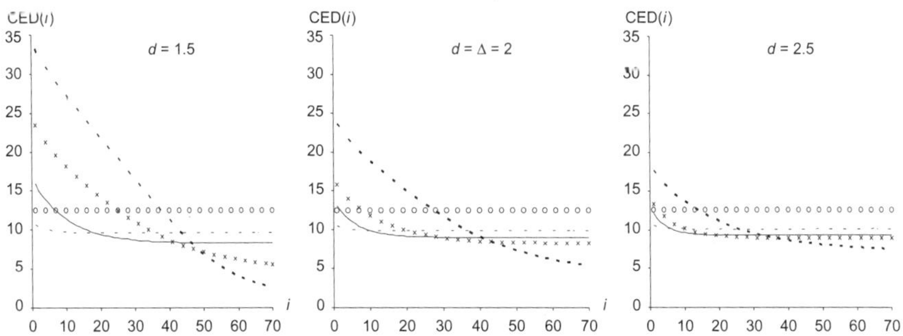

Figure 4.4 Conditional expected delay, CED(i). Shewhart(◯), CUSUM(− − −), SR(_____), LR(0.1)(×), LR(0.2)(—-).

among the methods. Though ARL^0 is not controlled here this illustrates the minimax optimality of the CUSUM method.

CUSUM is better in terms of CED but worse in terms of PSD with $m = 1$ (Figure 4.5) compared to Shewhart. The reason for this is that $P(t_A = i | t_A \geq$

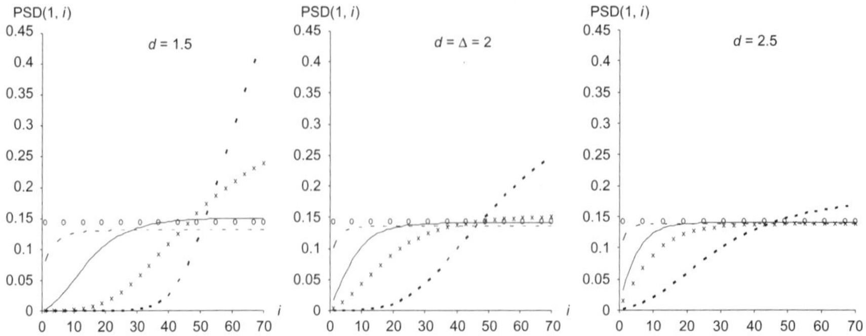

Figure 4.5 Probability of successful detection, PSD(m, i), with $m = 1$. Shewhart(◯), CUSUM(– – –), SR(_____), LR(0.1)(×), LR(0.2)(—-).

$\tau, \tau = i)$ that is PSD with $m = 0$, is higher for Shewhart compared to CUSUM which favours PSD with $m = 1$. The high $\mathrm{P}(t_A = i | t_A \geq \tau, \tau = i)$ of the Shewhart method is due to its optimality for $C(s) = \{\tau = s\}$ and $D(s) = \{\tau > s\}$ (see Section 4.4.2). The error spending behaviour of the LR methods explained in Section 4.5.1 influences the detection ability such that the methods have a large CED and a small PSD for early changes and the opposite for late τ.

4.5.3 The trust of alarms

The predictive value at time t, PV(t), reflects the trust you should have in an alarm. The predictive value at time point t is

$$\mathrm{P}(\tau \leq t | t_A = t) = \frac{\mathrm{PMA}(t)}{\mathrm{PMA}(t) + \mathrm{PFA}(t)} \tag{4.14}$$

where $\mathrm{PFA}(t) = \mathrm{P}(t_A = t | t < \tau) \cdot \mathrm{P}(\tau > t)$ and $\mathrm{PMA}(t) = \sum_{i=1}^{t} \mathrm{P}(\tau = i) \cdot \mathrm{P}(t_A = t | \tau = i)$ are probabilities of a false and a motivated alarm at time t, respectively. The PV is shown as a function of the time of the alarm in Figure 4.6 and 4.7.

Shewhart and CUSUM have high detection ability for early changes as seen in Figures 4.3–4.5, but at the same time a high false alarm probability (Figure 4.1). The result is a low predictive value of early alarms, i.e. these are not very trustworthy. The results get better for a large value of ν. The PV of SR and LR appear to be fairly robust to mis-specifications of Δ. For these methods PV is stable over time, which might be a desirable property as it simplifies matters if the same action can be used regardless of whenever an alarm occurs.

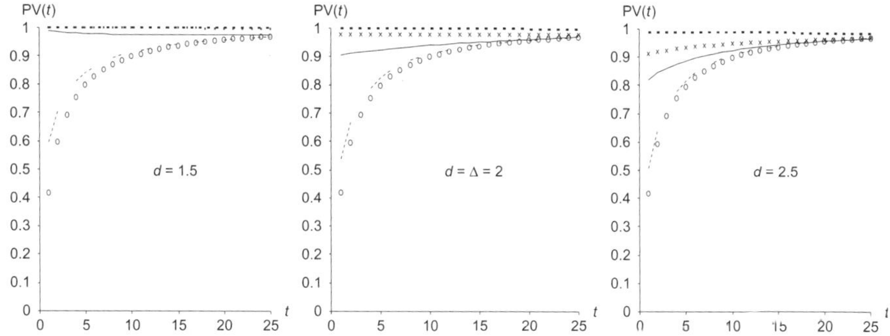

Figure 4.6 Predictive value, PV, with $\nu = 0.10$. Shewhart(○), CUSUM(- - -), SR(____), LR(0.1)(×), LR(0.2)(—-).

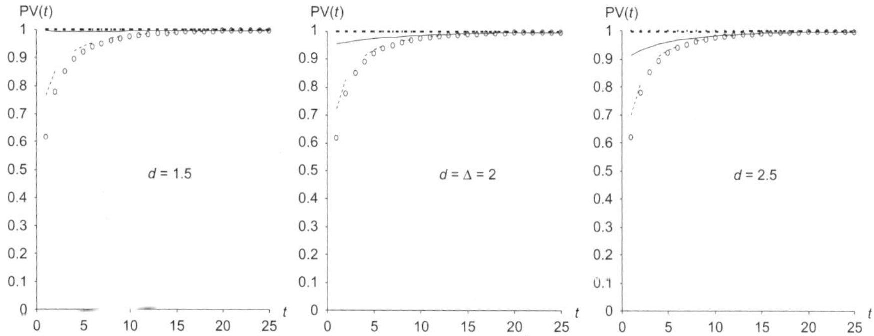

Figure 4.7 Predictive value, PV, with $\nu = 0.20$. Shewhart(○), CUSUM(- - -), SR(____), LR(0.1)(×), LR(0.2)(—-).

4.6 Illustrative example

The use of the methods is here illustrated by a simple example. The methods are monitoring the returns, denoted by r, of the stock market index Standard and Poor's 500 (S&P500).

Andreou and Ghysels (2002) applied a number of tests for homogeneity in the variance of the returns of S&P500 for the period 4 January 1989–19 October 2001 (3229 observations). The tests were made retrospectively, that is a historical data set of given length were analysed. Both tests for a single change point and for multiple change points were applied. For the latter the number of breaks was determined by the test of Kokoszka and Leipus (2000) applied to the squared returns and a sequential segmentation approach. It was concluded that changes in the volatility occurred at 31 December 1991, 18 December 1995 and 26 March 1997. Whether the second change in 18 December 1995 could have been detected online is investigated below.

The period of monitoring is 9 October 1995–25 March 1997 that is 370 observations and $\tau = 50$. Financial returns are known to be severely heteroscedastic. An in-control model of the heteroscedasticity is estimated by a historical set of data, 31 December 1991–6 October 1995 (954 observations). The returns of the historical period and the monitoring period are shown in Figure 4.8.

A common way of characterizing the heteroscedasticity is by ARCH processes. The portmanteau Q-test of the squared residuals (McLeod and Li 1983) and the Lagrange multiplier test by Engle (1982) for ARCH disturbances are applied to the returns of the historical period. The tests can be used to identify the order of an ARCH process. The p-values at different lags are shown in Table 4.1 and these indicate that there are high order ARCH effects which could be described by a first-order Gaussian GARCH process, GARCH(1, 1); $r(t) = \mu(t) + \varepsilon(t) \cdot h(t)^{1/2}$ where $h(t) = \omega + \alpha_1 \cdot (r(t-1) - \mu(t-1))^2 + \beta_1 \cdot h(t-1)$, $\omega > 0, \alpha_1 > 0, \beta_1 \geq 0, \alpha_1 + \beta_1 < 1$ and $\varepsilon \sim$ iid N(0, 1). We estimate the parameters of the Gaussian GARCH(1, 1) model with $\mu(t) = \mu$ (a constant) using the historical data set. The parameter estimates are given in Table 4.1. It should be pointed out that a proper modelling strategy requires a much more

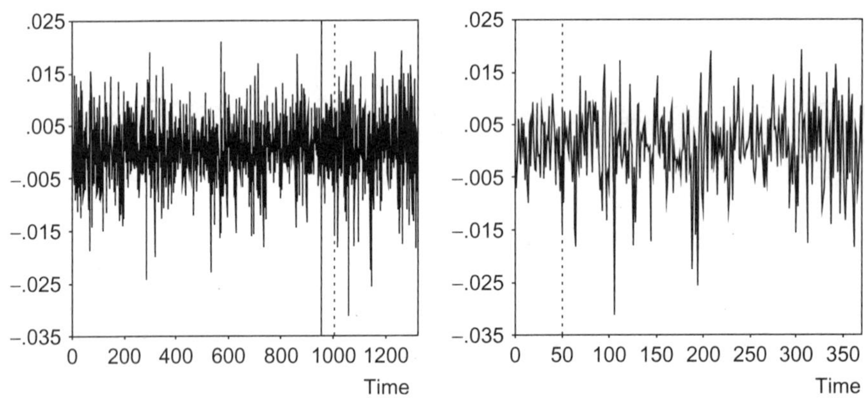

Figure 4.8 Daily returns of S&P500. The start of the monitoring period is marked with a solid vertical line. The time of the change τ is marked with a dashed vertical line. Left: 31 December 1991–25 March 1997. Right: 9 October 1995–25 March 1997.

thorough analysis than this. But for illustration the model is used as a rough approximation.

The statistic under surveillance is the squares of the standardized residuals $X(t) = (r(t) - \hat{\mu})/\hat{h}(t)^{1/2}$ where $\hat{h}(t)$ is the estimated conditional variance.

Since Δ is unknown the values of d, for which the methods are optimized for, used in Section 4.5 are used also here. The alarm limits used earlier are used also here. The alarm times are given in Table 4.1.

Table 4.1 Results from the modeling strategy and the alarm times of the methods.

P-values of the portmanteau Q-test and the Lagrange multiplier test at different lags.

Lag	1	2	3	4	5	6	7	8
Q test	0.442	0.670	0.233	0.087	0.044	0.025	0.036	0.054
LM test	0.440	0.671	0.237	0.099	0.058	0.041	0.049	0.077

Parameter estimates of the GARCH(1, 1) model. Standard errors within brackets.

μ	ω	α_1	β_1
0.000452	1.6561E−6	0.0356	0.9134
(0.000183)	(7.8792E−7)	(0.0125)	(0.0327)

Alarm times ($\tau = 50$).

LR(1.5; 0.2)	LR(2; 0.2)	LR(2.5; 0.2)
51	50	50
55	51	50
SR(1.5)	SR(2)	SR(2.5)
50	50	50
CUSUM(1.5)	CUSUM(2)	CUSUM(2.5)
51	50	50
Shewhart		
50		

All the methods give alarms at τ or immediately after. The variances of X before and after τ as estimated by

$$\hat{\sigma}^2 = \frac{1}{n-1} \cdot \sum_{t=1}^{n} (x(t) - \bar{x})^2$$

yields a shift of the size $\hat{\Delta} = 1.495$. At τ there is a highly negative return (see Figure 4.8, right) influencing the alarm statistics. For the model used in the simulations $P(t_A = \tau | \tau = 50)$ varies between 0.035 and 0.057 for the methods and the outcome in Table 4.1 is hence rather extreme. The residuals thus appear to deviate from the process of interest. The validity of the Gaussian GARCH(1, 1) model for describing financial time series is in fact frequently debated in the empirical finance literature. Alternative GARCH models are discussed in Chapter 2. This illustrates many of the difficulties encountered by case studies.

4.7 Concluding remarks

Different likelihood-based methods of statistical surveillance for detecting a change in the variance have been evaluated. The methods differ with respect to how the different observations available at each decision time are treated and the way the alarm limit change with the decision time. The methods differ with respect to the number of parameters they depend on. All methods but the Shewhart method can be optimized for the size of the change you want to detect and the LR methods can also be optimized for the intensity of the change-point time.

The robustness of the methods with respect to mis-specifications of the change has been examined. The results demonstrates the same behaviour that Frisén and Wessman (1999) found for a change in location: the larger the size of the change, d, for which the methods are optimized for, the more similar to the Shewhart method are the methods. Hence, if we optimize for a very large d all weight is allocated to the last observation. If, on the other hand, we optimize for a small d, more weight is given to earlier observations than for Shewhart because more observations are needed for having enough evidence for a change.

Differences in the weighting are reflected in the skewness of the run length densities. For a large d early alarms are more frequent compared to a small d. Consequences of these differences are that for the former situation PFA is high compared to the latter as early alarms have a great influence because of the geometric distribution.

The detection ability as measured by CED and PSD is rather constant and great at a large d. When the d is small the detection ability is worse at early changes but get better the later the change occurs. The price of the good detection ability of early changes of Shewhart and CUSUM is however that early alarms are not very reliable. LR and SR have better predictive values at early alarms.

The LR method has the parameter υ to optimize for the intensity for a change. This is avoided by the SR method. For the values of MRL^0 and υ used, LR seems in terms of PFA and PV (Figure 4.2, 4.6 and 4.7) to be robust against mis-specifications of the intensity and SR appear to be a good approximation of LR for small values of the intensity.

The way the methods differed in shapes of the run length densities noted by Frisén and Wessman (1999) is also seen here. It depends on the way the false alarms are controlled, as explained in Section 4.5.1.

In the illustration of the methods on the S&P500 data all the methods gave alarms very close to the change-point time. This would however be improbable if the residuals under surveillance were independent and Gaussian. This illustrates many of the difficulties and limitations encountered by case studies.

Acknowledgements

The author is grateful for valuable comments by Professor Marianne Frisén. The Bank of Sweden Tercentenary Foundation supported the research.

References

Acosta-Mejia, C. A. (1998). Monitoring reduction in variability with the range *IIE Transactions*, **30**, 515–523.

Acosta-Mejia, C. A., Pignatello, J. J. J. and Rao, B. V. (1999). A comparison of control charting procedures for monitoring process dispersion. *IIE Transactions*, **31**, 569–579.

Acosta-Mejia, C. A. and Pignatiello, J. J. (2000). Monitoring process dispersion without subgrouping. *Journal of Quality Technology*, **32**, 89–102.

Andersson, E., Bock, D. and Frisén, M. (2004). Detection of turning points in business cycles. *Journal of Business Cycle Measurement and Analysis*, **1**, 93–108.

Andersson, E., Bock, D. and Frisén, M. (2005). Statistical surveillance of cyclical processes. Detection of turning points in business cycles. *Journal of Forecasting*, **24**, 465–490.

Andersson, E., Bock, D. and Frisén, M. (2006). Some statistical aspects on methods for detection of turning points in business cycles, *Journal of Applied Statistics*, **33**, 257–278.

Andreou, E. and Ghysels, E. (2002). Detecting multiple breaks in financial market volatility dynamics. *Journal of Applied Econometrics*, **17**, 579–600.

Bock, D. (2008). Aspects on the control of false alarms in statistical surveillance and the impact on the return of financial decision systems. *Journal of Applied Statistics*, **35**.

Chen, G., Cheng, S. W. and Xie, H. (2004). A new EWMA control chart for monitoring both location and dispersion. *Quality Technology & Quantitative Management*, **1**, 217–231.

Chu, C.-S. J., Stinchcombe, M. and White, H. (1996). Monitoring structural change. *Econometrica*, **64**, 1045–1065.

Costa, A. F. B. and Rahim, M. A. (2004). Monitoring process mean and variability with one non-central chi-square chart. *Journal of Applied Statistics*, **31**, 1171–1183.

Crowder, S. V. and Hamilton, M. D. (1992). An EWMA for monitoring a process standard-deviation. *Journal of Quality Technology*, **24**, 12–21.

Domangue, R. and Patch, S. C. (1991). Some omnibus exponentially weighted moving average statistical process monitoring schemes. *Technometrics*, **33**, 299–313.

Engle, R. F. (1982). Autoregressive conditional heteroscedasticity with estimates of the variance of United Kingdom inflation. *Econometrica*, **50**, 987–1008.

Frisén, M. (1992). Evaluations of methods for statistical surveillance. *Statistics in Medicine*, **11**, 1489–1502.

Frisén, M. (1994). Statistical surveillance of business cycles 1994:1 (Revised 2000), Technical, Department of Statistics, Göteborg University, Sweden.

Frisén, M. (2003). Statistical surveillance. Optimality and methods. *International Statistical Review*, **71**, 403–434.

Frisén, M. and de Maré, J. (1991). Optimal surveillance. *Biometrika*, **78**, 271–280.

Frisén, M. and Sonesson, C. (2006). Optimal surveillance based on exponentially weighted moving averages methods. *Sequential Analysis*, **25**, 379–403.

Frisén, M. and Wessman, P. (1999). Evaluations of likelihood ratio methods for surveillance. Differences and robustness. *Communications in Statistics. Simulations and Computations*, **28**, 597–622.

Gan, F. (1993). An optimal-design of EWMA control chart based on median run-length. *Journal of Statistical Computation and Simulation*, **45**, 169–184.

Hawkins, D. L. (1992). Detecting shifts in functions of multivariate location and covariance parameters. *Journal of Statistical Planning and Inference*, **33**, 233–244.

Hawkins, D. M. (1981). A CUSUM for a scale parameter. *Journal of Quality Technology*, **13**, 228–231.

Hawkins, D. M. and Zamba, K. D. (2005). A change point model for a shift in the variance. *Journal of Quality Technology*, **37**, 21–37.

Järpe, E. and Wessman, P. (2000). Some power aspects of methods for detecting shifts in the mean. *Communications in Statistics. Simulations and Computations*, **29**, 633–646.

Knoth, S. and Schmid, W. (2002). Monitoring the mean and the variance of a stationary process. *Statistica Neerlandica*, **56**, 77–100.

Kokoszka, P. and Leipus, R. (2000). Change-point estimation in ARCH models. *Bernoulli*, **6**, 513–539.

MacGregor, J. F. and Harris, T. J. (1993). The exponentially weighted moving variance. *Journal of Quality Technology*, **25**, 106–118.

McLeod, A. I. and Li, W. K. (1983). Diagnostic checking ARMA time series models using squared-residual autocorrelations. *Journal of Time Series Analysis*, **4**, 269–273.

Moustakides, G. V. (1986). Optimal stopping times for detecting changes in distributions. *The Annals of Statistics*, **14**, 1379–1387.

Ncube, M. and Li, K. (1999). An EWMA–CUSCORE quality control procedure for process variability. *Mathematical and Computer Modelling*, **29**, 73–79.

Page, E. S. (1954). Continuous inspection schemes. *Biometrika*, **41**, 100–114.

Page, E. S. (1963). Controlling the standard deviation by CUSUMS and warning lines. *Technometrics*, **5**, 307–315.

Pollak, M. and Siegmund, D. (1975). Approximations to the expected sample size of certain sequential tests. *Annals of Statistics*, **3**, 1267–1282.

Reynolds, M. R. and Soumbos, Z. G. (2001). Monitoring the process mean and variance using individual observations and variable sampling intervals. *Journal of Quality Technology*, **33**, 181–205.

Rigdon, S. E., Cruthis, E. N. and Champ, C. W. (1994). Design strategies for individuals and moving range control charts. *Journal of Quality Technology*, **26**, 274–287.

Roberts, S. W. (1959). Control chart tests based on geometric moving averages. *Technometrics*, **1**, 239–250.

Roberts, S. W. (1966). A comparison of some control chart procedures. *Technometrics*, **8**, 411–430.

Schipper, S. and Schmid, W. (2001a). Control charts for GARCH processes. *Nonlinear Analysis*, **47**, 2049–2060.

Schipper, S. and Schmid, W. (2001b). Sequential methods for detecting changes in the variance of economic time series. *Sequential Analysis*, **20**, 235–262.

Severin, T. and Schmid, W. (1998). Statistical process control and its application in finance. In Contributions to Economics: *Risk Measurement, Econometrics and Neural Networks*, eds. G. Bol, G. Nakhaeizadeh and C.-H. Vollmer. Physica Verlag, Heidelberg, pp. 83–104.

Severin, T. and Schmid, W. (1999). Monitoring changes in GARCH processes. *Allgemeines Statistisches Archiv*, **83**, 281–307.

Shiryaev, A. N. (1963). On optimum methods in quickest detection problems. *Theory of Probability and its Applications*, **8**, 22–46.

Sonesson, C. and Bock, D. (2003). A review and discussion of prospective statistical surveillance in public health. *Journal of the Royal Statistical Society A*, **166**, 5–21.

Srivastava, M. S. (1997). CUSUM procedures for monitoring variability. *Communications in Statistics. Theory and Methods*, **26**, 2905–2926.

Srivastava, M. S. and Chow, W. (1992). Comparison of the CUSUM procedure with other procedures that detect an increase in the variance and a fast accurate approximation for the ARL of the CUSUM procedure. Technical Report 9122, Department of Statistics, University of Toronto.

5

Surveillance of univariate and multivariate linear time series

Yarema Okhrin and Wolfgang Schmid
Department of Statistics, European University Viadrina, Frankfurt (Oder), Germany

5.1 Introduction

Financial time series are subject to frequent structural changes. Changes in the management of companies or important announcements do have substantial impact on the price of the company stocks. Similarly, political changes or macroeconomic decisions may influence the exchange rate of the local currency with respect to foreign currencies. In terms of the time series model for the underlying process (return process or exchange rate process) this leads to possible shifts in the mean, in the variance, to outliers or drifts. For illustration purposes we take daily data on MSCI country indices for Germany, the US and Japan over the period from January, 1999 to June, 2006. Figure 5.1 plots rolling means and variances based on 20 last observations of the log-returns. As we observe, the average asset returns do not exhibit any substantial shifts in the mean. However, for the volatility we can clearly determine numerous shifts in the series followed by periods with relatively constant variance.

There are numerous standard tests for structural change in the econometric literature, however, it is usually assumed that the time point of the shift is known. This is frequently not a plausible assumption for practitioners, who prefer to monitor the process permanently and to detect the change as soon as possible after its occurrence. The approach discussed here is based on the

Financial Surveillance Edited by Marianne Frisén

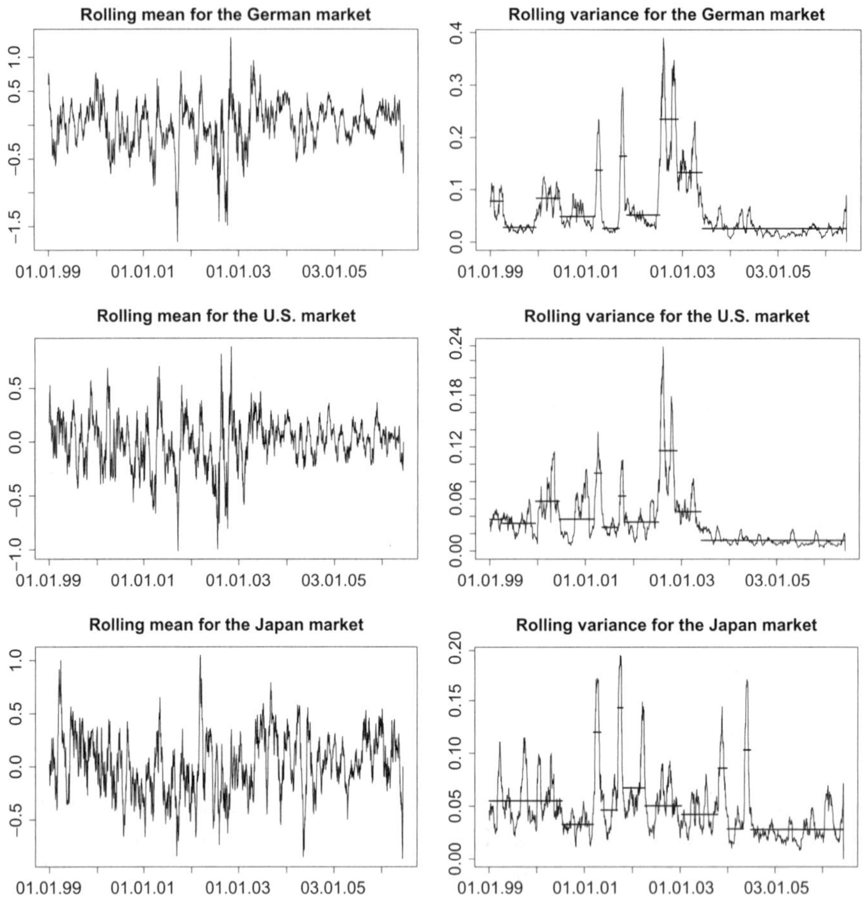

Figure 5.1 Rolling means and variances of MSCI index returns. Rolling means and variances are computed using 20 recent observations. The horizontal lines show the mean of rolling variances for periods with presumably constant variance.

sequential detection of the structural change in a linear time series. The monitoring is performed using control charts, which are the main tools of the statistical process control. Earlier these methods were mainly used in engineering, however, recently they become increasingly popular in economic sciences as well. For further discussion the reader may refer to Chapter 1 of this handbook which provides an overview of this development, while Chapters 3 and 7 contain insightful financial applications. The methods considered in this chapter are complemented by the case of nonlinear time series discussed in Chapter 6, where the authors concentrate on the changes in volatility. The alternative likelihood-based volatility surveillance is discussed in detail in Chapter 4. It also provides a discussion of the evaluation measures used in surveillance.

This chapter is structured as follows. The next section extends Chapter 2 of this handbook and provides a short review of multivariate time series models. Section 5.3 states the model. Section 5.4 and 5.5 discuss the control charts for the mean (Section 5.4) and for the variance (Section 5.5) of a univariate time series. These sections also contain extensive comparison studies and examples. The extension to the multivariate case is done in Section 5.6.

5.2 Foundations of time series analysis

The main tasks of time series analysis are the modelling and the forecasting of data observed at different points in time. Here we focus on discrete time series with equidistant observations. Let us consider a series of realizations of a stochastic process $\{Y_t\}$ with $t \in \mathbb{Z}$. First it is assumed that the observations are univariate.

Frequently we distinguish between linear and nonlinear processes. A linear process can be represented as an infinite linear combination of independent innovations (see Brockwell and Davis 1991). In this chapter we present some general results for stationary processes and a special family of linear processes, so-called stationary ARMA (AutoRegressive Moving Average) processes. Control procedures for nonlinear processes are discussed in detail in Chapter 6 of this book.

5.2.1 Stationary processes and ARMA processes

Stationary processes constitute the most widely discussed family of time series. A stochastic process $\{Y_t\}$ is called (weakly) stationary if

$$E(Y_t) = \mu_0, \qquad \text{Cov}(Y_t, Y_{t+h}) = \gamma_h \qquad \text{for all} \quad t, h \in \mathbb{Z}.$$

The mean of a stationary process is constant over time and its covariance depends only on the time lag and not on the time point itself. These assumptions appear to be valid for most financial time series or at least for some suitable transformations. The quantity γ_h is called the autocovariance of $\{Y_t\}$ at lag h. The autocorrelation $\{Y_t\}$ at lag h we denote by ρ_h and it is defined as $\rho_h = \gamma_h/\gamma_0$.

The most well studied and popular family of autocorrelated processes are ARMA processes. A stochastic process is called an ARMA process of order

(p, q) if it is a solution of the following stochastic difference equation

$$Y_t = \mu_0 + \sum_{i=1}^{p} \alpha_i (Y_{t-i} - \mu_0) + \epsilon_t + \sum_{j=1}^{q} \beta_j \epsilon_{t-j},$$

where $\{\epsilon_t\}$ is a white noise process, i.e. $E(\epsilon_t) = 0$, $\text{Var}(\epsilon_t) = \sigma^2$ and $\text{Cov}(\epsilon_t, \epsilon_s) = 0$ for $t \neq s$. If $q = 0$ then we refer to $\{Y_t\}$ as an AR(p) (AutoRegressive) process and we call it an MA(q) (Moving Average) process if $p = 0$.

The stationarity of an ARMA(p, q) process is determined only by the coefficients of the AR part. The process is stationary if all roots of the polynomial $1 - \sum_{i=1}^{p} \alpha_i z^i = 0$ lie outside the unit circle given by $|z| = 1$ Brockwell and Davis (1991). For an ARMA(1,1) process, for example, this means that $|\alpha_1| < 1$. If the stationarity condition is fulfilled then there is a unique solution of the difference equation given by

$$Y_t = \mu_0 + \sum_{i=0}^{\infty} \theta_i \epsilon_{t-i}.$$

The coefficients $\{\theta_i\}$ are equal to the coefficients of the power series obtained by dividing the polynomials $1 + \sum_{j=1}^{q} \beta_j z^j$ and $1 - \sum_{i=1}^{p} \alpha_i z^i$, i.e.

$$\sum_{i=0}^{\infty} \theta_i z^i = \frac{1 + \sum_{j=1}^{q} \beta_j z^j}{1 - \sum_{i=1}^{p} \alpha_i z^i}.$$

In this case the process $\{Y_t\}$ is linear. Frequently it is assumed that $\epsilon_t \sim \mathcal{N}(0, \sigma^2)$ (univariate normal distribution with zero mean and variance σ^2). Then $\{Y_t\}$ is a Gaussian process, i.e. all of its finite-dimensional marginal distributions are multivariate normal. In general it is difficult to give explicit expressions for the variances and the autocovariances of an ARMA(p, q) process, however, in some special cases explicit expressions are available. In Table 5.1 some results are summarized.

Table 5.1 Autocovariance functions of some ARMA processes

Process	Autocovariances
AR(1)	$\gamma_0 = \frac{\sigma^2}{1-\alpha_1^2}$, $\gamma_h = \alpha_1^h \gamma_0$ for $h \geq 1$
MA(1)	$\gamma_0 = \sigma^2(1 + \beta_1^2)$, $\gamma_1 = \beta_1 \sigma^2$, $\gamma_h = 0$ for $h \geq 2$
AR(2)	$\gamma_0 = \frac{\sigma^2}{1-\alpha_1^2-\alpha_2^2}$, $\gamma_1 = \frac{\alpha_1}{1-\alpha_2}\gamma_0$, $\gamma_h = \alpha_1\gamma_{h-1} + \alpha_2\gamma_{h-2}$ for $h \geq 2$
ARMA(1, 1)	$\gamma_0 = \frac{1+2\alpha_1\beta_1+\beta_1^2}{1-\alpha_1^2}\sigma^2$, $\gamma_1 = \frac{(1+\alpha_1\beta_1)(\alpha_1+\beta_1)}{1-\alpha_1^2}\sigma^2$ and $\gamma_h = \alpha_1^{h-1}\gamma_1$ for $h \geq 2$

5.2.2 Prediction theory

Prediction appears to be one of the key tasks in financial applications of time series models. In general, a predictor of Y_t can be any function of the observations $Y_1, \ldots, Y_{t-1}$. However, in practice simple and intuitive predictors are usually preferred. The theory of linear predictors provides an elegant solution (Brockwell and Davis 1991). For any stationary process $\{Y_t\}$ the best (in L^2 sense) linear predictor of Y_t in terms of $Y_1, \ldots, Y_{t-1}$ is given by

$$\hat{Y}_t = \mu_0 + \sum_{i=1}^{t-1} \phi_{ti}(Y_{t-i} - \mu_0),$$

where the coefficients ϕ_{ti} for $i = 1, \ldots, t-1$ solve the minimization problem

$$\min_{\psi_{t1},\ldots,\psi_{tt-1}} \mathrm{E}\left[Y_t - \mu_0 - \sum_{i=1}^{t-1} \psi_{ti}(Y_{t-i} - \mu_0)\right]^2 .$$

It is straightforward to show that the parameters $\boldsymbol{\phi}_t = (\phi_{t1}, \ldots, \phi_{tt-1})'$ are a solution of the system of linear equations

$$\boldsymbol{\Gamma}_t \boldsymbol{\phi}_t = \boldsymbol{\gamma}_t,$$

where $\boldsymbol{\Gamma}_t = (\gamma_{|i-j|})_{i,j=1,\ldots,t-1}$ and $\boldsymbol{\gamma}_t = (\gamma_1, \ldots, \gamma_{t-1})'$. The Durbin–Levinson method for recursive computing $\boldsymbol{\phi}_t$ is provided in Brockwell and Davis (1991, Chapter 5). In case Y_t follows an AR(p) process the forecast of Y_t for $t > p$ is given by

$$\hat{Y}_t = \mu_0 + \sum_{i=1}^{p} \alpha_i(Y_{t-i} - \mu_0).$$

For the starting p values of the process the forecast is more complicated.

5.3 Modeling

Using the tools of SPC the practitioner aims to decide at each moment of time whether the observed process still possesses the required characteristics or if there are some substantial deviations. Thus we need to distinguish between the observed process $\{X_t\}$ and the target process $\{Y_t\}$. The observations $x_1, x_2, \ldots$ are considered to be a realization of the stochastic process $\{X_t\}$. The target process satisfies some predetermined requirements or, as is usually the case in financial applications, it determines the true underlying data generating process. If the observed process coincides with the target process we say that the process in 'in control' and we say that the process is 'out of control' otherwise.

Stating the relationship between $\{X_t\}$ and $\{Y_t\}$ is a very important issue, since it explicitly determines what kind of changes we allow to occur. If we restrict the discussion only to changes in the mean and in the variance, then this relationship can be written as

$$X_t = \begin{cases} \mu_0 + \delta\sqrt{\gamma_0} + \Delta(Y_t - \mu_0) & \text{for } t \geq \tau \\ Y_t & \text{for } t < \tau. \end{cases} \tag{5.1}$$

Here $\{Y_t\}$ denotes the target process or in-control process. It is a univariate time series. This means that at each time point exactly one observation is available. Moreover, it is assumed to be stationary with mean μ_0 and autocovariance function $\{\gamma_h\}$. The parameter $\tau \in I\!N \cup \{\infty\}$ is assumed to be an unknown deterministic quantity. δ measures the shift in the mean in standard deviation units of $\{Y_t\}$ and $\Delta > 0$ measures the relative shift in the standard deviation itself. It is assumed that either $\delta \neq 0$ or $\Delta \neq 1$. Thus the process is called in control if $\tau = \infty$ and it is out of control if $\tau < \infty$. We can write the out-of-control characteristics of the process $\{X_t\}$ in terms of the in-control characteristics of the process $\{Y_t\}$. It holds that

$$E(X_t) = \begin{cases} \mu_0 + \delta\sqrt{\gamma_0} & \text{for} \quad t \geq \tau \\ \mu_0 & \text{for} \quad 1 \leq t < \tau, \end{cases}$$

$$\mathrm{Var}(X_t) = \begin{cases} \Delta^2 \mathrm{Var}(Y_t) & \text{for} \quad t \geq \tau \\ \mathrm{Var}(Y_t) & \text{for} \quad 1 \leq t < \tau, \end{cases}$$

$$\mathrm{Cov}(X_t, X_{t+h}) = \begin{cases} \Delta^2 \gamma_h & \text{for} \quad t \geq \max\{\tau, \tau - h\} \\ \Delta \gamma_h & \text{for} \quad \min\{\tau, \tau - h\} \leq t < \max\{\tau, \tau - h\} \\ \gamma_h & \text{for} \quad t < \min\{\tau, \tau - h\}, \end{cases}$$

$$\mathrm{Corr}(X_t, X_{t+h}) = \rho_h.$$

Note that the change does not influence the autocorrelation of $\{X_t\}$ and that the observed process $\{X_t\}$ is not stationary if $\tau < \infty$.

In the following we use the notation E_C, Var_C, Cov_C, etc. to denote that there is a change at position $\tau = 1$ as described in (5.1). If there is no change, i.e. $\tau = \infty$, we use the notation E_D, Var_D, etc.

5.4 Introduction to control charts for univariate time series

Walter Shewhart can be considered to be the founder of statistical process control (SPC). He invented the idea of a control chart which nowadays can be regarded to be one of the main tools in statistics. In the 1920s and 1930s he introduced various control charts for the mean and for the variance (e.g. Shewhart 1931). A disadvantage of the charts proposed by Shewhart consists in the fact that they do not have a memory. Control charts with a memory are, e.g. the Exponentially Weighted Moving Average (EWMA) chart of Roberts (1959) and the CUmulative SUM (CUSUM) control chart of Page (1954), respectively. In these attempts an additional parameter controls the impact of the past values. Initially all these control charts were proposed for independent data and have been extended to the more practically relevant case of dependent data during the last 20 years. Many papers have been published about control charts for time series within the last years. Most of the papers deal with the surveillance of the mean behaviour of a time series. An overview and discussion about newer developments can be found in Knoth and Schmid (2004). The optimality properties of EWMA charts are discussed in the recent paper of Frisén and Sonesson (2006).

In this section we consider two-sided control charts for the mean of a univariate time series. This means that we want to detect a change in the mean. It may be an increase or a decrease. The underlying model is

$$X_t = \begin{cases} \delta\sqrt{\gamma_0} + Y_t & \text{for} \quad t \geq \tau \\ Y_t & \text{for} \quad t < \tau, \end{cases} \tag{5.2}$$

where $\delta \neq 0$ and Y_t is a stationary process with mean μ and autocovariance function $\{\gamma_v\}$.

We will not consider one-sided schemes in the following. They can be derived in a straightforward way from the two-sided procedures.

5.4.1 Modified control charts

Classical control charts can be adapted to time series case in two ways. On the one hand the original time series can be transformed to an independent series and then the classical monitoring procedures are applied to the transformed quantities. Such charts were proposed by Alwan and Roberts (1988) and have been sequentially discussed by several authors like, e.g. Harris and Ross (1991), Lu and Reynolds (1999), Montgomery and Mastrangelo (1991) and Wardell *et al.* (1994a,b). This approach is discussed in Section 5.4.2. The alternative approach is based on the direct modification of the classical procedures to incorporate the dependency structure of the process (see, e.g. Nikiforov 1975; Schmid 1995, 1997a,b; Vasilopoulos and Stamboulis 1978; Yashchin 1993).

5.4.1.1 Modified Shewhart control chart

Shewhart charts are based on the use of the present observation. Previous observations are not taken into account within the decision rule. The modified Shewhart control chart gives a signal at time t if

$$|X_t - \mu_0| > c\sqrt{\gamma_0}\,.$$

The constant $c > 0$ is called the critical value and it defines how wide the control limits should be. The quantities μ_0 and γ_0 are assumed to be known. This control chart is the direct extension of the classical Shewhart control chart to time series. Note that the difference between the observed value and its in-control mean is normalized by the standard deviation of the target process, i.e. $\sqrt{\gamma_0}$. Thus the structure of the chart is the same as in the iid case. A thorough discussion of the modified Shewhart chart was given by Schmid (1995).

In order to assess the performance of a control chart it is necessary to define suitable performance measures (e.g. Frisén 2003; Frisén and de Maré 1991; Lorden 1971; Pollak and Siegmund 1975). All relevant criteria are based on the run length of the control chart which is defined as

$$t_A = \inf\{t \in \mathbb{N} : |X_t - \mu_0| > c\sqrt{\gamma_0}\}.$$

The run length is equal to the number of observations taken until the first alarm of the chart. The most popular measures of the performance are the average run length (ARL) and the expected delay (ED). The ARL measures the average number of observations until the first alarm. The in-control ARL is given by $\mathrm{E}_D(t_A)$. In the out-of-control state it is assumed that the shift occurs at the first time point. Thus the out-of-control ARL is equal to $\mathrm{E}_C(t_A)$. Sometimes we prefer the notation ARL$(c; \delta)$, ARL(δ), or ARL(c). Then ARL$(c; \delta) = \mathrm{E}_C(t_A)$ for $\delta \neq 0$ and ARL$(c; 0) = \mathrm{E}_D(t_A)$. The ARL suffers under the assumption that in the out-of-control case the shift already arises at the first position.

Another important measure of the control chart performance is the conditional expected delay defined as

$$\mathrm{CED}_\tau = \mathrm{E}_\tau(t_A - \tau + 1 | t_A \geq \tau).$$

It measures the average number of observations taken from the change point to the signal conditional on the fact that the alarm occurs not before the shift at time τ. In order to have a measure independent of τ usually the supremum or the limit of this quantity is considered.

The choice of the critical value c is of crucial importance for the setup of the chart because it determines the width of the decision interval. Many practitioners prefer to work with a constant value of c. Mainly c is chosen equal to 3. Using such an approach it is impossible to control the probability of a false alarm. This is a great disadvantage of such a procedure. For that reason it is better to choose c by taking into account the ARL of the chart. Then, however, it is necessary to know the distribution of the target process. In most cases $\{Y_t\}$ is assumed to be a Gaussian process. Now, first, a desired value of the in-control ARL ξ is fixed and then c is determined such that the in-control ARL at position c is equal to ξ. This means that c is the solution of the equation $\mathrm{ARL}(c; 0) = \xi$. The practical calculation of c may be difficult even in the iid case. Several numerical methods have been proposed for calculating the ARL of the EWMA and CUSUM chart (e.g. Brook and Evans 1972; Crowder 1987). In the time series case the situation is much harder and no explicit formulas of the ARLs are available. This is the reason why in most cases simulations are used to estimate the ARL or the CED.

Choosing c in that way we observe an important difference between the classical Shewhart chart and the modified Shewhart chart. For time series the critical value c depends not only on ξ but on all parameters of the in-control process as well. For that reason it is necessary to determine these values for each process individually. Because c depends on the correlation structure of the target process we are faced with the situation that the modified Shewhart chart depends in some sense on the history of the process over the critical value c.

5.4.1.2 Modified EWMA control chart

The EWMA chart for independent observations was introduced by Roberts (1959) and it was extended to time series by Schmid (1997b). Here we consider a more general approach. Let $L_{\mu,t} = L_{\mu,t}(X_t, X_{t-1}, \ldots)$ denote some local measure of location at time point t fulfilling the requirement that $E_D[L_{\mu,t}(X_t, X_{t-1}, \ldots)] = \mu_0$. We can use, e.g. $L_{\mu,t}(X_t, X_{t-1}, \ldots) = X_t$ or more generally

$$L_{\mu,t}(X_t, X_{t-1}, \ldots) = \frac{1}{k}\sum_{j=0}^{k-1} X_{t-j} \tag{5.3}$$

or

$$L_{\mu,t}(X_t, X_{t-1}, \ldots) = \sum_{j=0}^{k-1} w_j X_{t-j} \tag{5.4}$$

with nonnegative weights w_j satisfying $\sum_{j=0}^{k-1} w_j = 1$.

The EWMA recursion is applied to the local location measure

$$Z_t = (1-\lambda)Z_{t-1} + \lambda L_{\mu,t} \, . \tag{5.5}$$

In practice the computation of Z_t is usually hindered by a starting problem if not enough observations are available for the evaluation of Z_t for $t \geq 1$. Frequently this problem is ignored in the literature.

The EWMA scheme gives a signal at time $t \geq 1$ if $|Z_t - \mu_0| > c\sqrt{\mathrm{Var}_D(Z_t)}$. Then the run length is given by

$$t_A = \inf\{t \in \mathbb{N} : |Z_t - \mu_0| > c\sqrt{\mathrm{Var}_D(Z_t)}\}.$$

The use of reflecting barriers presents a further possible modification of the proposed scheme (see, e.g. Morais 2001). Here a chart is used to detect upper changes and another one for lower changes. Each chart is restarted if one observation lies below and above the target value, respectively. The run length of the joint scheme is equal to the minimum of the two one-sided schemes.

In the following we focus on the simple location measure $L_{\mu,t} = X_t$. Then the EWMA statistic is computed as

$$Z_t = (1 - \lambda)Z_{t-1} + \lambda X_t, \qquad t \geq 1, \qquad Z_0 = z_0. \tag{5.6}$$

The starting value z_0 is usually set equal to the target mean μ_0. This choice is always assumed in the following. The smoothing parameter $\lambda \in (0, 1]$ controls the impact of past observations. If $\lambda = 1$ we obtain the Shewhart control chart. Solving the difference equation (5.6) leads to the following presentation

$$Z_t = \mu_0(1 - \lambda)^t + \lambda \sum_{i=1}^{t} (1 - \lambda)^{t-i} X_i.$$

Thus large values of λ lead to large impacts of the current observations and lower impacts of past values. The inverse situation is observed for small values of λ. In the in-control case it is found that

$$\mathrm{E}_D(Z_t) = \mu_0, \quad \mathrm{Var}_D(Z_t) = \lambda^2 \sum_{|i| \leq t-1} \gamma_i \sum_{j=\max\{0,-i\}}^{\min\{t-1,t-1-i\}} (1 - \lambda)^{2j+i}.$$

This control chart was introduced by Schmid (1997b). An analysis of its ARL behaviour in the one-sided case is given in Schmid and Schöne (1997) and Schöne *et al.* (1999).

5.4.1.3 Modified CUSUM control chart

CUSUM control charts, introduced by Page (1954), can be linked to the sequential probability ratio test. In contrast to the modified EWMA approach, there are several possibilities to extend the CUSUM procedure to dependent data. They all lead to the same CUSUM procedure in case of independent data, however, they differ slightly in case of dependency. Here we apply a procedure proposed by Schmid (1997a).

Let $f_0(\mathbf{X}_t)$ denote the Gaussian density of the observed vector $\mathbf{X}_t = (X_1, \ldots, X_t)'$ at time t in the in-control state, i.e., if there there is no shift. Let $\boldsymbol{\mu}_t = \mu_0(1, \ldots, 1)'$, γ_0, and $\mathbf{C}_t$ denote the corresponding mean vector, variance, and correlation matrix. By $f_{\delta,m}(\mathbf{X}_t)$ we denote the density if there is a shift of size δ at time m. Then it can be shown that

$$-2 \log \frac{f_0(\mathbf{X}_t)}{\max_{0\le m\le t} f_{\delta,m}(\mathbf{X}_t)} = 2 \max\left\{0, \max_{1\le m\le t} \delta \frac{T_{mt} - \delta\sqrt{\gamma_0}K_{mt}/2}{\sqrt{\gamma_0}}\right\}$$

where $T_{mt} = \mathbf{e}'_{mt}\mathbf{C}_t^{-1}(\mathbf{X}_t - \boldsymbol{\mu}_t)$ and $K_{mt} = \mathbf{e}'_{mt}\mathbf{C}_t^{-1}\mathbf{e}_{mt}$. $\mathbf{e}_{mt}$ denotes a column vector with zeros in the first $m-1$ positions and ones in the remaining $t-m+1$ positions. The control charts gives a signal if

$$\max_{1\le m\le t} \left\{0, T_{mt} - K\sqrt{\gamma_0}K_{mt}, \ -(T_{mt} + K\sqrt{\gamma_0}K_{mt})\right\} > h\sqrt{\gamma_0}.$$

The reference value K is taken as in the case of independent observations equal to $|\delta|/2$. For large values of t the evaluation of these quantities may be time demanding. However, it is possible to compute the test statistic recursively (see Schmid 1997a). For an AR(1) model the run length is given by

$$t_A = \inf\left\{t \in \mathbb{N} : S_t^+ > h\sqrt{\gamma_0} \quad \text{or} \quad S_t^- < -h\sqrt{\gamma_0}\right\},$$

with

$$S_t^+ = \max\{0, \epsilon_t - K\sqrt{\gamma_0}, S_{t-1}^+ + (1-\alpha_1)(\epsilon_t - (1-\alpha_1)K\sqrt{\gamma_0})\}, \quad t \ge 2,$$
$$S_1^+ = \max\{0, (1-\alpha_1^2)(X_1 - \mu_0 - K\sqrt{\gamma_0})\},$$

$$S_t^- = \min\{0, \epsilon_t + K\sqrt{\gamma_0}, S_{t-1}^- + (1-\alpha_1)(\epsilon_t + (1-\alpha_1)K\sqrt{\gamma_0})\}, \quad t \ge 2,$$
$$S_1^- = \min\{0, (1-\alpha_1^2)(X_1 - \mu_0 + K\sqrt{\gamma_0})\}.$$

If the sign of the change is known, then just a single chart based either on S_t^+ or on S_t^- should be applied. If the sign is unknown then we implement a two-sided CUSUM chart by running simultaneously two one-sided charts with run lengths t_A^+ and t_A^- respectively. The run length of the two-sided chart is equal to $t_A = \min\{t_A^-, t_A^+\}$.

The SPRT approach applied to time series by Nikiforov (1975) compares the densities $f_0(\mathbf{X}_t)$ and $f_{\delta,1}(\mathbf{X}_t)$. The procedure is based on

$$T_t = \log \frac{f_{\delta,1}(\mathbf{X}_t)}{f_0(\mathbf{X}_t)} = \delta \frac{T_{1t} - \delta\sqrt{\gamma_0}K_{1t}/2}{\sqrt{\gamma_0}}.$$

The chart gives an alarm at time t if

$$\max_{1\le m\le t}\left\{T_{1m}-K\sqrt{\gamma_0}K_{1m},\ -(T_{1m}+K\sqrt{\gamma_0}K_{1m})\right\} > h\sqrt{\gamma_0}.$$

The computation can be simplified by using a recursion similar to that applied in Schmid (1997a).

The critical value h is chosen in the same way as for the modified EWMA chart. We demand the in-control average run length to be equal to a fixed value ξ and h is determined as the solution of this equation.

5.4.2 Residual control charts

Residual charts constitute an alternative approach to monitor time series data. They have been intensively discussed in the last few years (see Alwan and Roberts 1988; Harris and Ross 1991; Lu and Reynolds 1999; Montgomery and Mastrangelo 1991). The idea is to transform the original dependent observations X_t to an independent and identically distributed time series ϵ_t. Following the discussion of Section 5.2.2 we define the residuals by $\hat{\epsilon}_t = X_t - \hat{X}_t$, where $\hat{X}_t$ is the best linear forecast of X_t in the in-control state as discussed in Section 5.2.2. This implies that

$$\hat{\epsilon}_t = X_t - \mu_0 - \sum_{i=1}^{t-1}\phi_{ti}(X_{t-i}-\mu_0).$$

It is easy to see that for a Gaussian process $\{Y_t\}$ the variables $\{\hat{\epsilon}_t\}$ are independent. However, they are identically distributed only in the in-control state but not in the out-of-control situation.

In the next step the residuals are normalized to have the same variance. For example, if in the in-control state the target process follows an AR(1) process, then the best linear forecast is given by $\hat{X}_t = \mu_0 + \alpha_1(X_{t-1}-\mu_0)$ for $t \ge 2$ and $\hat{X}_1 = \mu_0$. This implies that the variance of the forecast error $\hat{\epsilon}_t = \hat{X}_t - X_t$ is given by $\mathrm{Var}_D(\hat{\epsilon}_t) = \sigma^2/(1-\alpha_1^2)$ for $t \ge 2$ and $\mathrm{Var}_D(\hat{\epsilon}_1) = \sigma^2$. The normalized residuals are then given by

$$\hat{\Delta}_t = \frac{X_t - \mu_0 - \alpha_1(X_{t-1}-\mu_0)}{\sigma}, \qquad \text{for} \quad t \ge 2,$$

$$\hat{\Delta}_1 = \frac{\sqrt{1-\alpha_1^2}(X_1-\mu_0)}{\sigma}.$$

Now the normalized residuals are independently and identically distributed if the observed process is in control. Thus standard control charts for independent data can be directly applied to $\{\hat{\Delta}_t\}$.

A great advantage of the residual charts is that the control limits do not depend on the parameters of the target process. Thus the determination of these values is much easier than for modified charts. A disadvantage lies in the fact that we monitor not the observations directly but the residuals and, thus, the interpretation of a signal may be difficult. Many practitioners favour monitoring the original process. A second disadvantage is that the residuals are no longer identically distributed if the process is out of control and thus they do not completely behave like the classical charts in the out-of-control state.

Residual control charts have been generalized to so-called ARMA charts (Jiang *et al.* 2000) where the transformation is aimed at detecting a shift as soon as possible.

5.4.3 Neglecting the time series structure

In this section we analyse the consequences of the application of standard control charts for iid data to dependent data. We assume that the true process is an AR(1) process with known coefficient α_1 and Gaussian white noise $\{\epsilon_t\}$. The variance of the target process γ_0 is assumed to be known. In practice the variance can be estimated using a pre-run, which is assumed to be generated from the target process.

In the out-of-control state we assume a shift in the process mean of size $\sqrt{\gamma_0}$, i.e. $\delta = 1$. The in-control ARL is set to $\xi = 500$. In the setup of the EWMA charts we set the smoothing parameter equal to $\lambda = 0.5$ and the parameter K of the CUSUM chart is set to 0.5. In this case we expect the fastest detection of the shift of size $\delta = 1$ (cf. Lucas and Saccucci 1990). This procedure leads to the following stopping times $t_{A,s}$, $t_{A,e}$ and $t_{A,c}$ for three different charts (Shewhart, EWMA, CUSUM)

$$
\begin{aligned}
t_{A,s} &= \inf\{t \in \mathbb{N} : |X_t| \geq 3.0902\sqrt{\gamma_0}\},\\
t_{A,e} &= \inf\{t \in \mathbb{N} : |Z_t| \geq 2.8143\sqrt{\lambda/(2-\lambda)}\sqrt{\gamma_0}\}, \quad \text{with} \quad Z_0 = 0, \quad \lambda = 0.1,\\
t_{A,c} &= \inf\{t \in \mathbb{N} : \max\{-S_t^-, S_t^+\} \geq 5.0707\sqrt{\gamma_0}\},\\
&\qquad \text{with} \quad S_0^- = S_0^+ = 0, \quad k = 0.5.
\end{aligned}
$$

Let ARL_s, ARL_e, and ARL_c denote the corresponding average run lengths. In the in-control $\delta = 0$ state all charts have the same ARL of 500. If, however, $\delta = 1$ then

$$\text{ARL}_s(c_s; \delta = 1) = 54.6, \ \ \text{ARL}_e(c_e; \delta = 1) = 10.3, \ \ \text{ARL}_c(c_c; \delta = 1) = 10.5.$$

Figure 5.2 illustrates the impact of the application of the classical Shewhart control chart to autocorrelated data. The graph shows the in-control ARL as a

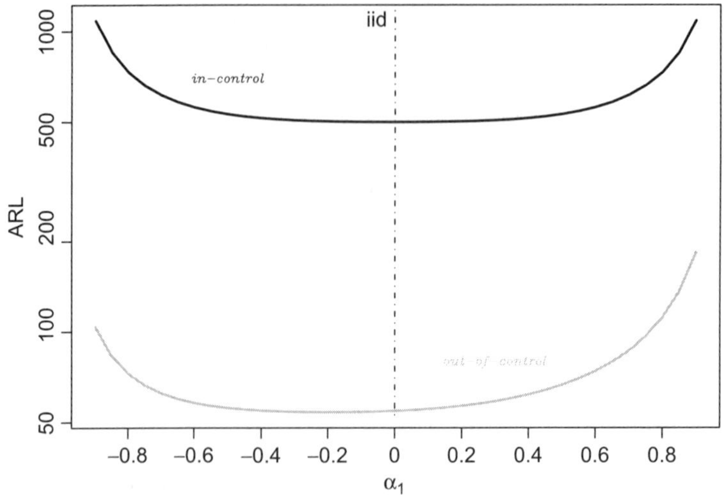

Figure 5.2 ARL of the standard Shewhart chart in the presence of autocorrelation. In the out-of-control state (grey lines) δ is set to one.

function of α_1. Below the graph of the the out-of-control ARL is given. We see that the chart is relatively robust and for $|\alpha_1| < 0.5$ the changes in the ARL are minor. Moreover, we see that the in-control ARL is symmetric with respect to α_1 and the out-of-control ARL is more sensitive to positive correlation than to a negative one. A completely different situation is observed in Figure 5.3 for the EWMA and CUSUM charts. Both charts are extremely sensitive even to a minor correlation in the data in the in-control state. Moreover, the in-control ARL decreases with positive correlation and increases with negative correlation. For example, for $\alpha_1 = 0.2$ the in-control ARL for CUSUM chart becomes 155 instead of the expected value of $\xi = 500$. For the Shewhart chart such correlation leads only to a marginal increase to 502.5. For negative autocorrelation the impact is even more dramatic. For $\alpha_1 = -0.1$ the ARL of the CUSUM chart increases to 1121. In the out-of-control state all charts appear to be more robust to dependencies. Similarly as for Shewhart chart the deviations are larger for positive correlation and smaller for negative correlation.

The fact that the Shewhart chart appears to be more robust to dependencies than the EWMA chart seems to be surprising. However, note that the the process gives a signal if the deviation of the control statistics exceeds some multiple of its standard deviation. This correct standard deviation is assumed to be known for the Shewhart chart, but this is not the case for EWMA scheme. This is the reason why the latter may show a worse performance.

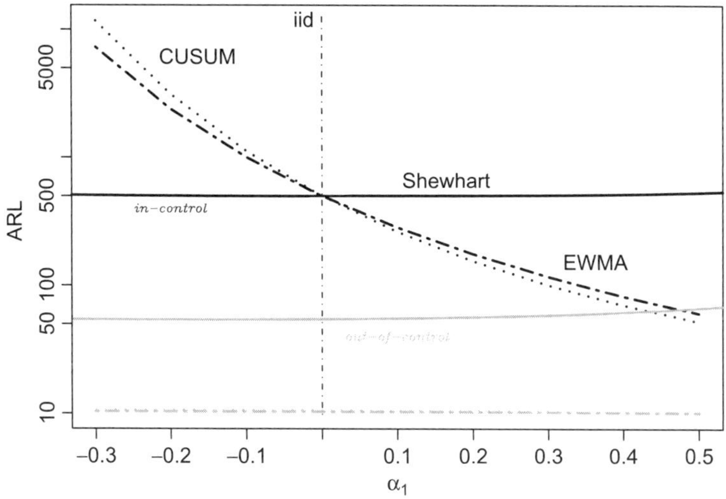

Figure 5.3 ARL of the standard Shewhart (solid line), EWMA (dashed line), and CUSUM (dotted line) charts in the presence of autocorrelation. In the out-of-control state (grey lines) δ is set to one.

5.4.4 Comparison study

Knoth and Schmid (2004) provide an extensive review and comparison of the introduced charts on the example of an AR(1) process. Here we only briefly restate the main findings and the interested reader is referred to the original paper. For negative α_1 the modified charts, especially based on the approach EWMA, tend to outperform in terms of the out-of-control ARL. However, for positive changes the residual charts become better. Moreover, the authors also documented the known fact fact that EWMA charts tends to detect small changes faster than CUSUM charts, while CUSUM outperforms for shifts $\delta > 1$. Concerning the design parameters it is concluded that the residual and the modified charts lead in general to similar parameters. Also it can be observed that for the residual charts the values are close to those of the classical charts for a shift of size $(1 - \alpha_1)\delta$.

5.5 Surveillance of the variance of a univariate linear time series

Some financial time series like asset returns rarely exhibit some shifts in the mean behaviour. However, changes in the risk of an asset are much more common. The importance of the variance as a risk measure leads to an increased attention to its modelling. Starting with the seminal results of Engle (1982) GARCH models and their modifications play a key role in volatility modeling. Further, the modelling and forecasting of stochastic volatility became extremely popular. The availability of high-frequency data led to new techniques of estimating the variance from intra-day data. This motivates the need for monitoring tools, that can detect a shift in the variance of an observed process as quickly as possible. This section deals with monitoring techniques for the variance of a linear process. The surveillance of the variance behaviour of a nonlinear process is the subject of Section 6.3.

For risk monitoring, similarly as for location charts, two different approaches are available in case of data dependency. One approach is to modify the classical charts, so that they can be directly applied to time series. Alternatively, we transform the original autocorrelated series to an uncorrelated series and the classical charts are applied.

As in Section 5.4 the target process $\{Y_t\}$ is assumed to be a stationary process with mean μ_0 and autocovariance function $\{\gamma_h\}$. The shift in the variance is modelled as in (5.1) but with $\delta = 0$ and $\Delta \neq 1$. This means that we completely focus on a change in the variance while the mean is supposed to stay in control.

5.5.1 Modified control charts for the variance

5.5.1.1 EWMA type charts

The first EWMA control chart for time series was introduced by MacGregor and Harris (1993). Schipper and Schmid (2001) introduced several one-sided variance charts for stationary processes, however, their main focus was in the area of nonlinear time series.

Our starting point is some local measure $L_{V,t} = L_{V,t}(X_t, X_{t-1}, \ldots)$ for the variation. It is assumed that in the in-control state it is unbiased. There are many possibilities to choose such a measure. The easiest way is to take $(X_t - \mu_0)^2$. Another possibility would be, e.g. $\sum_{j=0}^{k-1}(X_{t-j} - \mu_0)^2$. Then previous observations are taken into account as well. In both cases the in-control mean is equal to γ_0. It is possible to make use of other variation measures. This is done if we choose, e.g.

$$L_{V,t} = \frac{1}{k}\sum_{j=0}^{k-1} |X_{t-j} - \mu_0| \quad \text{or} \quad L_{V,t} = \underset{j=0,\ldots,k-1}{\text{MAD}} |X_{t-j} - \mu_0|$$

where MAD stands for the median of the absolute distances. Another popular approach is to use a variation measure for a transformed quantity. An example is, e.g. the quantity $\ln((X_t - \mu_0)^2/\gamma_0)$.

Applying the EWMA recursion to such a measure leads to

$$Z_t = (1 - \lambda)Z_{t-1} + \lambda L_{V,t}(X_t, X_{t-1}, \ldots).$$

As described for the mean charts there may be a starting problem in calculating Z_t for $t \geq 1$. Further we provide a detailed discussion of two of the most popular measures.

Chart based on squared observations MacGregor and Harris (1993) were the first to apply the EWMA recursion to the squared observations. Let

$$Z_t = (1 - \lambda)Z_{t-1} + \lambda(X_t - \mu_0)^2, \qquad t \geq 1, \qquad Z_0 = \gamma_0.$$

This scheme was analysed in detail by Schipper and Schmid (2001). They proved the following result.

Proposition 1 Let $\{Y_t\}$ be a stationary process with mean μ_0 and autocovariance function $\{\gamma_h\}$.

(a) Then

$$\begin{aligned} E(Z_t) &= \gamma_0 + \gamma_0\,(\Delta^2 - 1)\,[1 - (1 - \lambda)^{t-\tau+1}]\; I_{\{\tau,\tau+1,..\}}(t) \\ &\to \Delta^2\,\gamma_0 \qquad \text{as} \qquad t \to \infty\,. \end{aligned}$$

(b) Let $\{(Y_t - \mu_0)^2\}$ be stationary with autocovariance function $\{\delta_h\}$. Then it follows for $\tau = 1$ that

$$\mathrm{Var}(Z_t) \;=\; \Delta^4\,\lambda^2 \sum_{v=-(t-1)}^{t-1} \delta_v \sum_{i=\max\{0,-v\}}^{\min\{t-1-v,t-1\}} (1 - \lambda)^{2i+v}\,.$$

If additionally $\{\delta_h\}$ is absolutely summable then

$$\begin{aligned} \lim_{t\to\infty} \mathrm{Var}(Z_t) &= \Delta^4\,\frac{\lambda}{2 - \lambda} \sum_{v=-\infty}^{\infty} \delta_v\,(1 - \lambda)^{|v|} \\ &= \Delta^4\,\lambda^2 \int_{-\pi}^{\pi} f_{Y^2}(x)\,\frac{1}{1 - 2\,(1 - \lambda)\,\cos x + (1 - \lambda)^2}\,dx\,. \end{aligned}$$

f_{Y^2} denotes the spectral density of the stationary process $\{(Y_t - \mu_0)^2\}$.

(c) Let $\{Y_t\}$ be a two-sided moving average with mean μ_0, i.e. $Y_t = \mu_0 + \sum_{i=-\infty}^{\infty} a_i \, \varepsilon_{t-i}$ with $\{a_i\}$ absolutely summable. Let $\{\varepsilon_t\}$ be independent and normally distributed with mean 0 and variance σ^2. Then $\delta_h = \gamma_0^2 + 2\gamma_h^2$.

For a stationary ARMA(1,1) process it follows that for $\tau = 1$

$$\lim_{t\to\infty} \mathrm{Var}(Z_t) = \Delta^4 \frac{1}{2-\lambda} \left[2\gamma_0^2 + \lambda\gamma_0^2 + 4\gamma_1^2 \frac{\lambda(1-\lambda)}{1-(1-\lambda)\alpha_1^2} \right]$$

with

$$\gamma_0 = \sigma^2 \frac{1 + 2\alpha_1\beta_1 + \beta_1^2}{1-\alpha_1^2} \qquad \text{and} \qquad \gamma_1 = \sigma^2 \frac{(1+\alpha_1\beta_1)(\alpha_1+\beta_1)}{1-\alpha_1^2} .$$

In finance we are interested in detecting not only an increase of the risk measure but also to detect a decrease in the risk. For that reason we consider a two-sided control scheme. The two-sided chart gives an alarm at time t if

$$Z_t < c_1\gamma_0 \quad \text{or} \quad Z_t > c_2\gamma_0,$$

where $c_1,\ c_2 > 0$. Because the distribution of Z_t is skewed we do not have symmetric control limits area around the target value as in the case of mean charts for Gaussian processes. For that reason an asymmetric interval is chosen which is determined by the parameters c_1 and c_2. These quantities are determined such that the in-control ARL of the chart is equal to a specified value ξ, i.e.

$$\mathrm{E}_D(\inf\{t \in I\!N : Z_t < c_1\gamma_0 \quad \text{or} \quad Z_t > c_2\gamma_0\}) = \xi. \tag{5.7}$$

Because we have two parameters we add a side restriction in order to have a unique solution. It is demanded that the in-control ARL of both one-sided charts is the same. Consequently (5.7) is solved under the restriction that

$$\mathrm{E}_D(\inf\{t \in I\!N : Z_t < c_1\gamma_0\}) = E_D(\inf\{t \in I\!N : Z_t > c_2\gamma_0\}).$$

It is important to note that solving this problem numerically is computationally more demanding compared with the case with a single critical value. In our applications we used Broyden's version of the bivariate secant method. The method converges sufficiently quickly for large number of repetitions and not very large values of α_1.

A chart based on the logarithm of the squared observations By taking the logarithm of the squared quantities in (5.2) the model for a scale deviation is transformed to a model with a shift. We get

$$\ln \frac{(X_t - \mu_0)^2}{\gamma_0} = \ln \frac{(Y_t - \mu_0)^2}{\gamma_0} + 2\ln(\Delta) \qquad \text{for} \qquad t \geq \tau.$$

Consequently all control charts for the mean can be applied to $\ln((X_t - \mu_0)^2/\gamma_0)$. If we apply an EWMA recursion we get

$$Z_t = (1-\lambda)\, Z_{t-1} + \lambda\, L_{\mu,t}\left[\ln \frac{(X_t-\mu_0)^2}{\gamma_0}, \ln \frac{(X_{t-1}-\mu_0)^2}{\gamma_0}, \dots\right],$$

for $t \geq 1$ and with $Z_0 = E(\ln(Y_t - \mu_0)^2/\gamma_0) = \gamma_0^*$. For independent observations this chart was considered by Crowder and Hamilton (1992). The variance of the control statistic can be computed as for the mean chart, however, with θ_h^* instead of δ_h, which denotes the autocovariance function of $\ln(Y_t - \mu_0)^2/\gamma_0$. Explicit expressions are difficult to obtain even in the asymptotic case. The chart signals a shift if

$$Z_t < c_1\gamma_0^* \quad \text{or} \quad Z_t > c_2\gamma_0^* \,.$$

The quantities $c_1 > 0$ and $c_2 > 0$ are chosen in the same way as described above.

5.5.1.2 CUSUM type charts

Because of the complicated structure of the probability distribution of a stationary Gaussian process, variance charts for such processes up to now have not been derived directly over the LR approach or the SPRT. The starting point of the considerations is an independent Gaussian sample. The CUSUM scheme for independent variables is derived and after that the same recursion is applied to stationary processes as well.

For an iid Gaussian series the log likelihood ratio for fixed and known value of τ is given by

$$\ln \frac{f_{0,\Delta}(\mathbf{X}_t)}{f_0(\mathbf{X}_t)} = \frac{1-1/\Delta^2}{2\gamma_0}(S_t - t\gamma_0 K(\Delta) - (S_{\tau-1} - (\tau-1)\gamma_0 K(\Delta)))$$

for $t \geq \tau$ and with $S_t = \sum_{i=1}^{t}(X_i - \mu_0)^2$ and $K(\Delta) = 2\ln \Delta^{1/(1-1/\Delta^2)}$. We conclude that the process is out of control if

$$S_t^+ = S_t - t\gamma_0 K(\Delta) - \min_{0\leq m\leq t}(S_m - m\gamma_0 K(\Delta)) \geq c_2\gamma_0$$

or

$$S_t^- = S_t - t\gamma_0 K(\Delta) - \max_{0\leq m\leq t}(S_m - m\gamma_0 K(\Delta)) \leq -c_1\gamma_0 \,.$$

Note that we prefer to work with two different critical values $c_1 > 0$ and $c_2 > 0$ and not with one quantity as proposed in Hawkins and Olwell (1998). For computational convenience S_t^+ and S_t^- can also be computed recursively by

$$S_t^+ = \max\left\{0, S_{t-1}^+ + (X_t - \mu_0)^2 - K(\Delta)\gamma_0\right\}, \; t \geq 1, \quad S_0^+ = 0$$
$$S_t^- = \min\left\{0, S_{t-1}^- + (X_t - \mu_0)^2 - K(\Delta)\gamma_0\right\}, \; t \geq 1, \quad S_0^- = 0.$$

Note that in contrast to the mean charts, the expression for the reference value $K(\Delta)$ is complicated. The practitioner fixes the shift size Δ he wants to be protected against, and then computes the corresponding value of K.

For the above considerations the target process was assumed to be an independent normal sample. Now we assume that it is a stationary process. We apply the recursions from (5.8) directly to the observed time series. Because $K(\Delta)$ is unknown we replace it by a constant $K \geq 0$. Thus we obtain

$$S_t^+ = \max\left\{0, S_{t-1}^+ + (X_t - \mu_0)^2 - K\gamma_0\right\}, \; t \geq 1, \quad S_0^+ = 0$$
$$S_t^- = \min\left\{0, S_{t-1}^- + (X_t - \mu_0)^2 - K\gamma_0\right\}, \; t \geq 1, \quad S_0^- = 0.$$

The process is concluded to be out-of-control if $S_t^+ > c_2\gamma_0$ or $S_t^- < -c_1\gamma_0$. For each fixed value of K we determine the value of c such that the in-control ARL is equal to some predetermined quantity ξ.

Next we describe a chart based on the logarithm of the squared observations. As motivated above we can apply a CUSUM chart for the mean to the transformed data. We consider the CUSUM recursion of the mean chart in the iid case and apply it to a stationary process. Assuming that the distribution of Y_t is symmetric with respect to μ_0 we get

$$S_t^+ = \max\left\{0, S_{t-1}^+ + \ln\frac{(X_t - \mu_0)^2}{\gamma_0} - K\right\}, \; t \geq 1$$
$$S_t^- = \min\left\{0, S_{t-1}^- + \ln\frac{(X_t - \mu_0)^2}{\gamma_0} + K\right\}, \; t \geq 1$$

with $S_t^+ = 0$, $S_t^- = 0$, and $K \geq 0$. The chart gives an alarm if

$$S_t^+ > c \quad \text{or} \quad S_t^- < -c.$$

5.5.2 Residual control charts for the variance

In Section 5.4.2 we have seen that in the case of mean shifts the residuals $\hat{\varepsilon}_t = X_t - \hat{X}_t$ are still independent variables. They are identically distributed in the in-control state but are not iid in the out-of-control situation. Note that $\hat{X}_t$ stands for the best linear predictor of X_t given $X_{t-1}, \ldots, X_1, 1$. This quantity was calculated under the assumption that the process is in control. Here we are dealing with changes in the variances and the question arises how the residuals behave in the case of variance changes?

Of course in the in-control state the residuals are independent and identically distributed if the target process is a Gaussian process. However, in the out-of-control situation they are neither independent nor identically distributed. This shows that we can apply all of the classical control charts for the variance of independent variables to the residuals. We can use the same control limits as in the iid case. But we cannot expect that the out-of-control behaviour of the charts is the same as in the iid situation. Each residual chart has to be analysed separately with respect to the underlying dependency structure.

5.5.3 Neglecting the time series structure

Here we assess the impact of the time dependency on the performance of classical control charts as in Section 5.4.3, however, for variance charts. We consider the same process as in Section 5.4.3, i.e. AR(1) process with Gaussian residuals, and we make use of model (5.1) with $\delta = 0$. We apply the standard Shewhart, EWMA, and CUSUM variance charts for the iid case. In the out-of-control state we assume a jump in the process of size $\Delta = 2$. The smoothing parameter of the EWMA chart is set to $\lambda = 0.4$ and the reference value K of the CUSUM chart is set 1.5. This leads to the stopping times

$$\begin{aligned} t_{A,s} &= \inf\{t \in \mathbb{N} : X_t^2 \leq 1.573 \cdot 10^{-6}\gamma_0 \vee X_t^2 \geq 10.828\gamma_0\}, \\ t_{A,e} &= \inf\{t \in \mathbb{N} : Z_t \leq 8.271 \cdot 10^{-2}\gamma_0 \vee Z_t \geq 5.088\gamma_0\}, \\ &\quad \text{with} \quad Z_0 = \gamma_0, \ \ \lambda = 0.4, \\ t_{A,c} &= \inf\{t \in \mathbb{N} : S_t^- \leq -405.283\gamma_0 \vee S_t^+ \geq 13.373\gamma_0\}, \\ &\quad \text{with} \quad S_0^- = S_0^+ = 0, \ \ K = 1.5. \end{aligned}$$

Let ARL_s, ARL_e, and ARL_c denote the corresponding average run lengths. In the in-control state all charts have the same ARL of 500, however, in the out-of-control state with $\Delta = 2$ we obtain

$$\begin{aligned} \text{ARL}_s(c_s; \Delta = 2) &= 9.97, \\ \text{ARL}_e(c_e; \Delta = 2) &= 8.14, \\ \text{ARL}_c(c_c; \Delta = 2) &= 7.41. \end{aligned}$$

Our results are given in Figures 5.4 and 5.5. The performance of the Shewhart chart for the variance is similar to the performance of this chat for the mean as documented previously in Figure 5.2. The chart is not sensitive to modest autocorrelation with $|\alpha_1| < 0.5$. However, it heavily overestimates the target

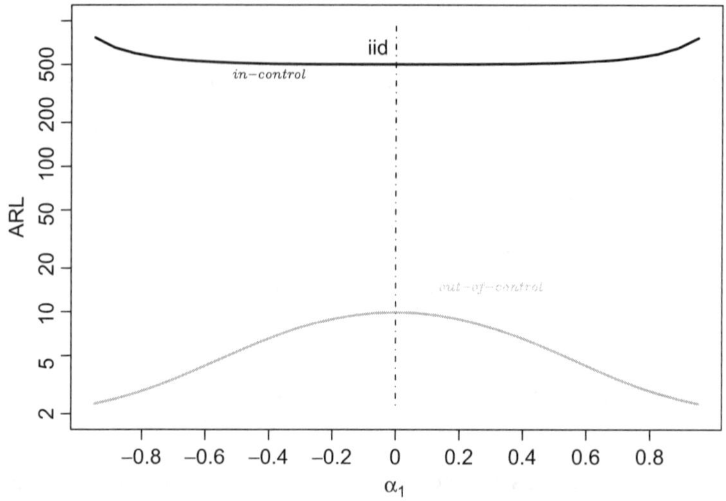

Figure 5.4 ARL of the standard Shewhart chart for the variance in the presence of autocorrelation. In the out-of-control state (the grey line) $\delta = 0$ and $\Delta = 2$.

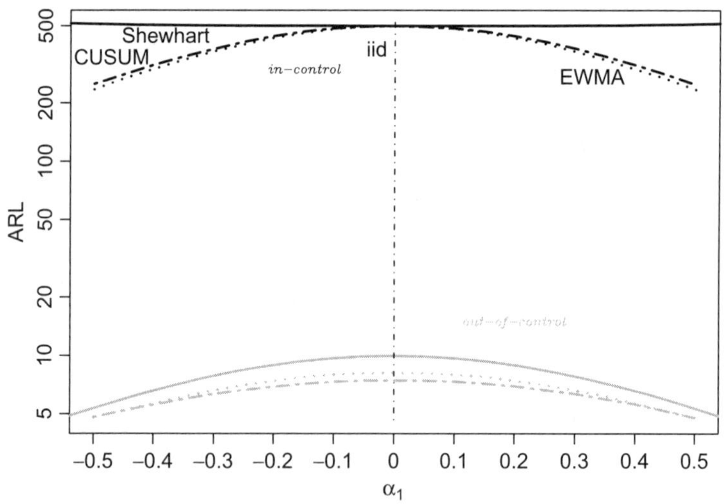

Figure 5.5 ARL of the standard Shewhart (solid line), EWMA (dashed line) and CUSUM (dotted line) charts for the variance in the presence of autocorrelation. In the out-of-control state (grey lines) $\delta = 0$ and $\Delta = 2$.

ARL for strong dependencies. Note, that in contrast to the mean chart, here the ARL is a symmetric function of α_1.

Figure 5.5 provides similar plots for EWMA and CUSUM charts. Both charts with the memory give an alarm for dependent data significantly faster than the Shewhart chart. For $\alpha_1 = 0.5$ the ARLs of the EWMA and the CUSUM charts are equal to 235 and 249 respectively, which are substantially lower than the ARL of the Shewhart chart which is equal to 510 or 500 in the in-control and out-of-control states, respectively. The same holds in the out-of-control state, however here the impact of the autocorrelation is smaller. A similar discussion as in Section 5.4.3 on the robustness of the Shewhart chart applies here as well.

5.5.4 Comparison study

In this section we perform a simulation study to compare the two-sided control charts for the variance introduced in the previous section. As above it is assumed that the target process $\{Y_t\}$ is an AR(1) process with the residuals following standard normal distribution.

We consider only charts based on the squared current observation. The average run length and the maximum of the average delay are taken as performance measures. To make the charts comparable we choose the critical values of all charts in such way that the in-control average run length is equal to some prespecified value, in our case to $\xi = 500$. The smoothing parameter λ is chosen from $\{0.05, 0.1, 0.2, 0.4, 0.6, 0.8, 1.0\}$ and the reference value K is chosen from $\{0.25, 0.5, 0.75, 1.0, 1.25, 1.5, 1.75, 2.0\}$. The parameter of the AR(1) process takes values from $\{0, 0.2, 0.4, 0.6, 0.8\}$ and the residuals are generated from the standard normal distribution. For given values of the parameter λ for the EWMA chart or of the reference value K for the CUSUM chart we determine the corresponding critical values c_1 and c_2 within a simulation study with 10^6 repetitions. The critical values are determined by numerical optimization using Broyden's version of the bivariate secant method. The methods showed good convergence for EWMA charts, but had some stability problems for CUSUM charts applied to processes with large α_1. The results of the simulation study are presented in Table 5.2. Note that the critical values strongly depend on the parameters of the underlying process and, thus, for each new process specification the critical values must be determined individually.

Using the computed critical values we determine the out-of-control ARLs for different sizes of the shift Δ. Table 5.3 summarizes our findings. For each value of α_1 the first two rows refer to the modified and residual EWMA charts, the third and the fourth rows to modified and residual CUSUM charts respectively. In parentheses we provide the optimal value of the parameter λ for EWMA and of K for CUSUM charts. For both residual charts we assumed

Table 5.2 Critical values of the two-sided EWMA and CUSUM charts for the variance*.

EWMA chart

λ		α_1 = 0.0	0.2	0.4	0.6	0.8
0.05	c_1	$5.410\cdot10^{-1}$	$5.329\cdot10^{-1}$	$5.069\cdot10^{-1}$	$4.557\cdot10^{-1}$	$3.563\cdot10^{-1}$
	c_2	1.730	1.768	1.888	2.127	2.620
0.1	c_1	$3.803\cdot10^{-1}$	$3.727\cdot10^{-1}$	$3.491\cdot10^{-1}$	$3.041\cdot10^{-1}$	$2.216\cdot10^{-1}$
	c_2	2.255	2.326	2.540	2.934	3.671
0.2	c_1	$2.170\cdot10^{-1}$	$2.119\cdot10^{-1}$	$1.957\cdot10^{-1}$	$1.655\cdot10^{-1}$	$1.133\cdot10^{-1}$
	c_2	3.223	3.336	3.672	4.243	5.145
0.4	c_1	$8.271\cdot10^{-2}$	$8.042\cdot10^{-2}$	$7.344\cdot10^{-2}$	$6.078\cdot10^{-2}$	$3.987\cdot10^{-2}$
	c_2	5.088	5.234	5.637	6.255	7.026
0.6	c_1	$3.058\cdot10^{-2}$	$2.968\cdot10^{-2}$	$2.697\cdot10^{-2}$	$2.214\cdot10^{-2}$	$1.443\cdot10^{-2}$
	c_2	6.970	7.087	7.408	7.873	8.283
0.8	c_1	$8.443\cdot10^{-3}$	$8.187\cdot10^{-3}$	$7.459\cdot10^{-3}$	$6.136\cdot10^{-3}$	$4.037\cdot10^{-3}$
	c_2	8.878	8.952	9.112	9.303	9.287
1.0	c_1	$1.573\cdot10^{-6}$	$1.579\cdot10^{-6}$	$1.572\cdot10^{-6}$	$1.568\cdot10^{-6}$	$1.569\cdot10^{-6}$
	c_2	10.828	10.824	10.781	10.657	10.173

CUSUM chart

K		α_1 = 0.0	0.2	0.4	0.6	0.8
0.25	c_1	1.169	1.189	1.265	1.450	2.050
	c_2	592.818	591.931	592.209	592.285	592.898
0.50	c_1	3.909	4.020	4.428	5.464	8.727
	c_2	398.253	397.135	397.372	398.054	400.107
0.75	c_1	10.391	10.807	12.321	15.958	26.201
	c_2	205.064	204.968	207.057	209.509	215.963
1.00	c_1	42.625	44.231	49.283	59.789	83.817
	c_2	42.605	44.116	49.065	59.365	82.135
1.25	c_1	211.280	212.512	215.675	222.363	239.406
	c_2	17.666	18.830	22.580	30.404	48.990
1.50	c_1	405.283	406.258	408.707	414.862	429.117
	c_2	13.373	14.330	17.377	23.695	38.461
1.75	c_1	602.052	602.510	604.872	610.004	622.293
	c_2	11.465	12.273	14.920	20.293	32.856
2.00	c_1	797.277	800.109	802.939	806.015	818.669
	c_2	10.332	11.040	13.325	18.058	28.899

* Critical values of the two-sided EWMA and CUSUM control charts for the variance based on squared observations for different parameters of the charts λ and K and different parameters α_1 of the true target $AR(1)$ process. The in-control ARL is set to 500. The simulation study is based on 10^6 repetitions.

Table 5.3 Out-of-control ARL of modified and residual EWMA and CUSUM control charts*.

α_1	Δ 0.50	0.75	1.00	1.25	1.50	1.75	2.00	2.50	3.00
0.0	(0.10)16.53	(0.05)47.56	EWMAm	(0.05)49.88	(0.05)18.38	(0.10)10.58	(0.10)7.28	(0.20)4.47	(0.20)3.31
	(0.10)16.53	(0.05)47.56	EWMAr	(0.05)49.88	(0.05)18.38	(0.10)10.58	(0.10)7.28	(0.20)4.47	(0.20)3.31
	(0.50)15.77	(0.75)49.18	CUSUMm	(1.25)48.46	(1.50)18.40	(1.75)10.49	(1.75)7.18	(2.00)4.41	(2.00)3.29
	(0.50)15.77	(0.75)49.18	CUSUMr	(1.25)48.46	(1.50)18.40	(1.75)10.49	(1.75)7.18	(2.00)4.41	(2.00)3.29
0.2	(0.20)16.59	(0.05)47.53	(0.60)499.93	(0.05)49.80	(0.05)18.17	(0.10)10.38	(0.10)7.07	(0.20)4.24	(0.20)3.08
	(0.10)16.63	(0.05)47.78	(0.10)499.20	(0.05)49.27	(0.05)18.00	(0.10)10.22	(0.10)6.95	(0.20)4.17	(0.20)3.02
	(0.50)15.81	(0.75)49.17	(0.25)498.72	(1.25)48.27	(1.50)18.20	(1.75)10.26	(2.00)6.95	(2.00)4.20	(2.00)3.07
	(0.50)15.88	(0.75)49.38	(0.25)499.33	(1.25)48.01	(1.50)18.01	(1.75)10.13	(2.00)6.85	(2.00)4.11	(2.00)3.01
0.4	(0.20)16.59	(0.05)47.38	(0.05)498.00	(0.05)47.52	(0.05)16.69	(0.10)9.29	(0.20)6.17	(0.60)3.54	(1.00)2.48
	(0.10)16.95	(0.05)48.62	(0.10)498.93	(0.05)47.32	(0.05)16.59	(0.10)9.00	(0.10)5.95	(0.20)3.42	(0.40)2.44
	(0.50)16.10	(0.75)49.81	(0.50)498.82	(1.25)46.11	(1.50)16.70	(2.00)9.16	(2.00)6.08	(2.00)3.62	(2.00)2.61
	(0.50)16.26	(0.75)50.24	(0.75)499.56	(1.25)46.33	(1.50)16.53	(1.75)8.92	(2.00)5.84	(2.00)3.40	(2.00)2.44
0.6	(0.20)16.62	(0.05)46.88	(0.40)497.24	(0.05)38.31	(0.10)12.77	(1.00)6.57	(1.00)4.23	(1.00)2.53	(1.00)1.91
	(0.10)17.56	(0.05)50.34	(0.10)499.13	(0.05)42.00	(0.10)13.20	(0.10)6.72	(0.20)4.31	(0.40)2.55	(0.40)1.90
	(0.25)16.22	(0.50)51.20	(0.50)499.03	(1.50)37.64	(2.00)12.58	(2.00)6.87	(2.00)4.66	(2.00)2.89	(2.00)2.16
	(0.50)16.96	(0.75)51.97	(1.25)499.49	(1.25)41.66	(1.50)13.12	(2.00)6.62	(2.00)4.28	(2.00)2.56	(2.00)1.93
0.8	(0.20)17.08	(0.05)46.12	(1.00)499.06	(1.00)20.60	(1.00)6.53	(1.00)3.86	(1.00)2.81	(1.00)1.93	(1.00)1.57
	(0.10)18.68	(0.05)54.38	(0.10)498.70	(0.05)24.96	(0.20)7.08	(0.20)4.01	(0.40)2.83	(0.60)1.92	(0.60)1.55
	(0.25)14.79	(0.50)44.66	(0.25)499.32	(2.00)20.63	(2.00)7.62	(2.00)4.79	(2.00)3.58	(2.00)2.46	(2.00)1.94
	(0.50)18.25	(0.75)55.93	(2.00)499.40	(1.50)25.00	(2.00)7.03	(2.00)4.00	(2.00)2.86	(2.00)1.95	(2.00)1.58

* Out-of-control ARL for different shift sizes and different values of α_1. The first two lines in each block contain values for the modified and residual EWMA charts, the third and the forth for the modified and residual CUSUM charts. The critical values are determined for the in-control ARL of 500. In parentheses the optimal value of the parameter λ for EWMA charts and K for CUSUM charts are given. The study is based on 10^6 repetitions.

that the true value of α_1 is known. Otherwise there arises the problem of estimating α_1 and this leads to non-Gaussian and non-iid residuals. Known α_1, however, may cause a slightly better performance of the residual charts compared with modified ones.

If the the variance decreases ($\Delta < 1$) this leads to the out-of-control ARL which is very robust to different values of α_1. For modified charts the ARL slightly decreases and for the residual charts it increases with α_1. For $\Delta > 1$ the out-of-control ARL for all charts decreases rapidly with increasing α_1. As also documented by other authors it is difficult to determine a chart which is the best for all parameter constellations. However, it is evident that EWMA charts perform better for larger values of α_1 while CUSUM tends to be better for smaller α_1's. The same holds when we compare either the modified or the residual charts. Concerning the optimal values of the chart parameters our results support the previous evidence. A small shift in the process leads to small optimal values of λ and of K. For large values of Δ the optimal parameters tend to be larger. The optimal parameters of the modified charts tend to be close to those of the residual charts.

Taking into account that the average run length is often criticized as a performance measure of control charts, we also run a study based on the maximum conditional expected delay. The results are summarized in Table 5.4. The time of the shift τ takes values between 1 and 30. The study is based on 10^5 repetitions. For each value of λ for EWMA chart (or K for CUSUM chart) we determine the value τ which leads to the largest delay. Further we choose such λ (or K) that provides the lowest delay for the given worst τ. Both optimal parameters are given in the table in parentheses. For $\Delta < 1$ the CUSUM type charts seem to provide a shorter expected delay, especially for larger values of α_1. For $\Delta > 1$ CUSUM also leads to shorter delays and only for stronger dependencies EWMA performs better. A similar behaviour is observed both for modified and for residual charts. Nevertheless, the difference between the delays is in most cases extremely small, which does not allow us to conclude that any of the charts exhibit systematically better performance than other charts. In the case of control charts for the mean of independent data, the CUSUM procedure with the known optimal K has optimal properties in terms of CED. In our case, however, we have a control procedure for the variance of dependent data with unknown optimal K. Therefore we cannot expect similar optimal performance of the CUSUM charts.

5.5.5 Example

We apply the developed control schemes to the daily MSCI indices for Germany, the US and Japan considered in the introduction. Since in financial

Table 5.4 Maximum of CED of modified and residual EWMA and CUSUM control charts*.

α_1	Δ 0.50	0.75	1.00	1.25	1.50	1.75	2.00	2.50	3.00
0.0	(1,0.10)16.39	(1,0.05)47.43	EWMAm	(6,0.05)49.76	(1,0.05)18.41	(2,0.10)10.59	(18,0.10)7.31	(10,0.20)4.50	(6,0.20)3.32
	(1,0.10)16.39	(1,0.05)47.43	EWMAr	(6,0.05)49.76	(1,0.05)18.41	(2,0.10)10.59	(18,0.10)7.31	(10,0.20)4.50	(6,0.20)3.32
	(1,0.50)15.21	(1,0.75)48.39	CUSUMm	(1,1.25)47.96	(1,1.50)18.25	(2,1.75)10.35	(1,1.75)7.09	(1,2.00)4.36	(1,2.00)3.26
	(1,0.50)15.21	(1,0.75)48.39	CUSUMr	(1,1.25)47.96	(1,1.50)18.25	(2,1.75)10.35	(1,1.75)7.09	(1,2.00)4.36	(1,2.00)3.26
0.2	(1,0.20)16.39	(1,0.05)47.43	(1,0.20)499.07	(5,0.05)49.92	(1,0.05)18.30	(6,0.10)10.51	(3,0.10)7.23	(16,0.20)4.45	(28,0.20)3.31
	(1,0.10)16.50	(1,0.05)47.66	(1,0.10)496.52	(5,0.05)49.42	(1,0.05)18.12	(1,0.10)10.34	(14,0.10)7.10	(16,0.20)4.36	(28,0.20)3.23
	(1,0.50)15.25	(1,0.75)48.45	(1,0.75)498.20	(1,1.25)47.97	(1,1.50)18.17	(1,1.75)10.31	(1,2.00)7.03	(1,2.00)4.35	(1,2.00)3.26
	(1,0.50)15.34	(1,0.75)48.65	(1,2.00)498.18	(1,1.25)47.80	(1,1.50)17.96	(1,1.75)10.16	(1,1.75)6.91	(1,2.00)4.24	(1,2.00)3.18
0.4	(1,0.20)16.35	(1,0.05)47.27	(2,0.05)497.92	(1,0.05)47.73	(30,0.05)17.10	(4,0.10)9.77	(2,0.20)6.68	(2,0.60)4.13	(11,1.00)3.06
	(1,0.10)16.88	(1,0.05)48.54	(2,0.10)497.97	(1,0.05)47.57	(9,0.05)16.96	(29,0.10)9.43	(4,0.10)6.41	(2,0.20)3.93	(18,0.40)2.95
	(1,0.50)15.58	(1,0.75)49.15	(1,0.50)497.71	(1,1.25)46.00	(2,1.50)16.82	(2,2.00)9.54	(2,2.00)6.54	(2,2.00)4.17	(1,2.00)3.19
	(1,0.50)15.74	(1,0.75)49.48	(2,1.00)498.64	(1,1.25)46.16	(1,1.50)16.62	(1,1.75)9.20	(1,2.00)6.22	(2,2.00)3.86	(1,2.00)2.92
0.6	(1,0.20)16.49	(1,0.05)46.99	(1,0.20)497.46	(22,0.05)38.83	(30,0.10)13.54	(27,1.00)7.44	(25,1.00)5.08	(11,1.00)3.31	(8,1.00)2.61
	(1,0.10)17.67	(1,0.05)50.49	(2,0.05)499.15	(3,0.05)42.33	(1,0.10)13.83	(2,0.10)7.42	(25,0.20)5.03	(11,0.40)3.23	(24,0.40)2.51
	(1,0.25)15.86	(1,0.50)50.78	(1,0.50)498.34	(1,1.50)37.94	(1,2.00)13.28	(2,2.00)7.64	(1,2.00)5.47	(1,2.00)3.71	(2,2.00)2.96
	(1,0.50)16.68	(1,0.75)51.38	(2,1.00)498.17	([illegible],1.25)41.56	(1,1.50)13.60	(1,2.00)7.27	(1,2.00)4.96	(1,2.00)3.20	(1,2.00)2.52
0.8	(1,0.20)17.06	(1,0.05)46.24	(2,0.20)496.49	(10,1.00)21.68	(22,1.00)7.53	(24,1.00)4.80	(26,1.00)3.71	(26,1.00)2.74	(11,1.00)2.29
	(1,0.10)19.24	(1,0.05)54.74	(1,0.10)498.40	(6,0.05)25.74	(6,0.20)7.91	(20,0.20)4.77	(7,0.40)3.55	(26,0.40)2.52	(11,0.60)2.07
	(1,0.25)14.69	(1,0.50)44.28	(1,0.25)497.12	(2,2.00)21.51	(6,2.00)8.56	(3,2.00)5.72	(2,2.00)4.52	(1,2.00)3.39	(2,2.00)2.85
	(1,0.50)18.42	(1,0.75)55.37	(1,1.00)498.08	(2,1.50)25.53	(1,2.00)7.80	(1,2.00)4.74	(1,2.00)3.55	(1,2.00)2.54	(1,2.00)2.09

*Maximum of the conditional expected delay for different shift sizes and different values of α_1. The first two lines in each block contain values for the modified and residual EWMA charts, the third and the forth for the modified and residual CUSUM charts. The critical values are determined for the in-control ARL of 500. In parentheses we provide the optimal value of τ from the interval 1 to 30 as well as the optimal parameter λ for EWMA charts and K for CUSUM charts. The study is based on 10^5 repetitions.

application we are mainly interested in changes in the riskiness of assets, we restrict the discussion to control charts for the variance. For the daily returns from January 1998 to December 1998 we fit an AR(1) process and estimate the initial in-control value of the parameter α_1. On the 1st of January 1999 we start EWMA and CUSUM control charts with $\lambda = 0.1$ and $K = 1$. respectively. After an alarm we re-estimate the parameter α_1 using the 50 last observations and compute new control limits. This is repeated after each alarm. The control statistics of the EWMA and the CUSUM charts together with control bounds is plotted in Figures 5.6 and 5.7. Comparing the volatility plots in Figure 5.1 with plotted control schemes, we observe that the charts manage to detect shifts in the variance rather quickly. We obtained between 20 and 30 alarms for around 2000 observations, which implies one alarm per three or four months. This is a reasonable quantity which should allow the practitioner to react quickly. This shows that the control techniques can be used by financial analysts for monitoring risk changes in the assets.

5.6 Surveillance of the covariance matrix of a multivariate linear time

In this section we deal with the surveillance of multivariate linear time series. First, we present a brief introduction into multivariate time series analysis. After that we discuss the relationship between the observed and the target process. Note that we focus on the surveillance of the covariances between the time series components at the same time point. We do not consider mean charts which are discussed in detail by, e.g. Bodnar and Schmid (2007). Moreover, we focus on EWMA type charts. The considered charts were proposed by Sliwa and Schmid (2005).

5.6.1 Introduction to multivariate time series

In many cases in practice we want to consider several univariate interdependent time series as a single quantity of interest. This leads to multivariate time series models. In financial applications such models are of special importance. To model financial spillovers we need a joint model for the indices or asset returns on several markets. Similarly, the presence of unobserved components in a factor model is usually solved using state-space models, which can be seen as a generalization of multivariate time series. Such models also play a crucial role in portfolio selection problems with autocorrelated asset returns.

In statistical process control we may be interested in simultaneous monitoring of several interdependent and autocorrelated quantities of interest. For

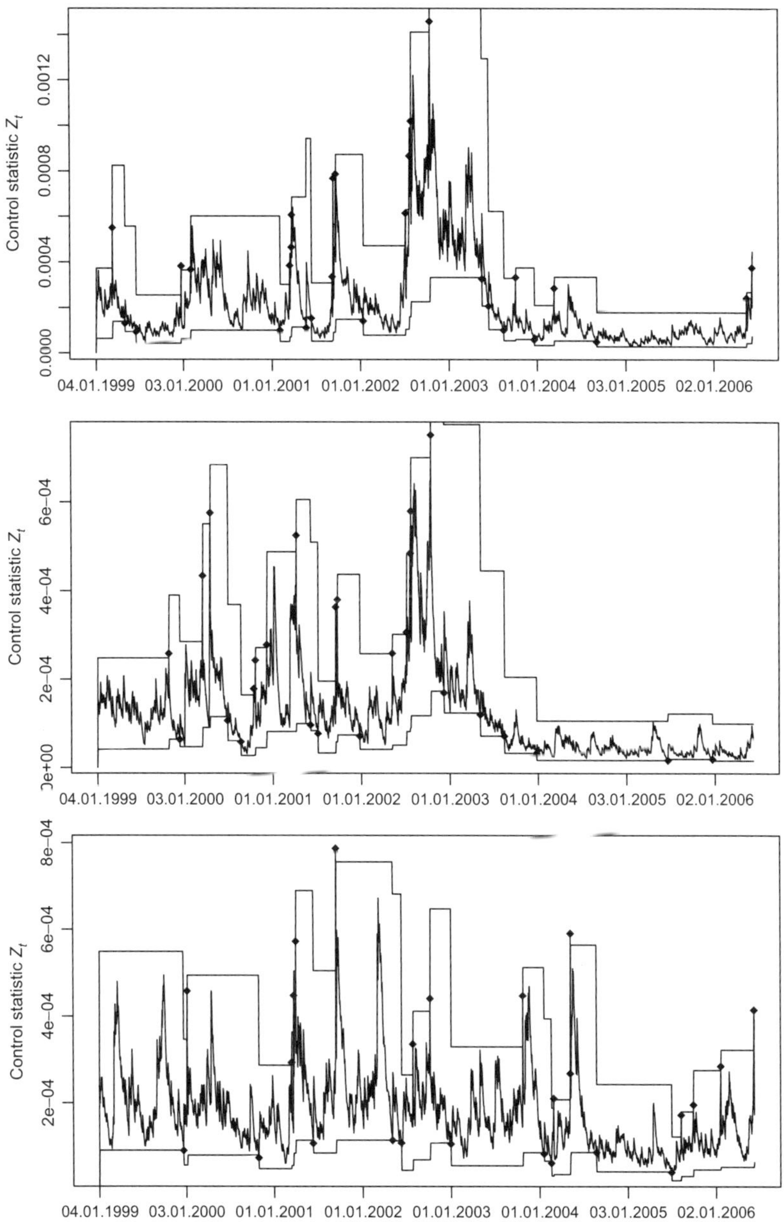

Figure 5.6 EWMA control chart for the volatility applied to daily returns on MSCI country indices for Germany, US and Japan. $\lambda = 0.1$ and 50 last observations are used to reestimate α_1 after an alarm.

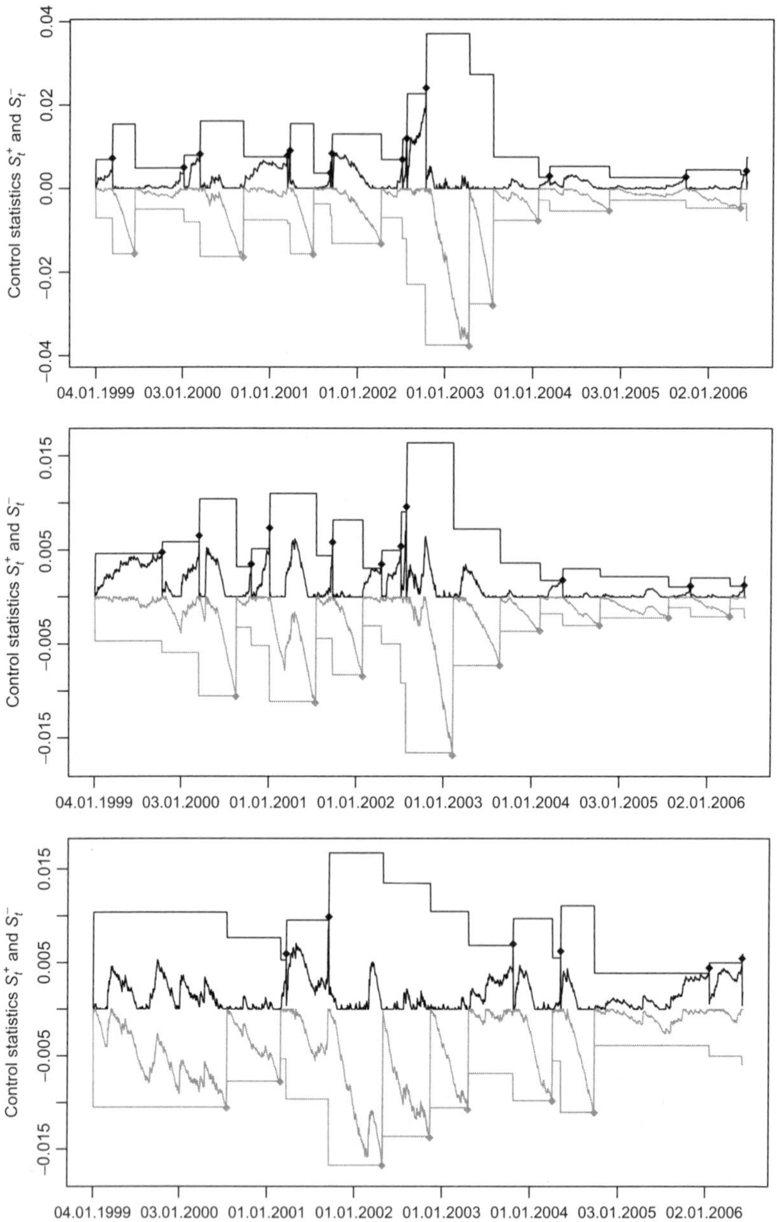

Figure 5.7 CUSUM control chart for the volatility applied to daily returns on MSCI country indices for Germany, US and Japan. $K = 1$ and 50 last observations are used to reestimate α_1 after an alarm. Black and grey lines show S_t^+ and S_t^- statistics and their bound respectively.

example, an investor can use the tools of SPC for active trading based on monitoring optimal portfolio weights. This problem is discussed in detail in Chapter 8 of this handbook. Also economic examples with applications to business cycles in Andersson *et al.* (2005) or examples in medicine in Frisén (1992) can be potentially extended to the multivariate framework. This requires an adoption of the above described univariate tools to multidimensional case.

Let $\mathbf{Y}_t = (Y_{1t}, \ldots, Y_{mt})'$ denote an m-variate vector of observations at time point $t \in \mathbb{Z}$. The second order properties of $\{\mathbf{Y}_t\}$ are then characterized by the mean vector $\mathrm{E}(\mathbf{Y}_t)$ and by the cross-covariance matrices $\mathbf{\Gamma}_t(h) = \mathrm{Cov}(\mathbf{Y}_{t+h}, \mathbf{Y}_t)$. The process $\{\mathbf{Y}_t\}$ is called stationary if the mean and the cross-covariance matrices are independent of the time point t. Thus $\mathrm{E}(\mathbf{Y}_t) = \boldsymbol{\mu}$ and $\mathbf{\Gamma}_t(h) = \mathbf{\Gamma}(h) = \{\gamma_{ij}(h)\}_{i,j=1,\ldots,m}$ for all t.

Next we extend the univariate models in Section 5.2.1 to the multivariate case. A multivariate VARMA process of order (p, q) is a solution of the difference equation

$$\mathbf{Y}_t = \boldsymbol{\mu}_0 + \sum_{i=1}^{p} \mathbf{A}_i(\mathbf{Y}_{t-i} - \boldsymbol{\mu}_0) + \boldsymbol{\varepsilon}_t + \sum_{j=1}^{q} \mathbf{B}_j \boldsymbol{\varepsilon}_{t-j},$$

where $\{\boldsymbol{\varepsilon}_t\}$ is a multivariate white noise process, i.e. $\mathrm{E}(\boldsymbol{\varepsilon}_t) = \mathbf{0}$, $\mathrm{Var}(\boldsymbol{\varepsilon}_t) = \mathbf{\Sigma}$ and $\mathrm{Cov}(\boldsymbol{\varepsilon}_t, \boldsymbol{\varepsilon}_s) = \mathbf{0}$ for $t \neq s$. It is demanded that $\mathbf{\Sigma}$ is positive definite. The coefficients α_i and β_j in the univariate model are now replaced by the matrix coefficients $\mathbf{A}_i$ and $\mathbf{B}_j$ of dimension $m \times m$. The stationarity depends as in the univariate case only on the coefficients of the AR part. The ARMA(p, q) process is stationary if $\det(\mathbf{I} - \sum_{i=1}^{p} \mathbf{A}_i z^i) = 0$ has no roots within the unit circle $|z| \leq 1$, $z \in \mathbb{C}$. In case of stationarity there is a unique solution of the difference equation written by

$$\mathbf{Y}_t = \boldsymbol{\mu}_0 + \sum_{i=0}^{\infty} \mathbf{\Theta}_i \boldsymbol{\varepsilon}_{t-i},$$

where the matrix coefficients are equal to the coefficients of the expanded product $(\mathbf{I} - \sum_{i=1}^{p} \mathbf{A}_i z^i)^{-1}(\mathbf{I} - \sum_{j=1}^{q} \mathbf{B}_j z^j)$. If we assume that $\boldsymbol{\varepsilon}_i \sim \mathcal{N}_m(\mathbf{0}, \mathbf{\Sigma})$ then the process $\{\mathbf{Y}_t\}$ is Gaussian, meaning that all finite marginal distributions follow a matrix variate normal distribution.

The best (in L^2 sense) linear predictor of $\mathbf{Y}_t$ given $\mathbf{Y}_{t-1}, \ldots, \mathbf{Y}_1, 1$ for a stationary VARMA process is given by

$$\hat{\mathbf{Y}}_t = \boldsymbol{\mu}_0 + \sum_{i=1}^{t-1} \boldsymbol{\psi}_{ti}(\mathbf{Y}_{t-i} - \boldsymbol{\mu}_0) .$$

It minimizes for every nonzero vector of constants $\mathbf{d}$ the mean squared error $E[(\mathbf{d}'\mathbf{Y}_t - \mathbf{d}'\hat{\mathbf{Y}}_t)^2]$. The optimal matrix coefficients can be determined from the system of equations (see Brockwell and Davis 1991, p. 421)

$$\sum_{j=1}^{t-1} \boldsymbol{\psi}_{tj}\boldsymbol{\Gamma}(i-j) = \boldsymbol{\Gamma}(i), \qquad i = 1, \ldots, t-1.$$

The multivariate extension of the Durbin–Levinson algorithm provides a convenient technique for a recursive computation of the weighting matrices $\boldsymbol{\psi}_{ti}$, $i = 1, \ldots, t-1$. If, however, $q = 0$, then the optimal linear forecasts are completely determined by the coefficients of the VAR(p) process. It holds for $t \geq p+1$ that

$$\hat{\mathbf{Y}}_t = \boldsymbol{\mu}_0 + \sum_{i=1}^{p} \mathbf{A}_i(\mathbf{Y}_{t-i} - \boldsymbol{\mu}_0).$$

For $t \leq p$ the forecasting is complicated due to unknown starting observations of the process.

5.6.2 Modelling

Our aim is to construct tools for the sequential detection of structural changes in the target process $\{\mathbf{Y}_t\}$ which is assumed to be driven by an m-dimensional stationary process with mean $\boldsymbol{\mu}_0$ and cross-covariance matrices $\{\boldsymbol{\Gamma}(h)\}$. For the purpose of our study, we assume that the actually observed process $\{\mathbf{X}_t\}$ is linked to the target process $\{\mathbf{Y}_t\}$ by the following relationship

$$\mathbf{X}_t = \begin{cases} \boldsymbol{\mu}_0 + \boldsymbol{\Delta}(\mathbf{Y}_t - \boldsymbol{\mu}_0) & \text{for} \quad t \geq \tau \\ \mathbf{Y}_t & \text{for} \quad t < \tau. \end{cases}$$

The $m \times m$ matrix $\boldsymbol{\Delta} \neq \mathbf{I}$ is unknown. It is assumed to be positive definite. If $\tau < \infty$ then the process is said to be out of control, else, if $\tau = \infty$, it is in control.

The first moments of the observed process are given by

$$\mathrm{E}(\mathbf{X}_t) = \boldsymbol{\mu}_0,$$

$$\mathrm{Cov}(\mathbf{X}_t) = \begin{cases} \boldsymbol{\Delta}\boldsymbol{\Gamma}(0)\boldsymbol{\Delta}' & \text{for} \quad t \geq \tau \\ \boldsymbol{\Gamma}(0) & \text{for} \quad t < \tau, \end{cases}$$

$$\mathrm{Cov}(\mathbf{X}_{t+h}, \mathbf{X}_t) = \begin{cases} \boldsymbol{\Delta}\boldsymbol{\Gamma}(h)\boldsymbol{\Delta}' & \text{for} \quad t \geq \max\{\tau, \tau - h\} \\ \boldsymbol{\Delta}\boldsymbol{\Gamma}(h) & \text{for} \quad \tau - h \leq t < \tau \\ \boldsymbol{\Gamma}(h)\boldsymbol{\Delta}' & \text{for} \quad \tau \leq t < \tau - h \\ \boldsymbol{\Gamma}(h) & \text{for} \quad t < \min\{\tau, \tau - h.\} \end{cases}$$

If $\boldsymbol{\Delta}$ is a diagonal matrix then $\mathrm{Corr}(\mathbf{X}_{t+h}, \mathbf{X}_t) = \mathrm{Corr}(\mathbf{Y}_{t+h}, \mathbf{Y}_t)$. Thus the cross-correlation matrix does not depend on the change. Note that the change influences not only the variances of the components of $\mathbf{X}_t$ but also its covariances.

In the next section we present techniques for detecting such type of changes.

5.6.3 Modified EWMA control charts

We consider two approaches to monitor multivariate time series. On the one hand we can use directly the multidimensional data. Alternatively we may appropriately transform the original data to one-dimensional quantities and then apply monitoring techniques to them. First we consider the modified control charts.

5.6.3.1 A chart based on a multivariate EWMA statistic

In this section all control schemes are based on a multivariate EWMA recursion. The starting point is an $m(m+1)/2$-dimensional local measure $\mathbf{L}_{\boldsymbol{\Gamma},t} = \mathbf{L}_{\boldsymbol{\Gamma},t}(\mathbf{X}_t, \mathbf{X}_{t-1}, \ldots)$ of the covariances of the components of the observed process at the same point of time. In the in-control state its expectation should stay constant over time, i.e. $\mathrm{E}_D(\mathbf{L}_{\boldsymbol{\Gamma},t}) = \mathrm{vech}(\boldsymbol{\Gamma}(0))$. The vech operator transforms a matrix to a vector. For an $m \times m$ matrix $\mathbf{A}$ with elements a_{ij} it is defined as

$$\mathrm{vech}(\mathbf{A}) = (a_{11}, .., a_{m1}, a_{22}, .., a_{m2}, .., a_{m-1,m-1}, a_{m,m-1}, a_{mm})' .$$

We apply a multivariate EWMA recursion to such a measure. This leads to

$$\mathbf{Z}_t = (\mathbf{I} - \boldsymbol{\Lambda})\mathbf{Z}_{t-1} + \boldsymbol{\Lambda}\mathbf{L}_{\boldsymbol{\Gamma},t}(\mathbf{X}_t, \mathbf{X}_{t-1}, \ldots), \quad t \geq 1, \quad \mathbf{Z}_0 = \mathrm{vech}(\boldsymbol{\Gamma}(0)),$$

where $\boldsymbol{\Lambda}$ is a diagonal matrix with the elements on the main diagonal taking values on the interval $(0, 1]$. They are smoothing parameters for each component of the local covariance measure. Note that the starting value $\mathbf{Z}_0$ is a known deterministic quantity. By recursive substitution and using the fact that $(\mathbf{I} - \boldsymbol{\Lambda})^t + \boldsymbol{\Lambda}\sum_{i=0}^{t-1}(\mathbf{I} - \boldsymbol{\Lambda})^i = \mathbf{I}$ we obtain that

$$\begin{aligned}
\mathbf{Z}_t &= (\mathbf{I} - \boldsymbol{\Lambda})^t\mathbf{Z}_0 + \boldsymbol{\Lambda}\sum_{i=0}^{t-1}(\mathbf{I} - \boldsymbol{\Lambda})^i\mathbf{L}_{\boldsymbol{\Gamma},t-i} \\
&= \boldsymbol{\Lambda}\sum_{i=0}^{t-1}(\mathbf{I} - \boldsymbol{\Lambda})^i[\mathbf{L}_{\boldsymbol{\Gamma},t-i} - \mathrm{vech}[\boldsymbol{\Gamma}(0)]] + \mathrm{vech}[\boldsymbol{\Gamma}(0)].
\end{aligned}$$

This shows that $\mathrm{E}_D(\mathbf{Z}_t) = \mathrm{vech}(\boldsymbol{\Gamma}(0))$.

Now we need a measure of the distance between the multivariate control statistic $\mathbf{Z}_t$ and its target value $\mathrm{vech}(\boldsymbol{\Gamma}(0))$. For this purpose we can use the

Mahalanobis distance. We conclude that there is a shift in the covariance matrix of the time series at time t if

$$\{\mathbf{Z}_t - \mathrm{vech}[\mathbf{\Gamma}(0)]\}'\mathrm{Cov}_D(\mathbf{Z}_t)^{-1}\{\mathbf{Z}_t - \mathrm{vech}[\mathbf{\Gamma}(0)]\} > c,$$

where c is a predetermined critical value. The covariance matrix of the control statistic $\mathbf{Z}_t$ in the in-control state is given by

$$\mathrm{Cov}_D(\mathbf{Z}_t) = \mathbf{\Lambda}\sum_{i=0}^{t-1}\sum_{j=0}^{t-1}(\mathbf{I}-\mathbf{\Lambda})^i\mathrm{Cov}_D(\mathbf{L}_{\Gamma,t-i},\mathbf{L}_{\Gamma,t-j})(\mathbf{I}-\mathbf{\Lambda})^j\mathbf{\Lambda}.$$

The computation of this matrix at each moment of time can be computationally expensive. In practice to avoid this, the asymptotic covariance matrix is used. It appears that the precision loss due to the use of the asymptotic matrix is minor (see, e.g. Knoth 2004). Thus this control chart gives an alarm if

$$\{\mathbf{Z}_t - \mathrm{vech}\,[\mathbf{\Gamma}(0)]\}'\,[\lim_{t\to\infty}\mathrm{Cov}_D(\mathbf{Z}_t)]^{-1}\,\{\mathbf{Z}_t - \mathrm{vech}\,[\mathbf{\Gamma}(0)]\} > c.$$

The transformation of the multivariate control statistic to a univariate quantity leads of course to information loss. For example, after an alarm the practitioner has no information in which component of the multivariate time series the shift occurred.

Here we focus on one example of a possible local covariance measure. Because $\mathbf{\Gamma}(0) = E_D[(\mathbf{X}_t-\boldsymbol{\mu}_0)(\mathbf{X}_t-\boldsymbol{\mu}_0)']$ the simplest measure is given by $\mathbf{L}_{\Gamma,t} = \mathrm{vech}\left[(\mathbf{X}_t-\boldsymbol{\mu}_0)(\mathbf{X}_t-\boldsymbol{\mu}_0)'\right] = \boldsymbol{\tau}_t$. Then only the actual observation at time point t is used. In the out-control state $\mathrm{vech}\left[(\mathbf{X}_t-\boldsymbol{\mu}_0)(\mathbf{X}_t-\boldsymbol{\mu}_0)'\right]$ can also be given in terms of the target process. Let $\mathbf{D}_m$ denote the duplication matrix (see Magnus and Neudecker 1999), $\mathbf{D}_m^+ = (\mathbf{D}_m'\mathbf{D}_m)^{-1}\mathbf{D}_m'$ and $\otimes$ denotes the Kronecker product of matrices. Then

$$\mathrm{vech}\left[(\mathbf{X}_t-\boldsymbol{\mu}_0)(\mathbf{X}_t-\boldsymbol{\mu}_0)'\right] = \left\{\mathbf{I}+\left[\mathbf{D}_m^+(\mathbf{\Delta}\otimes\mathbf{\Delta})\mathbf{D}_m\right]-\mathbf{I}\right\}\,\mathrm{vech}\left(\mathbf{Y}_t\mathbf{Y}_t'\right).$$

This implies that

$$\mathrm{E}\left\{\mathrm{vech}\left[(\mathbf{X}_t-\boldsymbol{\mu}_0)(\mathbf{X}_t-\boldsymbol{\mu}_0)'\right]\right\} = \left\{\mathbf{I}+\left[\mathbf{D}_m^+(\mathbf{\Delta}\otimes\mathbf{\Delta})\mathbf{D}_m\right]-\mathbf{I}\right\}\mathrm{vech}(\mathbf{\Gamma}(0))$$

and subsequently $E_D(\mathbf{L}_{\Gamma,t}) = \mathrm{vech}\,[\mathbf{\Gamma}(0)]$. The covariance matrix of the EWMA recursion and its limit are derived in Śliwa and Schmid (2005).

5.6.3.2 A chart based on a mahalanobis distance

Alternatively to the previous section we may first transform the original observed process to a univariate quantity and after that apply a monitoring procedure to the one-dimensional characteristic. For example, let

$$L_{\Gamma,t} = \{\mathbf{L}_{\Gamma,t} - \mathrm{vech}[\mathbf{\Gamma}(0)]\}'\mathrm{Cov}_D(\mathbf{L}_{\Gamma,t})^{-1}\{\mathbf{L}_{\Gamma,t} - \mathrm{vech}[\mathbf{\Gamma}(0)]\}.$$

As the multivariate measure of variance $\mathbf{L}_{\Gamma,t}$ we can also take the products of current observations denoted above by $\boldsymbol{\tau}_t$. Similarly as in the previous case we can use the asymptotic variance instead of the exact one. The univariate EWMA recursion is then defined by

$$Z_t = (1-\lambda)Z_{t-1} + \lambda L_{\Gamma,t}(\mathbf{X}_t), \qquad t \geq 1, \qquad Z_0 = m(m+1)/2.$$

The smoothing parameter λ plays the same role as for EWMA charts for univariate time series. The chart gives an alarm if the deviation of Z_t from the target value is sufficiently large.

5.6.4 EWMA residual control charts

Residual charts are based on a transformation of the original observations to a sequence which is independent in the in-control state. Next we will analyse how this can be done in the multivariate case. It is assumed that $\{\mathbf{Y}_t\}$ is a stationary VARMA process.

The starting point is the best (in L^2 sense) linear predictor of $\mathbf{Y}_t$ given $\mathbf{Y}_{t-1}, .., \mathbf{Y}_1, 1$ which is given by (see Section 5.6.1)

$$\hat{\mathbf{Y}}_t = \boldsymbol{\mu}_0 + \sum_{i=1}^{t-1} \boldsymbol{\psi}_{ti}(\mathbf{Y}_{t-i} - \boldsymbol{\mu}_0) \ .$$

Because we have a realization of the process $\{\mathbf{X}_t\}$ we replace the target values by the observed values. This leads to

$$\hat{\mathbf{X}}_t = \boldsymbol{\mu}_0 + \sum_{i=1}^{t-1} \boldsymbol{\psi}_{ti}(\mathbf{X}_{t-i} - \boldsymbol{\mu}_0) \ .$$

The matrices $\boldsymbol{\psi}_{ti}$ are determined as described in Section 5.6.1. Note that this quantity is only equal to the best linear predictor if the process is in control. If, e.g. $\{\mathbf{Y}_t\}$ is a stationary VARMA(1,1) process with mean $\boldsymbol{\mu}_0 = 0$ then the best linear predictor is given by (cf. Brockwell and Davis 1991)

$$\hat{\mathbf{X}}_t = \mathbf{A}_1\mathbf{X}_{t-1} + \tilde{\mathbf{B}}_t(\mathbf{X}_{t-1} - \hat{\mathbf{X}}_{t-1}) \qquad \text{for} \quad t \geq 2$$

with $\hat{\mathbf{X}}_1 = \mathbf{0}$. The matrix $\tilde{\mathbf{B}}_t$ is computed recursively using

$$\tilde{\mathbf{B}}_t = \mathbf{B}_1 \boldsymbol{\Sigma} \mathbf{V}_{t-1}^{-1} \qquad \text{with} \qquad \mathbf{V}_t = \begin{cases} \boldsymbol{\Gamma}(0) & \text{for} \quad t = 1 \\ \boldsymbol{\Sigma} + \mathbf{B}_1 \boldsymbol{\Sigma} \mathbf{B}_1' - \tilde{\mathbf{B}}_t \mathbf{V}_{t-1} \mathbf{B}_t' & \text{for} \quad t \geq 2. \end{cases}$$

When $\{\mathbf{Y}_t\}$ is invertible then it holds in the in-control case that $\tilde{\mathbf{B}}_t \to \mathbf{B}_1$ and $\mathbf{V}_t \to \boldsymbol{\Sigma}$ as $t \to \infty$. Note that $\mathbf{V}_t = \mathrm{E}_D((\mathbf{X}_t - \hat{\mathbf{X}}_t)(\mathbf{X}_t - \hat{\mathbf{X}}_t)')$.

Next we consider the residuals $\hat{\boldsymbol{\varepsilon}}_t = \mathbf{X}_t - \hat{\mathbf{X}}_t$. If the target process is a Gaussian process then the residuals are independent and normally distributed in the in-control state. Note that in the out-of-control state they are neither independent nor normally distributed.

The normalized residuals are given by $\hat{\boldsymbol{\Delta}}_t = \mathbf{V}_t^{-1/2}\hat{\boldsymbol{\varepsilon}}_t$. It holds that $E_D(\hat{\boldsymbol{\Delta}}_t) = \mathbf{0}$, $\text{Cov}_D(\hat{\boldsymbol{\Delta}}_t) = \mathbf{I}$, and $\text{Cov}_D(\hat{\boldsymbol{\Delta}}_s, \hat{\boldsymbol{\Delta}}_t) = \mathbf{0}$ for $s \neq t$. Now we monitor the covariances of the normalized residual process. We take $\mathbf{L}_{\Gamma,t} = \text{vech}(\hat{\boldsymbol{\Delta}}_t\hat{\boldsymbol{\Delta}}_t') = \boldsymbol{\tau}_{r,t}$. It is possible to make use of the same procedure as described in the previous section, however, we replace $\boldsymbol{\tau}_t$ by $\boldsymbol{\tau}_{r,t}$. A detailed analysis of these charts is given in Śliwa and Schmid (2005).

5.6.5 Comparison study

The comparison of the control procedures in the multivariate is a very difficult task because there are many parameters which influence the behaviour of the charts. A very intensive comparison was given by Śliwa and Schmid (2005). They assumed that the target process is a VARMA(1,1) process and the observed process is obtained by changing the coefficients of the target process. They analyse how this procedure influences the covariance structure. As a measure of the performance they used the ARL. In principle it turns out that the residual chart based on the multivariate EWMA recursion leads to the smallest out-of-control ARL. Because the determination of the control limits is much easier for the residual charts they should be favoured in a multivariate situation.

5.7 Summary

In this chapter we provided a review of the methods used for monitoring univariate and multivariate linear time series. We discuss the EWMA and the CUSUM type charts with memory. To adopt the classical monitoring techniques to linear processes we consider the residual and the modified charts. Special attention is paid to the monitoring the variance of financial series. As local volatility measures we consider the squared observations as well the the logarithm of the squared observations. Within an extensive simulation study we assess the impact of neglecting the time series structure and compare the charts using the out-of-control ARL and maximum average delay as performance measures. The discussed techniques are illustrated by monitoring the volatility of MSCI index returns for several countries.

Acknowledgements

The authors are grateful to Professor Marianne Frisén and the participants of the workshop on Financial Surveillance at the University of Göteborg, Sweden for helpful discussions and suggestions.

References

Alwan, L. and Roberts, H. (1988). Time-series modeling for statistical process control. *Journal of Business and Economic Statistics*, **6**, 87–95.

Andersson, E., Bock, D. and Frisén, M. (2005). Statistical surveillance of cyclical processes with application to turns in business cycles. *Journal of Forecasting*, **24**, 465–490.

Bodnar, O. and Schmid, W. (2007). Surveillance of the mean behaviour of multivariate time series. *Statistica Neerlandica* (to appear).

Brockwell, P. and Davis, R. (1991). *Time Series: Theory and Methods* Springer,New York

Brook, D. and Evans, D. (1972) An approach to the probability distribution of CUSUM run length. *Biometrika*, **59**, 539–549.

Crowder, S. (1987). A simple method for studying run-length distributions of exponentially weighted moving average charts. *Technometrics*, **29**, 401–407.

Crowder, S. and Hamilton, M. (1992). An EWMA for monitoring a process standard deviation. *Journal of Quality Technology*, **24**, 12–21.

Engle, R. (1982). Autoregressive conditional heteroscedasticity with estimates of the variance of UK inflation. *Econometrica*, **50**, 987–1008.

Frisén, M. (1992). Evaluations of methods for statistical surveillance. *Statistics in Medicine*, **11**, 1489–1502.

Frisén (2003). Statistical surveillance. Optimality and methods. *International Statistical Review*, **71**, 403–434.

Frisén, M. and de Maré, J. (1991). Optimal surveillance. *Biometrika*, **78**, 271–280.

Frisén, M. and Sonesson, C. (2006). Optimal surveillance based on exponentially weighted moving averages. *Sequential Analysis*, **25**, 379–403.

Harris, T. and Ross, W. (1991). Statistical process control procedures for correlated observations. *Canadian Journal of Chemical Engineering*, **69**, 48–57.

Hawkins, D. and Olwell, D. (1998). *Cumulative Sum Charts and Charting for Quality Improvement*. Springer, New York.

Jiang, W., Tsui, K.-L. and Woodall, W. (2000). A new SPC monitoring method: The ARMA chart. *Technometrics*, **42**, 399–410.

Knoth, S. (2004). Fast initial response features for EWMA control charts. *Statistical Papers*, **46**(1), 47–64.

Knoth, S. and Schmid, W. (2004). Control charts for time series: A review. In *Frontiers in Statistical Quality Control*, eds. Lenz, H.-J. and Wilrich, P.-T. Physica-Verlag, Heidelberg, vol. 7, pp. 210–236.

Lorden, G. (1971). Procedures for reacting to a change in distribution. *Annals of Mathematical Statistics*, **41**, 1897–1908.

Lu, C.-W. and Reynolds, Jr, M. (1999). EWMA control charts for monitoring the mean of autocorrelated processes. *Journal of Quality Technology*, **31**, 166–188.

Lucas, J. and Saccucci, M. (1990). Exponentially weighted moving average control schemes: Properties and enhancements. *Technometrics*, **32**, 1–29.

MacGregor, J. and Harris, T. (1993). The exponentially weighted moving variance. *Journal of Quality Technology*, **25**, 106–118.

Magnus, J. and Neudecker, H. (1999). *Matrix Differential Calculus with Applications in Statistics and Econometrics*. John Wiley & Sons Ltd, New York.

Montgomery, D. and Mastrangelo, C. (1991). Some statistical process control methods for autocorrelated data. *Journal of Quality Technology*, **23**, 179–204.

Morais, M. (2001). Stochastic ordering in the performance analysis of quality control schemes. Ph.D. thesis, Universidade Técnica de Lisboa, Lisbon, Portugal.

Nikiforov, I. (1975). Sequential analysis applied to autoregression processes. *Automation and Remote Control*, **36**, 1365–1368.

Page, E. (1954). Continuous inspection schemes. *Biometrika*, **41**, 100–115.

Pollak, M. and Siegmund, D. (1975). Approximations to the expected sample size of certain sequential tests. *Annals of Statistics*, **3**, 1267–1282.

Roberts, S. (1959). Control chart tests based on geometric moving averages. *Technometrics*, **1**, 239–250.

Schipper, S. and Schmid, W. (2001). Sequential methods for detecting changes in the variance of economic time series. *Sequential Analysis*, **20**(4), 235–262.

Schmid, W. (1995). On the run length of a Shewhart chart for correlated data. *Statistical Papers*, **36**, 111–130.

Schmid, W. (1997a). CUSUM control schemes for Gaussian processes. *Statistical Papers*, **38**, 191–217.

Schmid, W. (1997b). On EWMA charts for time series In *Frontiers in Statistical Quality Control*, eds. Lenz, H.-J. and Wilrich, P.-T. Physica-Verlag, Heidelberg, pp. 115–137.

Schmid, W. and Schöne, A. (1997). Some properties of the EWMA control chart in the presence of autocorrelation. *Annals of Statistics*, **25**, 1277–1283.

Schöne, A., Schmid, W. and Knoth, S. (1999). On the run length of the EWMA chart: A monotonicity result for normal variables. *Journal of Statistical Planning and Inferences*, **79**, 289–297.

Shewhart, W. (1931). *Economic Control of Quality of Manufactured Product*. van Nostrand, Toronto.

Śliwa, P. and Schmid, W. (2005). Monitoring the cross-covariances of multivariate time series. *Metrika*, **61**, 89–115.

Vasilopoulos, A. and Stamboulis, A. (1978). Modification of control chart limits in the presence of data correlation. *Journal of Quality Technology*, **10**, 20–30.

Wardell, D., Moskowitz, H. and Plante, R. (1994a). Run length distributions of residual control charts for autocorrelated processes. *Journal of Quality Technology*, **26**, 308–317.

Wardell, D., Moskowitz, H. and Plante, R. (1994b). Run-length distributions of special-cause control charts for correlated processes. (With discussion). *Technometrics*, **36**, 3–27.

Yashchin, E. (1993). Performance of CUSUM control schemes for serially correlated observations. *Technometrics*, **35**, 37–52.

6

Surveillance of univariate and multivariate nonlinear time series

Yarema Okhrin and Wolfgang Schmid

Department of Statistics, European University Viadrina, Frankfurt (Oder), Germany

6.1 Introduction

Linear time series processes like ARMA processes are widely applied in practice. They have been extensively analysed in theory. Many procedures are available to fit such a process to a data set and to determine its goodness of fit (e.g. Brockwell and Davis 1991).

If we consider a financial time series like, e.g. the daily returns of a stock or the daily exchange rate between two currencies, we find a very specific behaviour which was first described by Mandelbrot (1963a,b). Frequently economic time series exhibit periods of unusually large oscillations followed by periods of relative tranquility. This means that the conditional variance of the process, the so-called volatility, is not constant over time. Because ARMA processes have a constant volatility they are not suitable to describe such a process. The problem of modeling stochastic volatility and conditional heteroscedasticity in the financial context has been an active area of research over the past decade (see Cox et al. 1996; Fan and Yao 2003; Mills 2004). Among all these procedures the most widely applied approach in econometrics and finance are

Financial Surveillance Edited by Marianne Frisén

ARCH (autoregressive conditional heteroscedasticity) processes. Engle (1982) introduced the idea of an ARCH process and he was honoured for this contribution by the Nobel Prize 2003 in economics. Bollerslev (1986) extended this model to the so-called GARCH (generalized ARCH) processes. Chapter 2 of this handbook provides a review of these processes.

In this chapter we will consider surveillance procedures for univariate and multivariate GARCH processes. Because we are again exclusively interested in the weakly stationary solution of a GARCH process the main ideas behind the control schemes are similar to those given in Chapter 5. Nevertheless, it is necessary to determine the control design for the nonlinear processes once again.

6.2 Modelling

6.2.1 Modelling the observed process

In the previous chapter we have seen that there may occur many possible types of changes in the course of a time series. In finance, however, we are mainly confronted with changes in the risk behaviour. For that reason we focus exclusively on the surveillance of the variance of the underlying process. Control charts for the mean behavior were proposed by, for example, Severin and Schmid (1998, 1999).

Let $\{Y_t\}$ denote the target process. It is assumed to be a weakly stationary process with mean μ_0 and autocovariance function $\{\gamma_t\}$. Then we have the following relationship between $\{Y_t\}$ and the observed process $\{X_t\}$

$$X_t = \begin{cases} Y_t & \text{for} \quad t < \tau \\ \mu_0 + \Delta(Y_t - \mu_0) & \text{for} \quad t \geq \tau \end{cases} \tag{6.1}$$

with $\Delta > 0$, $\Delta \neq 1$, and $\tau \in \mathbb{N} \cup \{\infty\}$. If $\tau < \infty$ then a change in the scale occurs at time τ. It is said that $\{X_t\}$ is out of control. If $\tau = \infty$ the process $\{X_t\}$ is said to be in control. For the observed process it holds that $E(X_t) = \mu_0$, $\text{Var}(X_t) = \Delta^2\gamma_0$ for $t \geq \tau$ and $\text{Var}(X_t) = \gamma_0$ for $t < \tau$. This shows that in the case $\Delta > 1$ the variance of the process increases and thus the asset is getting more risky. Otherwise, for $\Delta < 1$, the variance decreases and the asset becomes more safer.

We use the notation E_D, Var_D, Cov_D, etc. to denote that these quantities are calculated with respect to the target process, i.e., assuming that $\tau = \infty$.

In the following $\{Y_t\}$ is a GARCH process. Without restriction it is assumed that its mean μ_0 is equal to 0.

6.2.2 Univariate GARCH processes

In this section we provide a short presentation of univariate GARCH processes. The key idea behind GARCH models is to describe the conditional variance as a function of previous observations. The volatility is modelled by an ARMA process and the original process is equal to the product of a white noise process with the volatility. Thus the structure of a GARCH process is very simple. However, due to the multiplicative decomposition it is harder to analyse than a linear process.

Following Engle (1982) and Bollerslev (1986) a stochastic process $\{Y_t\}$ with $t \in \mathbb{Z}$ is called a GARCH (Generalized AutoRegressive Conditional Heteroscedasticity) process of order (p, q) with mean $\mu_0 = 0$ if

$$Y_t = \varepsilon_t h_t \tag{6.2}$$

with

$$h_t^2 = \alpha_0 + \sum_{i=1}^{q} \alpha_i Y_{t-i}^2 + \sum_{j=1}^{p} \beta_j h_{t-j}^2. \tag{6.3}$$

We assume that $\alpha_0 > 0$ and $\alpha_i, \beta_j \geq 0$ for $i = 1, \dots, q$ and $j = 1, \dots, p$. These assumptions guarantee that h_t^2 is positive. The variables $\{\varepsilon_t\}$ are assumed to be independent with $\mathrm{E}(\varepsilon_t) = 0$ and $\mathrm{Var}(\varepsilon_t) = 1$. If the coefficients satisfy the assumption $\sum_{i=1}^{q} \alpha_i + \sum_{j=1}^{p} \beta_j < 1$ then there is a unique weakly stationary solution of the stochastic equation (6.2). In this case it holds that

$$\mathrm{E}(Y_t) = 0, \qquad \mathrm{Cov}(Y_t, Y_s) = 0 \quad \text{for} \quad t \neq s\,, \tag{6.4}$$

$$\mathrm{Var}(Y_t) = \frac{\alpha_0}{1 - \sum_{i=1}^{q} \alpha_i - \sum_{j=1}^{p} \beta_j} \tag{6.5}$$

and $\mathrm{Var}(Y_t | Y_{t-1}, \dots) = h_t^2$. Consequently the variables of a GARCH process are uncorrelated. Moreover, the conditional variance is given by h_t^2 and it is time dependent.

It is an interesting fact that $\{Y_t^2\}$ can be presented as an ARMA($\max\{p, q\}$, p) process with mean γ_0. To obtain this we square Equation (6.2), substitute the expression for the conditional variance given in (6.3) and present the right-hand side in terms of the deviation $Y_{t-i}^2 - h_{t-i}^2$. This leads to

$$Y_t^2 - \gamma_0 = \sum_{i=1}^{q} \alpha_i (Y_{t-i}^2 - \gamma_0) + \sum_{j=1}^{p} \beta_j (Y_{t-j}^2 - \gamma_0) + \nu_t - \sum_{j=1}^{p} \beta_j \nu_{t-j} \tag{6.6}$$

with $\nu_t = Y_t^2 - h_t^2 = (\varepsilon_t^2 - 1) h_t^2$. It can be shown that $\mathrm{E}(\nu_t) = 0$ and that $\mathrm{Cov}(\nu_t, \nu_s) = 0$ for $t \neq s$ provided that the fourth moments of $\{Y_t\}$ exist. Note that the variables $\{\nu_t\}$ are uncorrelated but not independent. Equation (6.6) shows that the squared values of a GARCH process are correlated while, as seen above, the original observations are uncorrelated. This fact is frequently used for model identification.

It is a well-known fact from prediction theory (e.g. Brockwell and Davis 1991; Pourahmadi 2001) that the conditional variance h_t^2 is the best linear predictor in L^2 sense of Y_t^2 in terms of $Y_{t-1}^2, Y_{t-2}^2, \ldots$. Because in practice, however, only a finite number of observations is available, we have to determine the best linear predictor of Y_t^2 in terms of $Y_{t-1}^2, \ldots, Y_1^2, 1$. This quantity is denoted by $\hat{h}_t^2$ and is equal to

$$\hat{h}_t^2 = a_{tt} + \sum_{i=1}^{t-1} a_{ti} Y_{t-i}^2 \tag{6.7}$$

with some constants $a_{ti}, i = 1, \ldots, t$ depending on the autocovariance structure $\{\gamma_t\}$ of the process $\{Y_t\}$. Because $\mathrm{E}(Y_t^2) = \mathrm{E}(\hat{h}_t^2) = \gamma_0$ we obtain that $a_{tt} = \gamma_0(1 - \sum_{i=1}^{t-1} a_{ti})$. Moreover,

$$(a_{t1}, \ldots, a_{t,t-1})' = \mathbf{\Gamma}(0)^{-1}(\gamma_1, \ldots, \gamma_{t-1})' \quad \text{with} \quad \mathbf{\Gamma}(0) = \{\gamma_{i-j}\}_{i,j=1,\ldots,t-1}\,.$$

Using the projection theorem (see Brockwell and Davis 1991) it follows that

$$\mathrm{E}(Y_t^2 - \hat{h}_t^2)^2 = \mathrm{E}(Y_t^2 - \gamma_0)^2 - \mathrm{E}(\hat{h}_t^2 - \gamma_0)^2 = \mathrm{Var}(Y_t^2) - \mathrm{Var}(\hat{h}_t^2).$$

This implies that $\{\hat{h}_t^2\}$ is not weakly stationary since the left-hand side is not constant over time. However, $\mathrm{E}(Y_t^2 - \hat{h}_t^2)^2$ converges to γ_0 if t tends to infinity and, thus, $\{\hat{h}_t^2\}$ is at least asymptotically stationary.

Since $\{Y_t^2\}$ is an ARMA process we can determine the forecasts $\hat{h}_t^2$ recursively. For example, for an GARCH(1,1) process we obtain by applying the Durbin–Levinson algorithm (see Brockwell and Davis 1991) that for $t \geq 1$

$$\begin{aligned}
\hat{h}_t^2 &= \gamma_0 + (\alpha_1 + \beta_1)(Y_{t-1}^2 - \gamma_0) - \beta_1(Y_{t-1}^2 - \hat{h}_{t-1}^2)/r_{t-1}, \\
r_t &= 1 + \beta_1^2 - \beta_1^2/r_{t-1}
\end{aligned}$$

with

$$\hat{h}_1^2 = \gamma_0, \qquad r_1 = 1 + \frac{\alpha_1^2}{1 - (\alpha_1 + \beta_1)^2}.$$

It has to be noted that since the introduction of a GARCH process many other nonlinear time series have been introduced to model the returns of stocks like, e.g. EGARCH, IGARCH, GARCH-M, threshold GARCH processes. In the meantime we can speak about a GARCH zoo (Ruppert 2004). An overview about recent developments is given in Tsay (2005). Nonparametric nonlinear methods are discussed in Fan and Yao (2003). We will not consider these approaches here. In this chapter, we will focus on the most prominent member of nonlinear time series, the GARCH family.

6.3 Surveillance of the variance of a univariate nonlinear time series

In this section we introduce EWMA and CUSUM type charts for the variance of a nonlinear time series. The target process is always assumed to be a weakly stationary GARCH process. Furthermore it is assumed that (6.1) holds. Note that in this case the moments of the observed process for $t \geq \tau$ can be given explicitly by

$$E(X_t) = 0, \quad \mathrm{Var}(X_t) = \Delta^2 \gamma_0 \quad \text{and} \quad \mathrm{Cov}(X_t, X_s) = 0 \ \text{ for } \ t \neq s.$$

We focus on two-sided charts, i.e. we are interested in detecting an increase or a decrease of the variance. The control charts presented in this section were introduced by Schipper and Schmid (2001), however, only for the one-sided case of an increasing variance.

6.3.1 EWMA type control charts

Next we propose a class of EWMA control charts aimed to detect shifts in the scale of the observed process. As in Chapter 5 the starting point of our considerations is a local measure of the variance or of a suitable function of it which is denoted by $L_{V,t}(X_t, X_{t-1}, ..)$. Then the EWMA recursion is given by

$$Z_t = (1-\lambda)Z_{t-1} + \lambda L_{V,t}(X_t, X_{t-1}, ..), \quad \lambda \in (0, 1].$$

The smoothing parameter λ reflects the impact of the current value of the local measure on the control statistic. Large values of λ lead to a higher impact and small values to a lower impact.

6.3.1.1 A chart based on the volatility

The most interesting quantity for a practitioner within a GARCH process is the volatility. For that reason our first chart is based on the volatility.

Let $\hat{h}_t$ denote the best linear predictor of Y_t^2 given $Y_{t-1}, \ldots, Y_1, 1$. As given in (6.7) it is equal to $\hat{h}_t^2 = a_{tt} + \sum_{i=1}^{t-1} a_{ti} Y_{t-i}^2$ with some constants a_{ti}, $i = 1, \ldots, t$. Because we have a realization of the process $\{X_t\}$ and we do not know whether a change arises or not we cannot determine $\hat{h}_t^2$. Instead of this quantity we use

$$\hat{\sigma}_t^2 = \gamma_0 + \sum_{i=1}^{t-1} a_{ti}(X_{t-i}^2 - \gamma_0)$$
$$= \begin{cases} \Delta^2 \hat{h}_t^2 - (\Delta^2 - 1)(a_{tt} + \sum_{i=t-\tau+1}^{t-1} a_{ti} Y_{t-i}^2) & \text{for} \quad t > \tau \\ \hat{h}_t^2 & \text{for} \quad t \le \tau. \end{cases}$$

This means that the target process is replaced by the original process. As shown in the previous section it is possible to determine $\hat{\sigma}_t^2$ recursively.

To understand the behaviour of $\hat{\sigma}_t^2$ in the out-of-control state we calculate the first two moments. They are summarized in the following proposition. For the proof and a more detailed discussion see Schipper and Schmid (2001).

Proposition 1 *(a) Let $\{Y_t\}$ be a stationary GARCH process. Then*

$$E(\hat{\sigma}_t^2) = \begin{cases} \Delta^2 \gamma_0 - (\Delta^2 - 1)(a_{tt} + \gamma_0 \sum_{i=t-\tau+1}^{t-1} a_{ti}) & for \quad t > \tau \\ \gamma_0 & for \quad t \le \tau. \end{cases}$$

If the fourth moments of $\{Y_t\}$ are finite then it follows for $\tau = 1$ that

$$\text{Cov}(\hat{\sigma}_t^2, \hat{\sigma}_s^2) = \begin{cases} \Delta^4 \text{Cov}(\hat{h}_t^2, \hat{h}_s^2) & for \quad t, s \ge 2 \\ 0 & for \quad t = 1 \; or \; s = 1. \end{cases}$$

(b) Let $\{Y_t\}$ be a stationary GARCH(1,1) process. Then for $t > \tau$,

$$\text{E}(\hat{\sigma}_t^2) - \Delta^2 \gamma_0 = \frac{\beta_1}{r_{t-1}} [E(\hat{\sigma}_{t-1}^2) - \Delta^2 \gamma_0] - \alpha_0 (\Delta^2 - 1).$$

Moreover,

$$\lim_{t \to \infty} E(\hat{\sigma}_t^2) - \text{Var}(X_t) = -(\Delta^2 - 1) \frac{\alpha_0}{1 - \beta_1}.$$

Now we choose $L_{V,t}(X_t, X_{t-1}, \ldots) = \hat{\sigma}_{t+1}^2$. At the first look it seems to be strange to have the index $t + 1$. Because we want to detect a change at position t the statistic for the decision rule has to depend on X_t and it may depend on the previous values $X_{t-1}, \ldots, X_1$. This is the reason to take $\hat{\sigma}_{t+1}^2$. Then the EWMA recursion based on $\hat{\sigma}_{t+1}^2$ is given by

$$Z_t = (1 - \lambda) Z_{t-1} + \lambda \hat{\sigma}_{t+1}^2 \quad \text{for} \quad t \ge 1. \tag{6.8}$$

The starting value Z_0 is set equal to the in-control value of $E(\hat{\sigma}_t^2)$, i.e. to $E_\infty(\hat{\sigma}_{t+1}^2) = \gamma_0$. Using the results of Proposition 1 we can show that for a shift at the first moment of time ($\tau = 1$) it holds that

$$\mathrm{E}(Z_t) = \Delta^2\gamma_0 - (\Delta^2 - 1)\gamma_0(1-\lambda)^2 - \lambda(\Delta^2 - 1)\sum_{i=0}^{t-1}(1-\lambda)^i a_{t-i+1,t-i+1}$$

$$\underset{t\to\infty}{\longrightarrow} \Delta^2\gamma_0 - (\Delta^2 - 1)a_\infty,$$

where a_∞ stands for $\lim_{t\to\infty} a_{tt}$ provided that the limit exists. For the special case of a GARCH(1,1) it holds that $a_\infty = \alpha_0/(1-\beta_1)$ and

$$\lim_{t\to\infty} Var(Z_t) = (\kappa - 1)\Delta^4 \frac{\lambda}{2-\lambda}\gamma_0^2 \frac{\alpha_1^2}{1-\beta_1^2 - 2\alpha_1\beta_1 - \kappa\alpha_1^2}$$
$$\times \frac{1 + (1-\lambda)(\alpha_1 + \beta_1)}{1 - (1-\lambda)(\alpha_1 + \beta_1)},$$

assuming that E $(\varepsilon_t^4)\kappa < \infty$ for all t and $k\alpha_1^2 + 2\alpha_1\beta_1 + \beta_1^2 < 1$. The exact variance of the control statistic is difficult to derive due to the complexity of the process.

The monitoring procedure is based on the decision rule with the asymptotic variance and the process is concluded to be out of control if

$$Z_t < c_1\gamma_0 \quad \text{or} \quad Z_t > c_2\gamma_0,$$

where $c_1 > 0$ and $c_2 > 0$. This is a similar decision rule as in Chapter 5 for linear processes. Because the distribution of Z_t is skewed we have two control limits. The values c_1 and c_2 are determined in such a way that the in-control ARL of the chart is equal to a desired given value ξ. This equation is solved under the side condition that the in-control ARLs of both one-sided charts is the same. This means that

$$\mathrm{E}_D(\inf\{t \in I\!N : Z_t < c_1\gamma_0 \text{ or } Z_t > c_2\gamma_0\}) = \xi$$

under the restriction that

$$\mathrm{E}_D(\inf\{t \in I\!N : Z_t < c_1\gamma_0\}) = \mathrm{E}_D(\inf\{t \in I\!N : Z_t > c_2\gamma_0\}) \ .$$

Note that the control limits are multiples of γ_0. This seems to be a natural choice because the in-control mean of Z_t is γ_0. However, here the distribution of the control statistic is not symmetric and thus the interpretation of the mean is difficult. It is desirable that the control limits are multiples of α_0 because then the ARL of the chart does not depend on α_0. Consequently it is easier to determine the critical values. In spite of this property it has to be emphasized that they depend on all other parameters of the GARCH process.

6.3.1.2 A chart based on squared observations

In the same way as in the previous chapter we choose $L_{V,t}(X_t, X_{t-1}, ..) = X_t^2$. This chart was studied in Chapter 5 in the case of an arbitrary stationary process. We have seen that it is able to track the change. For the special case of a GARCH(1,1) process with $\varepsilon_t \sim F$, $E(\varepsilon_t^4) = \kappa < \infty$ for all t, and $\kappa\alpha_1^2 + 2\alpha_1\beta_1 + \beta_1^2 < 1$ it holds that for $\tau = 1$

$$\lim_{t\to\infty} Var(Z_t) = (\kappa - 1)\, \Delta^4 \, \frac{\lambda}{2-\lambda}\, \gamma_0^2$$

$$\times \frac{1 - \beta_1^2 - 2\alpha_1\beta_1 + (1-\lambda)(\alpha_1 - \beta_1 + \beta_1^2(\alpha_1 + \beta_1))}{(1-(1-\lambda)(\alpha_1+\beta_1))(1-\beta_1^2 - 2\alpha_1\beta_1 - \kappa\alpha_1^2)} .$$

The chart signals an alarm if

$$Z_t < c_1\gamma_0 \quad \text{or} \quad Z_t > c_2\gamma_0 .$$

The values c_1 and c_2 are determined as above. The in-control ARL is set equal to a prespecified value ξ. This equation is solved under the assumption that the one-sided charts have the same in-control ARL.

6.3.1.3 A chart based on the logarithm of the squared observations

By taking the logarithm of the squared observations we have seen in Chapter 5 that the scale change is transferred to a shift in the model. In this case we have $L_{V,t}(X_t, X_{t-1}, \ldots) = \ln\ X_t^2/\gamma_0$. The EWMA recursion is given by

$$Z_t = (1-\lambda)Z_{t-1} + \lambda \ln(X_t^2/\gamma_0)$$

for $t \geq 1$ with $Z_0 = E_D[\ln(X_t^2/\gamma_0)] = \gamma_0^*$. The chart gives a signal if

$$|Z_t - \gamma_0^*| > c\sqrt{\lim_{t\to\infty} \mathrm{Var}_D(Z_t)} .$$

Because the distribution of $\ln(Y_t^2/\gamma_0)$ is nearly symmetric a symmetric control interval is taken. Thus we have to determine only one critical value $c > 0$. This is done by setting the in-control ARL equal to a specified value ξ.

6.3.1.4 Residual charts

The residual charts are based on the consideration of the estimated residuals $X_t/\hat{\sigma}_t$. Here we have $L_{V,t}(X_t, X_{t-1}, ..) = X_t^2/\hat{\sigma}_t^2$. The EWMA recursion is given by

$$Z_t = (1-\lambda)Z_{t-1} + \lambda\ X_t^2/\hat{\sigma}_t^2$$

for $t \geq 1$ with $Z_0 = \mathrm{E}_D(X_t^2/\hat{\sigma}_t^2)$. As a starting value we take Z_0 = E $(\varepsilon_t^2) = 1$. Note that due to the starting problem it holds that in general $\mathrm{E}(\varepsilon_t^2) \neq \mathrm{E}_D(X_t^2/\hat{\sigma}_t^2)$. The chart gives a signal at time t if

$$Z_t < c_1 \quad \text{or} \quad Z_t > c_2$$

where $c_1 > 0$ and $c_2 > 0$ are determined in the usual way. Another possibility would be to consider the logarithm of the squared residuals and to apply an EWMA recursion to it. We will not discuss this procedure here.

6.3.2 CUSUM type charts

In the previous chapter we have seen that the derivation of CUSUM charts turns out to be much harder in the case of correlated data. For that reason in many cases a CUSUM recursion for independent data is applied to the correlated data. We have also made use of this procedure in section 5.5.1.2 where we introduced control charts for a linear process. This procedure leads to the following recursion in the case $\mu_0 = 0$

$$S_t^+ = \max\left\{0, S_{t-1}^+ + X_t^2 - K\gamma_0\right\}, t \geq 1, \quad S_0^+ = 0$$
$$S_t^- = \min\left\{0, S_{t-1}^- + X_t^2 - K\gamma_0\right\}, t \geq 1, \quad S_0^- = 0.$$

The process is concluded to be out of control if $S_t^+ > c_2\gamma_0$ or $S_t^- < -c_1\gamma_0$.

In the following we replace the squared observations by an arbitrary local measure for the variance or a transformed quantity $L_{V,t}(X_t, X_{t-1}, ..)$. This leads to

$$S_t^+ = \max\left\{0, S_{t-1}^+ + L_{V,t}(X_t, X_{t-1}, ..) - K\mathrm{E}_D(L_{V,t}(X_t, X_{t-1}, ..))\right\},$$
$$S_t^- = \min\left\{0, S_{t-1}^- + L_{V,t}(X_t, X_{t-1}, ..) - K\mathrm{E}_D(L_{V,t}(X_t, X_{t-1}, ..))\right\}$$

for $t \geq 1$ with $S_0^+ = S_0^- = 0$. Note that not in all cases is it advisable to subtract $\mathrm{E}_\infty(L_{V,t}(X_t, X_{t-1}, ..)$. The main aim is to use a quantity such that the in-control ARL does not depend on α_0.

Taking $L_{V,t}(X_t, X_{t-1}, \ldots) = \hat{\sigma}_{t+1}^2 - \alpha_0/(1 - \sum_{j=1}^p \beta_j)$ we obtain a chart based on the volatility. We use that $\mathrm{E}_D(L_{V,t}(X_t, X_{t-1}, \ldots) = \gamma_0 - \alpha_0/(1 - \sum_{j=1}^p \beta_j)$. Note that $\hat{\sigma}_{t+1}^2$ never attains zero. Since $h_t^2 \geq \alpha_0/(1 - \sum_{j=1}^p \beta_j)$, we subtract this quantity as a proxy for the lower bound of the forecast of the conditional volatility $\hat{\sigma}_{t+1}^2$. The chart gives a signal if $S_t^+ > c_2\gamma_0$ or $S_t^- < -c_1\gamma_0$.

The choice $L_{V,t}(X_t, X_{t-1}, \ldots) = X_t^2$ leads to the chart based on squared observations. It has the same structure and the same decision rule as in the iid case. The chart based on the logarithm of the squared observations is obtained by the choice $L_{V,t}(X_t, X_{t-1}, \ldots) = \ln(X_t^2/\gamma_0)$.

The above recursions can be applied to the residuals as well. In that case we have $L_{V,t}(X_t, X_{t-1}, \ldots) = X_t^2/\hat{\sigma}_t^2$ and $L_{V,t}(X_t, X_{t-1}, \ldots) = \ln(X_t^2/\hat{\sigma}_t^2)$, respectively.

6.3.3 Comparison of the control charts

In this section we compare the control charts for the variance introduced in the previous sections. The target process is a GARCH (1, 1) process with Gaussian residuals and the parameters set to

$$\alpha_0 = 10^{-6}, \quad \alpha_1 = 0.07 \quad \text{and} \quad \beta_1 = 0.92.$$

The values seem to be similar to those usually obtained for financial time series. The control limits of all charts are determined in such way that their in-control ARL is equal to a prespecified constant, here $\xi = 500$. Note that in most cases the control limits depend on the parameters of the target process. Because no explicit formula for the in-control ARL is available, simulations have been used to determine the control limits. The iterative procedure is based on Broyden's approach to the bivariate secant method with 10^6 replications. With preliminary calibration, we achieved good convergence within a few iterations. The smoothing parameter λ for the EWMA charts is taken from $\{0.05, 0.1, 0.2, 0.4, 0.6, 0.8, 1.0\}$ and the reference value K for the CUSUM charts is taken from $\{0.25, 0.5, 0.75, 1.0, 1.25, 1.5, 1.75, 2.0\}$. The results are presented in Table 6.1.

Using these limits the out-of-control behaviour of the charts is compared with each other on the basis of the out-of-control ARL and the maximum average delay. This is the same procedure as in Chapter 5. The size of the shift Δ takes values from $\{0.5, 0.75, 1.0, 1.25, 1.5, 1.75, 2.0, 2.5, 3.0\}$. The study is based on 10^6 replications. The out-of-control ARLs for each λ, K and Δ are given in Tables 6.2 and 6.3 for EWMA and CUSUM charts respectively. The first line in each block refers to the charts based on squared observations, the second line refers to the charts based on conditional volatility and the third line contains the results for the residual charts.

If $\Delta = 1$ we have no shift and the corresponding values of the ARL are close to $\xi = 500$. For minor changes, the charts based on squared observations leads to the optimal value of λ between 0.1 and 0.2. This is consistent with the evidence in the literature. The case of $\lambda = 1$ also performs well, however, note that the extremely small lower critical value of this chart is difficult to determine precisely. Thus the good performance of this chart can be due to

Table 6.1 Critical values of the two-sided EWMA and CUSUM charts for the variance*

EWMA chart				
λ		squared	volatility	residual
0.05	c_1	$2.461 \cdot 10^{-1}$	$2.644 \cdot 10^{-1}$	$5.425 \cdot 10^{-1}$
	c_2	2.595	1.853	1.729
0.10	c_1	$1.693 \cdot 10^{-1}$	$2.235 \cdot 10^{-1}$	$3.812 \cdot 10^{-1}$
	c_2	3.394	2.095	2.256
0.20	c_1	$1.003 \cdot 10^{-1}$	$1.978 \cdot 10^{-1}$	$2.177 \cdot 10^{-1}$
	c_2	4.655	2.301	3.223
0.40	c_1	$4.160 \cdot 10^{-2}$	$1.814 \cdot 10^{-1}$	$8.288 \cdot 10^{-2}$
	c_2	6.837	2.481	5.100
0.60	c_1	$1.651 \cdot 10^{-2}$	$1.747 \cdot 10^{-1}$	$3.058 \cdot 10^{-2}$
	c_2	8.925	2.574	6.989
0.80	c_1	$4.875 \cdot 10^{-3}$	$1.710 \cdot 10^{-1}$	$8.446 \cdot 10^{-3}$
	c_2	11.039	2.651	8.912
1.00	c_1	$1.100 \cdot 10^{-6}$	$1.683 \cdot 10^{-1}$	$1.538 \cdot 10^{-6}$
	c_2	12.930	2.744	10.874

CUSUM chart				
		squared	volatility	residual
0.25	c_1	2.126	0.471	1.167
	c_2	571.260	474.190	592.986
0.50	c_1	17.720	20.659	3.896
	c_2	357.119	256.705	397.354
0.75	c_1	76.345	99.226	10.350
	c_2	177.769	111.407	205.349
1.00	c_1	196.560	245.936	42.469
	c_2	98.712	56.645	42.574
1.25	c_1	363.577	428.878	210.781
	c_2	61.033	28.483	17.697
1.50	c_1	549.573	624.131	405.801
	c_2	39.859	13.571	13.400
1.75	c_1	744.916	826.738	603.342
	c_2	28.244	5.730	11.509
2.00	c_1	943.266	1033.834	802.539
	c_2	21.538	1.975	10.379

* Critical values of the twosided EWMA and CUSUM control charts based on squared observations, volatility and residual charts for different values of the parameters λ and K. The incontrol ARL is set to 500. The simulation study is based on 10^6 repetitions.

Table 6.2 Out-of-control ARL of the EWMA charts*

Δ	λ = 0.05	0.1	0.2	0.4	0.6	0.8	1.0
0.50	35.656	25.861	21.430	22.543	26.990	35.073	254.764
	37.650	27.201	21.295	19.027	18.611	18.463	18.408
	23.225	21.857	24.339	36.077	55.009	88.179	509.744
0.75	61.130	55.432	56.855	66.219	80.010	103.472	458.254
	62.624	55.304	52.844	52.322	52.236	52.192	52.209
	58.262	68.655	96.931	161.784	233.383	318.571	753.256
1.00	500.011	501.041	501.596	500.027	498.478	499.194	498.743
	498.847	501.154	500.052	500.902	499.764	499.617	500.771
	499.611	500.242	499.755	499.390	498.855	499.490	500.164
1.25	42.647	41.771	41.514	41.621	41.791	41.948	40.683
	45.064	42.901	41.193	40.040	39.401	39.026	38.717
	54.544	63.049	77.480	96.010	106.179	111.828	111.019
1.50	18.671	17.583	16.990	16.849	16.936	17.073	16.803
	20.869	18.709	16.975	15.761	15.176	14.822	14.514
	21.848	22.811	25.950	31.343	35.023	37.521	38.605
1.75	11.760	10.816	10.209	9.967	9.978	10.055	9.938
	13.656	11.856	10.330	9.212	8.688	8.372	8.111
	12.945	12.811	13.809	16.054	17.775	19.019	19.735
2.00	8.504	7.717	7.168	6.898	6.869	6.905	6.837
	10.158	8.682	7.389	6.402	5.943	5.671	5.456
	8.839	8.480	8.799	9.945	10.923	11.657	12.140
2.50	5.448	4.893	4.480	4.239	4.186	4.189	4.152
	6.756	5.716	4.770	4.017	3.671	3.469	3.321
	5.167	4.833	4.801	5.152	5.539	5.863	6.108
3.00	4.016	3.609	3.297	3.104	3.051	3.042	3.015
	5.092	4.307	3.583	3.003	2.737	2.585	2.478
	3.601	3.360	3.278	3.398	3.570	3.734	3.866

*The table contains the out-of-control ARLs of the EWMA charts for different shifts Δ and smoothing parameters λ. The first line in each block contains the values for the EWMA chart based on squared observations, the second line for the chart based on volatility and the third line for the EWMA residual chart. The number of replications is set to 10^6.

precision problems. With increasing Δ the case of $\lambda = 1$ becomes dominant. This value of λ is also optimal for the charts based on the conditional volatility for all sizes of the shift Δ. An opposite situation is observed for the residual charts, where 0.1 and 0.2 always lead to shorter ARLs. For mean charts it appears to be only a local optimum as shown by Frisén (2003) and Frisén and Sonesson (2006). In general for the considered type of the shift, the chart based on the volatility with $\lambda = 1$ appears to outperform other charts in all cases. The conclusion applies both to decreases and to increases in the variance of the process.

For the CUSUM type charts we observe that small values of K are optimal for decreases in the variance and large K's are optimal for increases. For

Table 6.3 Out-of-control ARL of the CUSUM charts*

Δ	K 0.25	0.50	0.75	1.00	1.25	1.50	1.75	2.00
0.50	19.348	45.274	112.402	208.510	303.571	379.073	437.962	483.363
	19.760	59.984	149.761	263.143	359.670	431.656	486.944	530.449
	33.607	22.182	25.707	61.437	214.587	327.783	404.925	460.977
0.75	51.002	71.322	140.512	239.862	336.425	411.413	468.911	512.567
	49.665	87.490	180.483	296.540	393.085	463.513	516.816	558.284
	155.584	81.672	59.749	103.984	311.816	436.765	510.809	559.915
1.00	500.208	500.692	500.386	499.482	500.047	499.343	500.275	500.309
	500.279	500.236	500.332	499.533	500.118	500.124	500.273	501.686
	500.400	500.204	500.497	499.422	500.037	499.687	500.190	500.415
1.25	103.246	87.721	67.722	55.179	47.904	43.208	40.540	38.970
	110.174	89.993	67.746	55.848	48.003	42.869	39.381	36.964
	422.654	368.727	256.817	80.729	53.278	57.986	64.938	71.277
1.50	49.532	42.784	33.605	27.126	22.851	19.799	17.864	16.696
	56.162	47.061	35.982	28.933	23.580	19.625	16.754	14.797
	296.326	231.973	143.168	40.935	23.237	22.135	23.110	24.503
1.75	31.988	27.825	22.047	17.800	14.863	12.691	11.253	10.341
	37.712	31.922	24.607	19.668	15.723	12.656	10.328	8.693
	215.306	161.073	95.487	27.199	14.341	12.801	12.770	13.171
2.00	23.322	20.359	16.212	13.102	10.912	9.257	8.137	7.416
	28.232	24.008	18.595	14.842	11.771	9.321	7.412	6.044
	163.753	119.855	69.999	20.041	10.016	8.629	8.374	8.476
2.50	14.879	13.043	10.452	8.474	7.054	5.964	5.217	4.728
	18.576	15.876	12.387	9.912	7.838	6.136	4.769	3.782
	104.993	75.566	44.038	12.477	5.927	5.009	4.749	4.688
3.00	10.823	9.508	7.642	6.217	5.189	4.396	3.855	3.501
	13.730	11.768	9.232	7.421	5.887	4.607	3.564	2.821
	73.921	53.095	31.251	8.539	4.099	3.497	3.305	3.238

*The table contains the out of-control ARLs of the CUSUM charts for different shifts Δ and the reference velue K. The first line in each block contains the values for the CUSUM chart based on squared observations, the second line for the chart based on volatility and the third line for the CUSUM residual chart. The number of replications is set to 10^6.

nonoptimal values of K the chart based on the squared observations usually leads to shorter out-of-control ARL. However, for the optimal K the chart based on the volatility performs better for $\Delta > 1$. The residual chart performs better only for nonoptimal K's with $\Delta < 1$ and $K = 1.5$ or 1.75 and large shifts Δ.

Following the criticism of the ARL as a measure of control chart performance we also consider the maximum expected delay. The results are summarized in Table 6.4. The shift occurs at the time point τ, which takes values between 1 and 30. The study is based on 10^5 repetitions. For each value of λ for the EWMA charts and each value of K for the CUSUM charts we

Table 6.4 Maximum of CED for EWMA and CUSUM control charts*

	Δ			
	0.50	0.75	1.00	1.25
EWMA sq.	(1,0.20) 21.83	(1,0.10) 56.05	(4,0.60) 497.83	(29,1.00) 42.28
EWMA vol.	(16,1.00) 25.17	(1,1.00) 53.17	(2,0.05) 499.12	(23,1.00) 43.75
EWMA res.	(1,0.10) 21.20	(1,0.05) 57.28	(6,0.80) 497.42	(1,0.05) 53.96
CUSUM sq.	(1,0.25) 19.68	(1,0.25) 51.61	(2,1.50) 498.49	(29,2.00) 42.03
CUSUM vol.	(21,0.25) 25.91	(9,0.25) 50.80	(2,0.50) 499.55	(23,2.00) 44.32
CUSUM res.	(1,0.50) 20.88	(1,0.75) 58.16	(1,1.50) 498.36	(1,1.25) 52.38

Δ				
1.50	1.75	2.00	2.50	3.00
(17,1.00) 18.18	(13,1.00) 11.01	(12,1.00) 7.74	(9,1.00) 4.83	(6,1.00) 3.56
(21,1.00) 19.04	(19,1.00) 11.91	(21,1.00) 8.60	(17,1.00) 5.52	(15,1.00) 4.12
(1,0.05) 21.44	(1,0.10) 12.75	(1,0.10) 8.63	(1,0.20) 5.13	(1,0.20) 3.64
(29,2.00) 18.58	(30,2.00) 11.72	(30,2.00) 8.53	(30,2.00) 5.56	(28,2.00) 4.20
(21,2.00) 21.11	(19,2.00) 14.12	(21,2.00) 10.74	(18,2.00) 7.38	(16,2.00) 5.74
(1,1.50) 21.51	(1,1.50) 12.54	(1,1.75) 8.42	(1,2.00) 4.98	(1,2.00) 3.57

* Maximum of the conditional expected delay for different sizes of the shift Δ. The first three lines in each block contain values for the EWMA charts based on the squared observations and on the volatility, as well as the EWMA residual chart. The next three lines contain the corresponding results for the CUSUM type charts. The critical values are determined for the in-control ARL of 500. In parentheses we provide the optimal value of τ from the interval 1 to 30 as well as the optimal parameter λ for EWMA charts and K for CUSUM charts. The study is based on 10^5 repetitions.

determine the value of τ which leads to the largest delay. Further the optimal parameter λ or K is chosen to provide the lowest delay for the given worst value of τ. The parameters are provided in the table in parentheses. The EWMA chart based on the squared observations provides the best results, while the residual chart appears to be in most cases the worst. A similar situation is observed for CUSUM charts, however, here the residual chart outperforms other charts for very large shifts. As expected, the optimal parameters λ and K are lower for small shifts and larger for increasing shifts.

6.3.4 Example

To illustrate the developed control charts, we use the daily data on the MSCI country index for Germany from January 1998 to June 2006. The estimated parameters of the fitted GARCH(1, 1) process are close to those consider in the previous subsection. Thus we use the critical values from Table 6.1. The smoothing parameter λ is set to 1 and the reference value K to 1. Figures 6.1 and 6.2 illustrate the application of the discussed charts. Here we assume that

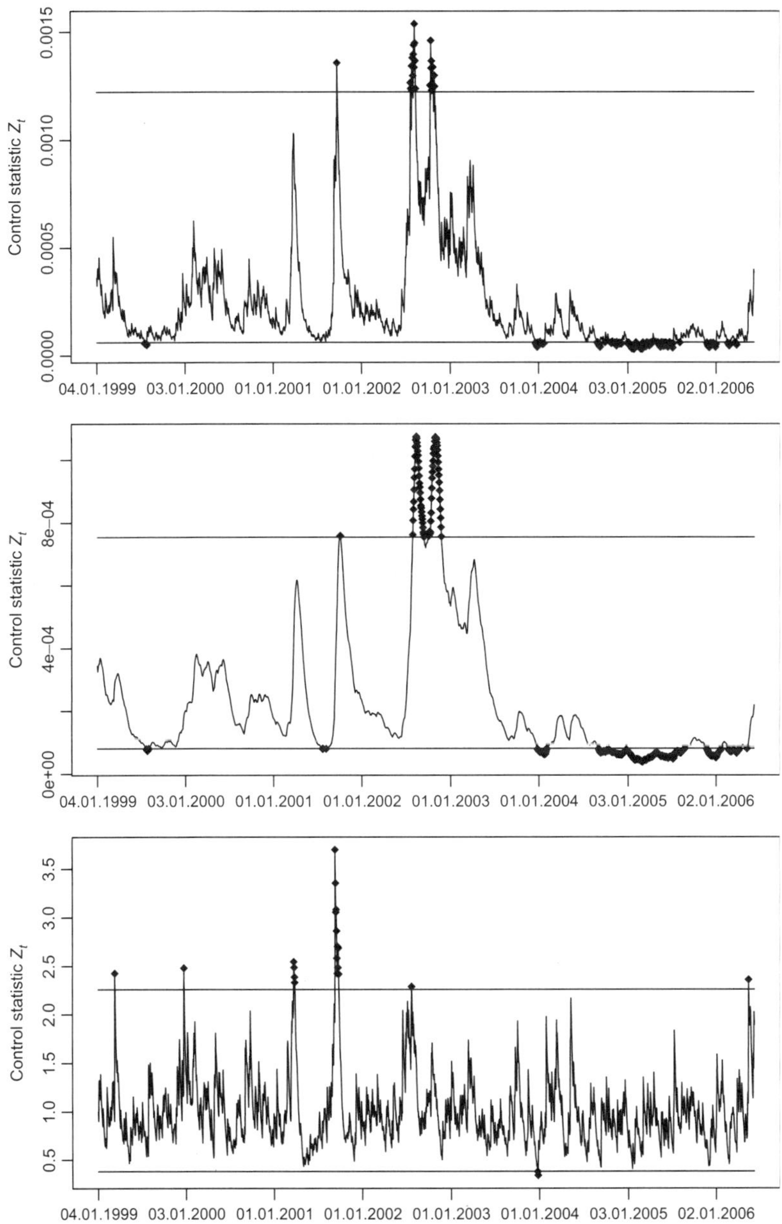

Figure 6.1 Modified EWMA control charts based on squared observations, volatility and the residual EWMA chart applied to daily returns on MSCI country index for Germany. The value of λ is set to 0.1.

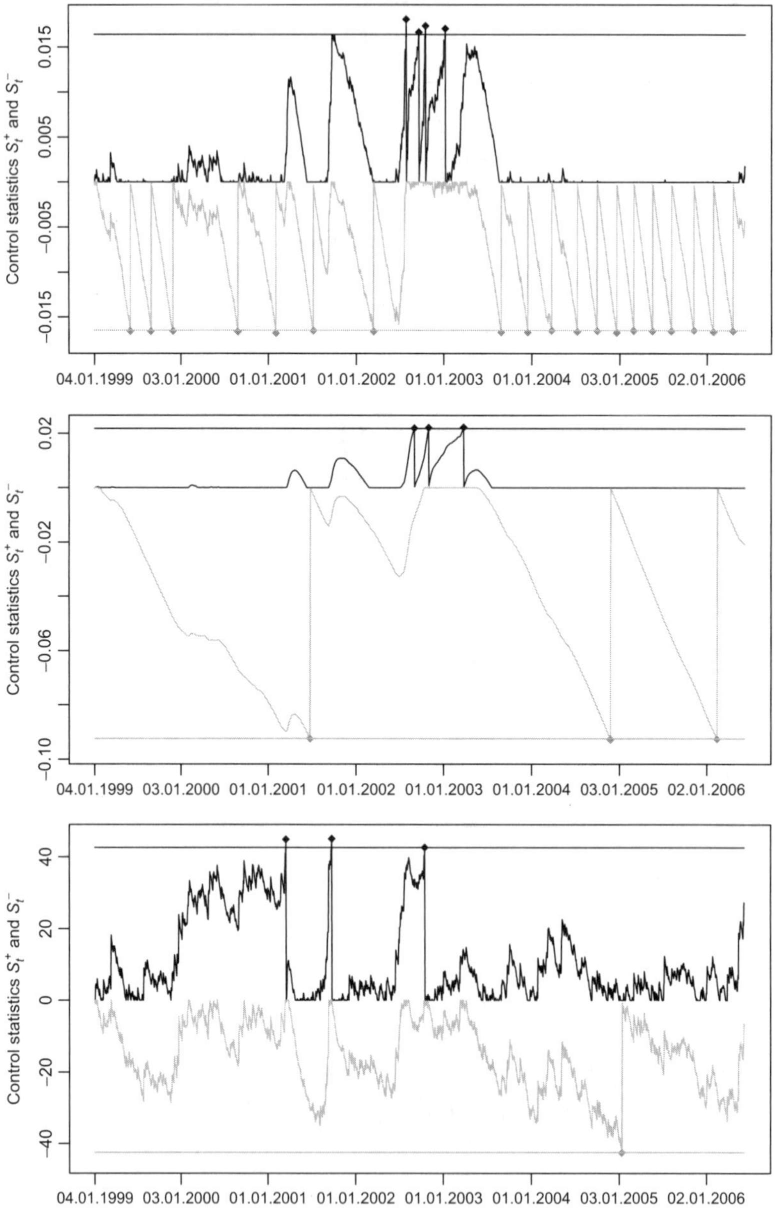

Figure 6.2 Modified CUSUM control chart based on squared observations, volatility and the residual CUSUM chart applied to daily returns on MSCI country index for Germany. The value of K is set to 1.

the initially determined control limits stay unchanged during the whole monitoring period. For the CUSUM type charts, the control statistic is reset to zero after each alarm.

The EWMA type charts based on squared observations and on the volatility exhibit similar performance. The periods with very high and very low volatilities are determined very precisely. This is due to the fact that both control statistics react similarly to shifts in the observed process. For the residual chart the deviations in the observations and in the conditional volatility cancel out by construction of the control statistic. Therefore, this chart aims to detect a specific shifts in the residuals of the process. The first two CUSUM type charts exhibit similar behaviour as the EWMA charts and correctly specify the periods with higher and lower volatility. The chart based on the residuals seems to be more sensitive to the deviations as the volatility chart. Concerning the residual chart a similar discussion as for the EWMA charts applies here as well.

6.4 Surveillane of the covariance matrix of a multivariate nonlinear time series

Up to now we have considered only univariate time series. In most applications, however, we are interested in the behaviour of assets of several stocks at the same point of time. Consequently a multivariate time series is considered.

As we have seen in the previous sections there are many possible types of changes but in finance we mainly find changes in the risk behaviour. Here we monitor not only the variances of each time series but their covariances as well.

The control charts presented in this section were introduced by Śliwa and Schmid (2005). They proposed several EWMA type control schemes. We will not consider CUSUM charts in this section. First attempts in this direction have been done by Bodnar and Schmid (2007).

6.4.1 Modelling the observed process

We use the same notation and the same modelling as in Chapter 5. The target process is denoted by $\{\mathbf{Y}_t\}$. It is assumed to be an m-dimensional weakly stationary process with mean $\boldsymbol{\mu}_0$. The observed process is denoted by $\{\mathbf{X}_t\}$. Both processes are related as follows

$$\mathbf{X}_t = \begin{cases} \boldsymbol{\Delta}\mathbf{Y}_t & \text{for} \quad t \geq \tau \\ \mathbf{Y}_t & \text{for} \quad t < \tau. \end{cases} \tag{6.9}$$

The shift matrix $\mathbf{\Delta} \neq \mathbf{I}$ is unknown, but it is assumed that it is positive definite. If $\tau = \infty$ then the observed process is said to in control. The in-control process (target process) is equal to the observed process. If, however, $\tau < \infty$ then the observed process is out of control. Then

$$\mathrm{Var}(\mathbf{X}_t) = \begin{cases} \mathrm{Cov}(\mathbf{Y}_t) & \text{for} \quad t < \tau \\ \mathbf{\Delta}\mathrm{Cov}(\mathbf{Y}_t)\mathbf{\Delta}' & \text{for} \quad t \geq \tau. \end{cases}$$

In the following the target process is equal to a multivariate GARCH process with $\boldsymbol{\mu}_0 = 0$. We introduce this family of processes in the next section.

6.4.2 Introduction to multivariate GARCH processes

There are several possibilities to generalize a univariate GARCH process to a multivariate one (see, e.g. Bollerslev 1990; Engle *et al.* 1990; Gourieroux 1997; Harvey *et al.* 1994). Here we make use of a model proposed by Engle and Kroner (1995).

Let $\{\mathbf{Y}_t\}$ be an m-dimensional process satisfying

$$\mathbf{Y}_t = \mathbf{H}_t^{1/2}\boldsymbol{\varepsilon}_t \tag{6.10}$$

for $t \in \mathbb{Z}$. The variables $\{\boldsymbol{\varepsilon}_t\}$ are assumed to be independent and identically distributed with zero mean and covariance matrix equal to the identity matrix $\mathbf{I}$. The matrix $\mathbf{H}_t$ is positive definite and symmetric. To simplify the presentation we vectorize $\mathbf{H}_t$ by applying the vech operator, i.e. let

$$\mathbf{h}_t = \mathrm{vech}(\mathbf{H}_t) = (h_{11}, \ldots, h_{m1}, h_{22}, \ldots, h_{m2}, \ldots, h_{m-1,m-1}, h_{m,m-1}, h_{mm})'.$$

The process $\{\mathbf{Y}_t\}$ follows an m-variate GARCH(p, q) process if the $m(m+1)/2$-dimensional vector $\mathbf{h}_t$ develops as

$$\mathbf{h}_t = \boldsymbol{\alpha}_0 + \sum_{i=1}^{q} \boldsymbol{\alpha}_i \boldsymbol{\eta}_{t-i} + \sum_{j=1}^{p} \boldsymbol{\beta}_j \mathbf{h}_{t-j}, \tag{6.11}$$

where $\boldsymbol{\eta}_t$ =vech$(\mathbf{Y}_t\mathbf{Y}_t')$. The matrices $\boldsymbol{\alpha}_i$ and $\boldsymbol{\beta}_j$ for $i = 1, \ldots, q$ and $j = 1, \ldots, p$ are $m \times m$-dimensional parameter matrices. The vector $\boldsymbol{\alpha}_0$ is a vectorized symmetric positive definite $m \times m$ matrix $\boldsymbol{\Omega}$. Engle and Kroner (1995) provide conditions for the stationary of a multivariate GARCH(p, q) process. They showed that $\{\mathbf{Y}_t\}$ is stationary if the following two conditions are satisfied

1. The matrix $\mathbf{H}_t$ is almost surely positive definite.

2. All eigenvalues of the matrix $\sum_{i=1}^{\max\{q,p\}} (\boldsymbol{\alpha}_i + \boldsymbol{\beta}_i)$ are smaller than one in absolute value where $\boldsymbol{\alpha}_i = \mathbf{0}$ for $i > q$ and $\boldsymbol{\beta}_j = \mathbf{0}$ for $j > p$.

Then it holds that

$$\mathrm{E}(\mathbf{Y}_t) = \mathbf{0}, \quad \mathrm{E}(\mathbf{Y}_t\mathbf{Y}_s') = \mathbf{0} \quad \text{for} \quad t \neq s$$

and the unconditional covariance matrix $\boldsymbol{\gamma}(0)$ is defined by its vectorization

$$\boldsymbol{\gamma} = \mathrm{vech}[\mathrm{E}(\mathbf{Y}_t\mathbf{Y}_t')] = \left[\mathbf{I} - \sum_{i=1}^{\max\{q,p\}} (\boldsymbol{\alpha}_i + \boldsymbol{\beta}_i)\right]^{-1} \boldsymbol{\alpha}_0.$$

Similarly as in the univariate case, it can be shown that $\mathbf{H}_t$ is the variance of $\mathbf{Y}_t$ conditional on all previous observations of the process, i.e.

$$\mathrm{E}(\mathbf{Y}_t\mathbf{Y}_t'|\mathbf{Y}_{t-1}, \mathbf{Y}_{t-2}, ...) = \mathbf{H}_t.$$

The conditional distribution of $\mathbf{Y}_t$ is determined by the distribution of $\boldsymbol{\varepsilon}_t$.

Using that $E(\mathbf{h}_t) = E(\boldsymbol{\eta}_t) = \boldsymbol{\gamma}$ we obtain from (6.11) that $\boldsymbol{\alpha}_0 = (\mathbf{I} - \sum_{i=1}^{q} \boldsymbol{\alpha}_i - \sum_{j=1}^{p} \boldsymbol{\beta}_j)\boldsymbol{\gamma}$. Substituting back into (6.11) and letting $\mathbf{v}_t = \boldsymbol{\eta}_t - \mathbf{h}_t$ we obtain a VARMA(max$\{p, q\}$, p) representation of the process $\boldsymbol{\eta}_t$ in the form

$$\boldsymbol{\eta}_t - \boldsymbol{\sigma} = \sum_{i=1}^{\max\{q,p\}} (\boldsymbol{\alpha}_i + \boldsymbol{\beta}_i)(\boldsymbol{\eta}_{t-i} - \boldsymbol{\sigma}) + \mathbf{v}_t - \sum_{i=1}^{p} \boldsymbol{\beta}_j \mathbf{v}_{t-j}. \tag{6.12}$$

It holds that $E(\mathbf{v}_t) = \mathbf{0}$ and $E(\mathbf{v}_t\mathbf{v}_s') = \mathbf{0}$ for $t \neq s$.

Assuming that the stationarity conditions are satisfied we may rewrite the process $\{\boldsymbol{\eta}_t\}$ as a linear process. Thus

$$\boldsymbol{\eta}_t = \boldsymbol{\sigma} + \sum_{i=0}^{\infty} \boldsymbol{\Phi}_l \mathbf{v}_{t-l},$$

where $\boldsymbol{\Phi}_0 = \mathbf{I}$ and $\boldsymbol{\Phi}_i$ for $i \geq 1$ are determined by the coefficients of the GARCH process. For example, in the case of a GARCH(1,1) process we have that $\boldsymbol{\Phi}_i = (\boldsymbol{\alpha}_1 + \boldsymbol{\beta}_1)^{i-1}\boldsymbol{\alpha}_1$ for $i \geq 1$. Further, denoting the covariance matrix of $\mathbf{v}_t$ by $\boldsymbol{\Sigma}_\mathbf{v}$ and the autocovariance matrices of $\boldsymbol{\eta}_t$ by $\boldsymbol{\Sigma}(i) = \mathrm{Cov}(\boldsymbol{\eta}_{t+i}, \boldsymbol{\eta}_t)$, it can be shown that

$$\boldsymbol{\Sigma}(0) - (\boldsymbol{\alpha}_1 + \boldsymbol{\beta}_1)\boldsymbol{\Sigma}(0)(\boldsymbol{\alpha}_1 + \boldsymbol{\beta}_1)' = \boldsymbol{\Sigma}_\mathbf{v} - (\boldsymbol{\alpha}_1 + \boldsymbol{\beta}_1)\boldsymbol{\Sigma}_\mathbf{v}\boldsymbol{\beta}_1' - \boldsymbol{\beta}_1\boldsymbol{\Sigma}_\mathbf{v}\boldsymbol{\alpha}_1'$$

$$\boldsymbol{\Sigma}(i) = (\boldsymbol{\alpha}_1 + \boldsymbol{\beta}_1)^{i-1}((\boldsymbol{\alpha}_1 + \boldsymbol{\beta}_1)\boldsymbol{\Sigma}(0) - \boldsymbol{\beta}_1\boldsymbol{\Sigma}_\mathbf{v}), \quad \text{for} \quad i \geq 1.$$

Defining $\boldsymbol{\Sigma}_\eta = \mathrm{E}(\boldsymbol{\eta}_t\boldsymbol{\eta}_t')$ and $\boldsymbol{\Sigma}_\mathbf{h} = \mathrm{E}(\mathbf{h}_t\mathbf{h}_t')$ it can be proved that

$$\boldsymbol{\Sigma}_\mathbf{v} = \boldsymbol{\Sigma}_\eta - \boldsymbol{\Sigma}_\mathbf{h}, \qquad \boldsymbol{\Sigma}(0) = \boldsymbol{\Sigma}_\eta - \boldsymbol{\sigma}\boldsymbol{\sigma}'.$$

Furthermore, following Hafner (2003) we can establish a direct relationship between $\boldsymbol{\Sigma}_\eta$ and $\boldsymbol{\Sigma}_\mathbf{h}$ in the form $\mathrm{vec}(\boldsymbol{\Sigma}_\eta) = \mathbf{G}_m \; \mathrm{vec}(\boldsymbol{\Sigma}_\mathbf{h})$, where $\mathbf{G}_m$ is an

$m^2 \times m^2$ matrix defined in Hafner (2003) and Śliwa and Schmid (2005). In the special case of a GARCH(1,1) process we obtain

$$\mathrm{vec}(\boldsymbol{\Sigma}_\eta) = \mathbf{G}_m(\mathbf{I} - \mathbf{B})^{-1}\mathrm{vec}\big(\boldsymbol{\alpha}_0\boldsymbol{\alpha}_0' + (\boldsymbol{\alpha}_1 + \boldsymbol{\beta}_1)\boldsymbol{\sigma}\boldsymbol{\alpha}_0' + \boldsymbol{\alpha}_0\boldsymbol{\sigma}'(\boldsymbol{\alpha}_1 + \boldsymbol{\beta}_1)'\big)$$

with

$$\mathbf{B} = (\boldsymbol{\alpha}_1 \otimes \boldsymbol{\alpha}_1)\mathbf{G}_m + \boldsymbol{\alpha}_1 \otimes \boldsymbol{\beta}_1 + \boldsymbol{\beta}_1 \otimes \boldsymbol{\alpha}_1 + \boldsymbol{\beta}_1 \otimes \boldsymbol{\beta}_1.$$

Note that for an $m \times m$ matrix $\mathbf{A}$ with elements a_{ij} the operator vec($\mathbf{A}$) is defined as

$$\mathrm{vec}(\mathbf{A}) = (a_{11}, .., a_{m1}, a_{12}, .., a_{m2}, .., a_{1m}, .., a_{mm})'\ .$$

6.4.3 Control charts based on a multivariate EWMA recursion

Let $\mathbf{L}_{\Gamma,t}(\mathbf{X}_t, \mathbf{X}_{t-1}, \ldots)$ be an $m(m+1)/2$-dimensional local volatility measure based on the historical observations up to time point t. $\mathbf{L}_{\Gamma,t}$ is obtained by applying the vech operator to a suitable $m \times m$ variance matrix. Its mean $\mathrm{E}_D(\mathbf{L}_{\Gamma,t})$ should be constant over time and it is denoted by $\boldsymbol{\gamma}^*$. In this section we consider surveillance methods of the quantity $\boldsymbol{\gamma}^*$. All charts are based on exponential smoothing.

The MEWMA recursion is defined by

$$\mathbf{Z}_t = (\mathbf{I} - \boldsymbol{\Lambda})\mathbf{Z}_{t-1} + \boldsymbol{\Lambda}\mathbf{L}_{\Gamma,t}(\mathbf{X}_t, \mathbf{X}_{t-1}, \ldots) \qquad t \geq 1 \tag{6.13}$$

with a constant starting value $\mathbf{Z}_0$. The diagonal matrix $\boldsymbol{\Lambda}$ contains smoothing parameters, which are chosen individually for each element of the variance measure $\mathbf{L}_{\Gamma,t}$. The elements of $\boldsymbol{\Lambda} = \mathrm{diag}(\lambda_{11}, \lambda_{12}, \ldots, \lambda_{n-1,n}\lambda_{nn})$ take values in the interval (0, 1]. As shown in Chapter 5 of this handbook it can be proved that

$$\mathrm{E}_D(\mathbf{Z}_t) = \boldsymbol{\Gamma}^* + (\mathbf{I} - \boldsymbol{\Lambda})^t(\mathbf{Z}_0 - \boldsymbol{\Gamma}^*),$$

$$\mathrm{Cov}_D(\mathbf{Z}_t) = \boldsymbol{\Lambda}\sum_{i=0}^{t-1}\sum_{j=0}^{t-1}(\mathbf{I} - \boldsymbol{\Lambda})^i\mathrm{Cov}(\mathbf{L}_{\Gamma,t-i}, \mathbf{L}_{\Gamma,t-j})(\mathbf{I} - \boldsymbol{\Lambda})^j\boldsymbol{\Lambda}.$$

Our aim is to get a signal if there is a change in any component of $\boldsymbol{\gamma}^*$. Therefore, we measure the distance between $\mathbf{Z}_t$ and the target in-control value $\boldsymbol{\gamma}^*$. A convenient way to do this is to use the Mahalanobis distance. It is concluded that there is a shift in the cross-covariance structure if

$$(\mathbf{Z}_t - \boldsymbol{\gamma}^*)'\big[\mathrm{Cov}_D(\mathbf{Z}_t)\big]^{-1}(\mathbf{Z}_t - \boldsymbol{\gamma}^*) \ > \ c.$$

Of course such a transformation leads to an information loss because after a signal it is not clear which component of the variance measure is responsible for it.

The implementation of this control chart is hindered by two facts. First, it is computationally demanding to recompute the covariance matrix $\text{Cov}_D(\mathbf{Z}_t)$ at each moment of time. Therefore, it is recommended to use the asymptotic covariance matrix.

The second problem is the choice of the critical value c. This problem is treated in the usual way. Let t_A denote the run length of the chart based on the asymptotic variance, i.e.

$$t_A = \inf\{t \in \mathbb{N} : (\mathbf{Z}_t - \boldsymbol{\gamma}^*)'\big[\lim_{t\to\infty} \text{Cov}_D(\mathbf{Z}_t)\big]^{-1}(\mathbf{Z}_t - \boldsymbol{\gamma}^*) \; > \; c\}.$$

The control limit c is determined such that the in-control average run length $\text{E}_D[t_A(c)]$ is equal to some predetermined value ξ.

6.4.3.1 A chart based on the conditional covariances

There are several possibilities to choose the local covariance measure $\mathbf{L}_{\Gamma,t}$. First, we take it equal to the observed multivariate conditional volatility.

In the previous section we have seen that the conditional variance of $\mathbf{Y}_t$ given all previous observations is equal to $\mathbf{H}_t$. In practice, however, only a finite number of observations is available. For that reason we determine the conditional variance of η_t given $\eta_{t-1}, .., \eta_1$. This quantity is denoted by $\hat{\eta}_t$. Applying the vech operator to this matrix expression we get the vector $\hat{\mathbf{h}}_t$. This quantity can be recursively calculated by applying a multivariate Durbin–Levinson algorithm (see Brockwell and Davis 1991). In the case of an GARCH(1,1) process we obtain $\hat{\mathbf{h}}_t$ as follows

$$\hat{\mathbf{h}}_t = \boldsymbol{\sigma} + (\boldsymbol{\alpha}_1 + \boldsymbol{\beta}_1)(\boldsymbol{\eta}_{t-1} - \boldsymbol{\sigma}) - \boldsymbol{\zeta}_t(\boldsymbol{\eta}_{t-1} - \hat{\mathbf{h}}_{t-1}) \qquad t \geq 2$$

with the starting value $\hat{\mathbf{h}}_1 = \boldsymbol{\sigma}$. The time varying matrix of coefficients $\boldsymbol{\gamma}_t$ is computed recursively using

$$\boldsymbol{\zeta}_t = \boldsymbol{\beta}_1 \boldsymbol{\Sigma}_{\mathbf{v}} \mathbf{V}_{t-1}^{-1} \qquad \text{with} \qquad \mathbf{V}_t = \begin{cases} \boldsymbol{\Sigma}_\eta - \boldsymbol{\sigma}\boldsymbol{\sigma}' & \text{for} \quad t = 1 \\ \boldsymbol{\Sigma}_{\mathbf{v}} + \boldsymbol{\beta}_1 \boldsymbol{\Sigma}_{\mathbf{v}} \boldsymbol{\beta}' - \boldsymbol{\zeta}_t \mathbf{V}_{t-1} \boldsymbol{\zeta}_t' & \text{for} \quad t \geq 2 \end{cases}$$

where the matrices $\boldsymbol{\Sigma}_{\mathbf{v}}$ and $\boldsymbol{\Sigma}_\eta$ are defined above. It holds that $\lim_{t\to\infty} \boldsymbol{\zeta}_t = \boldsymbol{\beta}_1$ and $\lim_{t\to\infty} \mathbf{V}_t = \boldsymbol{\Sigma}_{\mathbf{v}}$.

Note that η_t is defined through the observations of the target process which is unobservable. To make our procedure applicable we replace η_t by $\tau_t = \text{vech}(\mathbf{X}_t\mathbf{X}_t')$ in the expression of $\hat{\mathbf{h}}_t$. Thus the local variance measure at time

$t+1$ is the forecast of the conditional volatility for the time point $t+1$ based on the observations available up to time t. Note that this quantity is calculated under the assumption that the observed process is in control. It is given by

$$\hat{\boldsymbol{\xi}}_{t+1} = \boldsymbol{\gamma} + (\alpha_1 + \beta_1)(\boldsymbol{\tau}_t - \boldsymbol{\gamma}) - \zeta_t(\boldsymbol{\tau}_t - \hat{\boldsymbol{\xi}}_t) \qquad t \geq 2.$$

The starting value is set to the target value $\boldsymbol{\Gamma}$. The computation of the variance of $\hat{\boldsymbol{\xi}}_{t+1}$ is computationally demanding and, as above, the asymptotic variance is used. In the in-control state it holds that

$$\lim_{t\to\infty} \mathrm{Var}(\hat{\boldsymbol{\xi}}_t) = \lim_{t\to\infty} \mathrm{Var}(\hat{\mathbf{h}}_t) = \boldsymbol{\Sigma}_h - \boldsymbol{\gamma}\boldsymbol{\gamma}'.$$

Here we choose $L_{\boldsymbol{\Gamma},t} = \hat{\boldsymbol{\xi}}_{t+1}$. Then the control statistic is given by

$$\mathbf{Z}_t = (\mathbf{I} - \boldsymbol{\Lambda})\mathbf{Z}_{t-1} + \boldsymbol{\Lambda}\hat{\boldsymbol{\xi}}_{t+1} \qquad t \geq 1 \tag{6.14}$$

with the starting value equal to $\boldsymbol{\gamma}$. For the decision we further need the covariance matrix of $\mathbf{Z}_t$. In general it is difficult to obtain it even for simple processes. However, the asymptotic covariance matrix in case of an GARCH(1,1) process can be found in Theorem 3 of Śliwa and Schmid (2005).

6.4.3.2 A chart based on the present observations

While for the calculation of the local covariance measure based on the conditional covariances previous observations are taken into account an easier measure is obtained if only the present values are used. This leads to

$$L_{\boldsymbol{\Gamma},t} = \mathrm{vech}(\mathbf{X}_t\mathbf{X}_t') = \boldsymbol{\tau}_t.$$

It holds that $\boldsymbol{\gamma}^* = E_D(\boldsymbol{\tau}_t) = \boldsymbol{\gamma}$. An explicit expression of the in-control covariance matrix of the corresponding MEWMA recursion can be found in Śliwa and Schmid (2005).

6.4.3.3 Residual charts

Let $\hat{\boldsymbol{\xi}}_t$ denote the conditional volatility calculated with the observations of the observed process. Then $\hat{\boldsymbol{\xi}}_t$ is obtained from $\mathbf{h}_t$ by replacing $\boldsymbol{\eta}_{t-1}, \ldots, \boldsymbol{\eta}_1$ with $\boldsymbol{\tau}_{t-1}, \ldots, \boldsymbol{\tau}_1$. Moreover, let $\hat{\boldsymbol{\Xi}}_t$ represent the re-vectorized form of $\hat{\boldsymbol{\xi}}_t$, so that vech $(\hat{\boldsymbol{\Xi}}_t) = \hat{\boldsymbol{\xi}}_t$. Then the residuals are given by

$$\hat{\boldsymbol{\varepsilon}}_t = \hat{\boldsymbol{\Xi}}_t^{-1/2}\mathbf{X}_t, \quad t \geq 1\ . \tag{6.15}$$

An iterative procedure for computing $\hat{\boldsymbol{\Xi}}_t^{1/2}$ can be found in literature, e.g. in Harville (1997, p. 235).

Note that for a nonlinear process the residuals are defined in another way than for a linear process. The empirical residuals $\hat{\boldsymbol{\varepsilon}}_t$ fulfil the properties

$$\mathrm{E}_D(\boldsymbol{\varepsilon}_t) = \mathbf{0}\,, \qquad \lim_{t\to\infty} \mathrm{Cov}_D(\boldsymbol{\varepsilon}_t) = \mathbf{I}$$

and $\mathrm{E}_D(\boldsymbol{\varepsilon}_t\boldsymbol{\varepsilon}_s') = \mathbf{0}$ for any $t \neq s$.

Here we consider a control chart purely based on the present residuals at time t. This means that $L_{\Gamma,t} = \mathrm{vech}(\boldsymbol{\varepsilon}_t\boldsymbol{\varepsilon}_t')$ for $t \geq 1$. Here we have $\boldsymbol{\Gamma}^* = \mathrm{E}_D(\mathrm{vech}(\boldsymbol{\varepsilon}_t\boldsymbol{\varepsilon}_t') = \mathrm{vech}(\mathbf{I}_m)$. The in-control covariance matrix of $\hat{\boldsymbol{\varepsilon}}_t$ and of the corresponding MEWMA recursion is calculated in Śliwa and Schmid (2005). Thus the design of the MEWMA chart based on the residuals is given and the chart can be applied.

6.4.4 Control charts based on a univariate EWMA approach

In the previous section we considered a local measure of the covariance matrix. We applied a MEWMA recursion to this quantity and projected it to a one-dimensional statistic by calculating the Mahalanobis distance. Here we change the sequence of the procedure. First, we calculate the Mahalanobis distance between the local measure of the covariance matrix and its target value and after that we apply a univariate EWMA recursion to this expression.

Let

$$\begin{aligned} L_{\Gamma,t}(\mathbf{X}_t, \mathbf{X}_{t-1}, \ldots) &= (\mathbf{L}_{\Gamma,t}(\mathbf{X}_t, \mathbf{X}_{t-1}, \ldots) - \boldsymbol{\Gamma}^*)' \\ &\quad \times \left[\mathrm{Cov}_D(\mathbf{L}_{\Gamma,t}(\mathbf{X}_t, \mathbf{X}_{t-1}, \ldots))\right]^{-1} \\ &\quad (\mathbf{L}_{\Gamma,t}(\mathbf{X}_t, \mathbf{X}_{t-1}, \ldots) - \boldsymbol{\Gamma}^*). \end{aligned}$$

The univariate control statistic is then computed as

$$Z_t = (1-\lambda)Z_{t-1} + \lambda L_{\Gamma,t}(\mathbf{X}_t, \mathbf{X}_{t-1}, \ldots)$$

with the starting value $\lim_{t\to\infty} \mathrm{E}_D[L_{\Gamma,t}(\mathbf{X}_t, \mathbf{X}_{t-1}, \ldots)] = n(n+1)/2$.

As described in the previous section there are many possibilities to choose $L_{\Gamma,t}$. We refer to Śliwa and Schmid (2005) where this approach is treated in detail.

6.5 Summary

Financial time series often exhibit shifts in the volatility. In this chapter we provide a review of the monitoring techniques for the variance of nonlinear processes, with special attention paid to the GARCH processes. As monitoring tools we consider the EWMA and CUSUM control charts. These types of charts are developed to fully exploit memory and dependencies in the observed time series. Further we consider three local measure of the variance – based on the squared observations, the forecasts of the conditional variance and on the residuals. Within a simulation study we compare all charts using the out-of-control ARL and the maximum average delay as performance measures. In terms of the ARL both the EWMA and the CUSUM type charts based on the conditional volatility shows the best performance. However, the residual charts should be preferred if we use the maximum average delay as a performance measure. We also provide an illustration of the developed techniques on a real-world data. Finally, we discuss the generalization of the considered monitoring techniques to the multivariate GARCH processes.

Acknowledgements

The authors are grateful to Professor Marianne Frisén and the participants of the workshop on Financial Surveillance at the University of Göteborg, Sweden for helpful discussions and suggestions.

References

Bodnar, O. and Schmid, W. (2007). Surveillance of the mean behaviour of multivariate time series. *Statistica Neerlandica*, (to appear).

Bollerslev, T. (1986). Generalized autoregressive conditional heteroscedasticity. *Journal of Econometrics*, **31**, 307–327.

Bollerster, T. (1990). Modeling the coherence in short-run nominal exchange rates: A multivariate generalized ARCH approach. *Review of Economics and Statistics*, **72**, 498–505.

Brockwell, P. and Davis, R. (1991). *Time Series: Theory and Methods*. Springer-Verlag, New York.

Cox, D., Hinkley, D. and Barndorff-Nielsen. O. (1996). *Time Series Models in Econometrics, Finance and other Fields*. Chapman and Hall, London.

Engle, R. (1982). Autoregressive conditional heteroscedasticity with estimates of the variance of UK inflation. *Econometrica*, **50**, 987–1008.

Engle, R. and Kroner, K. (1995). Multivariate simultaneous generalized ARCH. *Econometric Theory*, **11**, 122–150.

Engle, R., Ng, V. and Rothschild, M. (1990). Asset pricing with a factor arch covariance structure: Empirical estimates for treasury bills. *Journal of Econometrics*, **45**(2), 213–237.

Fan, J. and Yao, Q. (2003). *Nonlinear Time-Series: Nonparametric and Parametric Methods*. Springer-Verlag, Berless.

Frisén, M. (2003). Statistical surveillance. Optimality and methods. *International Statistical Review*, **71**, 403–434.

Frisén, M. and Sonesson, C. (2006). Optimal surveillance based on exponentially weighted moving averages. *Sequential Analysis*, **25**, 379–403.

Gourieroux, C. (1997). *ARCH Models and Financial Applications*. Springer-Verlag, New York.

Hafner, C. (2003). Fourth moment structure of multivariate GARCH models. *Journal of Financial Econometrics*, **1**(1), 26–54.

Harvey, A., Ruiz, E. and Shephard, N. (1994). Multivariate stochastic variance models. *Review of Economic Studies*, **61**(2), 247–264.

Harville, D. (1997). *Matrix Algebra from a Statistician's Perspective*. Springer-Verlag, New York.

Mandelbrot, B. (1963a). New methods in statistical economics. *The Journal of Political Economy*, **71**, 421–440.

Mandelbrot, B. (1963b). The variation of certain speculative prices. *Journal of Business*, **36**, 394–419.

Mills, T. (2004). *The Econometric Modelling of Financial Time Series*. Cambridge University Press, Cambridge.

Pourahmadi, M. (2001). *Foundations of Time Series Analysis and Prediction Theory*. John Wiley & sons, Ltd, New York.

Ruppert, D. (2004). *Statistics in Finance: An Introduction*. Springer-Verlag, New York

Schipper, S. and Schmid, W. (2001). Sequential methods for detecting changes in the variance of economic time series. *Sequential Analysis*, **20**(4), 235–262.

Severin, T. and Schmid, W. (1998). Statistical process control and its application in finance. In *Contributions to Economics: Risk Measurement, Econometrics and Neural Networks*. Physica-Verlag, Hevdelbag. pp. 83–104.

Severin, T. and Schmid, w. (1999). Monitoring changes in GARCH models. *Journal of the German Statistical Society*, **83**, 281–307.

Śliwa, P. and Schmid, W. (2005). Surveillance of the covariance matrix of multivariate nonlinear time series. *Statistics*, **39**, 221–246.

Tsay, S. (2005), *Analysis of Financial Time Series*. Wiley-Interscience, New York

7

Sequential monitoring of optimal portfolio weights

Vasyl Golosnoy[a], Wolfgang Schmid[b] and Iryna Okhrin[b]

[a]*Institute of Statistics and Econometrics University of Kiel, Germany*
[b]*Department of Statistics, European University Viadrina, Frankfurt (Oder), Germany*

7.1 Introduction

The rapid development of financial markets in the second part of the twentieth century has initiated intensive research in the field of financial economics. Modern portfolio theory is based on the market efficiency hypothesis and no free arbitrage considerations. Market efficiency presumes quick and correct accommodation of new information into prices, while the no free arbitrage requirement implies the existence of equilibrium asset prices (Ingersoll 1987, pp. 51ff.). Portfolio analysis addresses all stages of the investment process, namely identifying an investor's objectives, constructing portfolios according to the investor's objectives, checking and revising obtained portfolios and evaluating the performance of the whole portfolio strategy.

On the one hand an investor wants to have a large expected return but on the other hand he should avoid risks associated with future payoffs. Since Markowitz (1952) the investor's objective function is often defined as a trade-off between the expected return and the risk of the portfolio. Portfolio variance

Financial Surveillance Edited by Marianne Frisén

serves as the most popular risk measure. The minimal attainable portfolio variance for a given level of the expected return of the portfolio defines the efficient frontier. A rational mean–variance investor should choose a portfolio lying on the efficient frontier portfolio by maximizing her objective function over the investment horizon. During the holding period, however, newly incoming information may change the optimal portfolio composition. Then the portfolio weights must be revised and adjusted to the new market situation. In the financial literature, there is overwhelming evidence about the presence of structural breaks in the parameters of interest (Banerjee and Urga 2005). In this chapter statistical methods are presented for determining the time points for a portfolio adjustment.

The important global minimum variance portfolio (GMVP) provides the smallest attainable portfolio variance (Section 7.2). The data from the empirical study in Section 7.6 is used for illustrating structural breaks in the GMVP composition. Figure 7.1 visualizes the time evolution of the sample estimators of the four GMVP weights. Here the true weights are estimated by a rolling estimation window using $n = 252$ previous observations. The straight lines denote the optimal portfolio composition estimated from the whole data set.

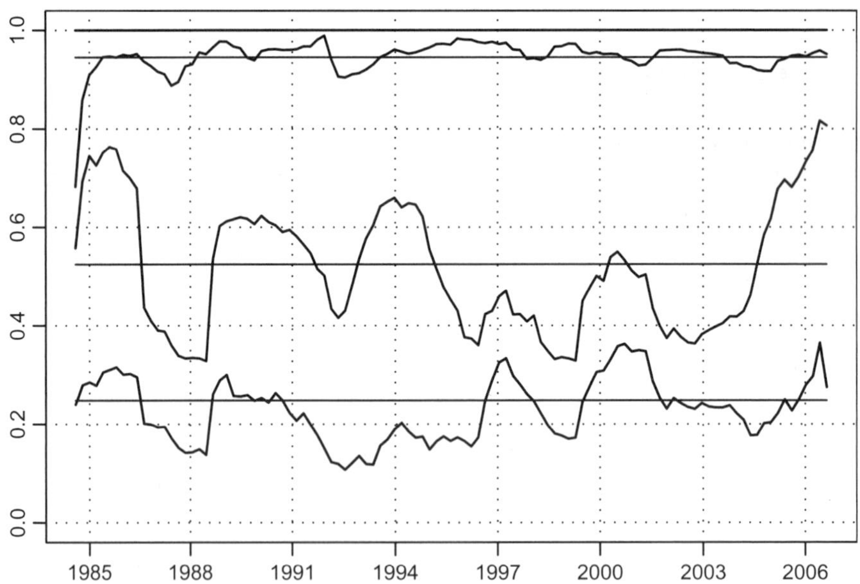

Figure 7.1 GMVP weights estimated by a rolling window consisting of $n =$ 252 observations.

Figure 7.1 shows that the optimal portfolio proportions are varying over time. The sudden large changes in the optimal portfolio proportions observed in the plot can be classified as structural breaks. The investor is interested in detecting these breaks as soon as possible in order to escape from a suboptimal portfolio holding. As in Chapters 5 and 6, the time points of the changes are assumed in this chapter to be unknown and deterministic. This is a different model compared to Chapters 3 and 4 where random change points are discussed.

The breaks should be detected in a sequential manner, because the investor has to decide at each time point about the optimal portfolio composition. Such problems are treated within statistical process control (SPC). The main tool of SPC are control charts (Montgomery 2005). In the present case the parameters of interest are the weights of the optimal portfolio because the investor requires only this information for her wealth allocation decisions. This chapter deals with the surveillance of the weights of the GMVP portfolio.

Golosnoy and Schmid (2007) derive several control charts for monitoring the GMVP weights. Their schemes are described and discussed later in this chapter. Differently to Golosnoy and Schmid (2007), here another model is used for the simulation study as well as another measure of the chart performance. Recommendations about the choice of the design are provided for the considered monitoring schemes. Moreover, the best control chart parameters are determined. Additionally, an example illustrates how the control charts can be applied to monitor a real portfolio.

The rest of the chapter is organized as follows. Section 7.2 gives a brief review of portfolio theory. Moreover, the estimation issues for the optimal portfolio weights are considered there. Section 7.3 shows the necessity to monitor the portfolio weights and provides a short introduction into SPC. Section 7.4 introduces the control charts for the GMVP weights. The performance of the control charts is evaluated within a Monte Carlo simulation study in Section 7.5. The empirical example in Section 7.6 illustrates the practical relevance of the described monitoring approach.

7.2 Optimal portfolio composition

The investor makes her decisions concerning wealth allocation in terms of the optimal portfolio proportions. In order to choose an appropriate algorithm for resolving this investment task, the form of the investor's utility function and the distribution of the asset returns are of crucial importance. The current portfolio theory (Brandt 2007) differentiates between discrete and continuous time portfolio problem settings, depending on the underlying assumption about the distribution of the asset returns. This study deals with a discrete-time model, which is attractive due to its simple implementation and practical relevance.

The theoretical solutions of a portfolio task are often based on restrictive model assumptions and should be used with caution. A practical investor is interested in relating the rather sophisticated methods to the data at hand. The major challenges of portfolio selection are time-varying distributions of the asset returns, uncertainty about the model and the distribution parameters as well as different kinds of frictions, present in the markets. Dealing with these problems complicates the portfolio solutions and optimal trading rules significantly. A recent state of the relevant literature for the portfolio selection is reviewed by Brandt (2007).

This study relates to a portfolio choice problem with nonconstant asset return distribution parameters. The impact of parameter changes on the optimal portfolio composition is investigated by means of on-line monitoring instruments. A myopic (one-period) investor with a mean–variance utility function is considered. There are k risky assets in the economy and no risk-free equivalent available. Moreover, the possible market frictions are neglected. Transaction costs are assumed to be zero. Such an approach is of importance both for theoretical and practical purposes because it is parsimonious and provides a working approximation for a broad class of the investor's objective functions.

7.2.1 Mean–variance portfolio selection

The mean–variance framework of Markowitz (1952) serves as a basic benchmark for numerous recent refinements of portfolio procedures (Brandt 2007). The Markowitz investor looks for the optimal trade-off between the expected portfolio return $E(X_p)$ and a portfolio risk, usually characterized by the portfolio variance $V(X_p)$. Then the vector of the optimal portfolio weights $\mathbf{w}$ is determined by maximizing the utility function under restriction $\boldsymbol{w}'\mathbf{1} = 1$

$$\mathbf{w} = \arg\max_{\boldsymbol{w}} \left[EU = E(X_p) - \frac{\gamma}{2} V(X_p) \right], \qquad X_p = \boldsymbol{w}'\mathbf{X}, \tag{7.1}$$

where $\mathbf{X}$ is the k-dimensional vector of the asset returns. The risk aversion coefficient $\gamma > 0$ measures the attitude towards uncertainty of future expected payoffs $E(X_p)$.

Let $\boldsymbol{\mu} = E(\mathbf{X})$ and $\boldsymbol{\Sigma} = \mathrm{Cov}(\mathbf{X})$. It is assumed that the matrix $\boldsymbol{\Sigma}$ is positive definite. Consequently, the portfolio expected return and variance can be written as $E(X_p) = \boldsymbol{w}'\boldsymbol{\mu}$ and $V(X_p) = \boldsymbol{w}'\boldsymbol{\Sigma}\boldsymbol{w}$, respectively.

The solution of problem (7.1) is given by the optimal proportions

$$\mathbf{w} = \frac{\mathbf{\Sigma}^{-1}\mathbf{1}}{\mathbf{1}'\mathbf{\Sigma}^{-1}\mathbf{1}} + \frac{1}{\gamma}\mathbf{Q}\boldsymbol{\mu}, \qquad \text{with}$$

$$\mathbf{Q} = \mathbf{\Sigma}^{-1} - \frac{\mathbf{\Sigma}^{-1}\mathbf{1}\mathbf{1}'\mathbf{\Sigma}^{-1}}{\mathbf{1}'\mathbf{\Sigma}^{-1}\mathbf{1}}. \tag{7.2}$$

The true parameters $\boldsymbol{\mu}$ and $\mathbf{\Sigma}$ are unknown in practice. In order to calculate the optimal portfolio weights it is necessary to estimate $\boldsymbol{\mu}$ and $\mathbf{\Sigma}$ by a random sample. In the following it is always presumed that a random sample $\mathbf{X}_1, \ldots \mathbf{X}_t$ of the values is available. The returns $\mathbf{X}_1, \ldots \mathbf{X}_t$ are assumed to be independent and multivariate normally distributed. At time t the parameters are estimated using the last n observations $\mathbf{X}_{t-n+1}, \ldots, \mathbf{X}_{t-1}, \mathbf{X}_t$. Here the sample mean and the unbiased sample covariance matrix are used to estimate the parameters. This leads to

$$\hat{\boldsymbol{\mu}}_{t,n} = n^{-1} \sum_{i=t-n+1}^{t} \mathbf{X}_i,$$

$$\hat{\mathbf{\Sigma}}_{t,n} = (n-1)^{-1} \sum_{i=t-n+1}^{t} (\mathbf{X}_i - \hat{\boldsymbol{\mu}}_{t,n})(\mathbf{X}_i - \hat{\boldsymbol{\mu}}_{t,n})'. \tag{7.3}$$

The estimator $\hat{\mathbf{w}}_{t,n}$ of the true weights $\mathbf{w}$ is obtained by inserting these estimators in (7.2). Okhrin and Schmid (2006) calculated the moments of the estimated weights $\hat{\mathbf{w}}_n$. They are given by

$$E(\hat{\mathbf{w}}_n) = \frac{\mathbf{\Sigma}^{-1}\mathbf{1}}{\mathbf{1}'\mathbf{\Sigma}^{-1}\mathbf{1}} + \frac{n-1}{n-k-1}\gamma^{-1}\mathbf{Q}\boldsymbol{\mu},$$

$$\mathrm{Cov}(\hat{\mathbf{w}}_n) = \frac{1}{n-k-1}\frac{\mathbf{Q}}{\mathbf{1}'\mathbf{\Sigma}^{-1}\mathbf{1}} + \frac{c_1}{\gamma^2}\mathbf{Q}\boldsymbol{\mu}\boldsymbol{\mu}'\mathbf{Q} + \frac{c_2}{\gamma^2}\boldsymbol{\mu}'\mathbf{Q}\boldsymbol{\mu}\mathbf{Q} + \frac{c_3}{\gamma^2 n}\mathbf{Q},$$

with

$$z_1 = \frac{(n-1)^2(n-k-1)}{(n-k)(n-k-1)^2(n-k-3)},$$

$$z_2 = \frac{(n-1)^2}{(n-k)(n-k-1)(n-k-3)},$$

$$z_3 = c_1 + c_2(k-1) + \frac{(n-1)^2}{(n-k-1)^2}.$$

The covariance matrix of the estimated optimal proportions $\mathrm{Cov}(\hat{\mathbf{w}}_n)$ represents the degree of estimation risk in the portfolio weights. In the case of

$n \to \infty$ the estimation risk reduces to zero, however, due to possible structural breaks in the data (Foster and Nelson 1996) it is reasonable to take a finite number of n previous values for estimation purposes.

The impact of the estimation risk on the portfolio composition has been recognized since Klein and Bawa (1976). Neglecting the estimation risk may have a damaging influence on the portfolio selection, thus a portfolio procedure should account for this type of risk as well. There are a number of methods for mitigating the impact of the estimation risk, such as constraining portfolio weights (Frost and Savarino 1988; Garlappi *et al.* 2007), factor models (Brandt *et al.* 2005), Bayesian estimation technique (Jorion 1986), and shrinkage estimators (Golosnoy and Okhrin 2007; Wang 2005). However, an overall best method does not seem to exist. Moreover, even these refinements of the initial Markowitz idea often provide a portfolio performance which is much worse than that of simple investment strategies, such as equal weight portfolios (DeMiguel *et al.* 2007). This happens primarily due to the difficulties in prediction of the expected asset returns (Michaud 1998). That is, the estimation risk cannot be reduced due to possible structural breaks in the $(\boldsymbol{\mu}, \boldsymbol{\Sigma})$ parameters, and, consequently, the estimated optimal portfolio composition $\hat{\mathbf{w}}_{t,n}$ may be very far from the true optimum.

7.2.2 The global minimum variance portfolio

The global minimum variance portfolio (GMVP) problem results from setting the risk aversion coefficient γ in (7.2) to infinity. Since Merton (1980) it has been known that the portfolio returns are hardly predictable. Best and Grauer (1991) show that the estimation errors in the mean returns may lead to strongly inefficient portfolio compositions. On the contrary, the well-documented presence of volatility clusters suggests a high degree of risk predictability, at least on short-term horizons. The predictability of the second moment suggests the use of the minimum-variance portfolio for investment purposes. The GMVP proportions are given by

$$\mathbf{w} = \frac{\boldsymbol{\Sigma}^{-1}\mathbf{1}}{\mathbf{1}'\boldsymbol{\Sigma}^{-1}\mathbf{1}}. \tag{7.4}$$

The GMVP is shown on the Markowitz mean–standard-deviation μ–σ space in Figure 7.2. It serves as a starting point of the efficient frontier (Ingersoll 1987, pp. 82ff.). Accordingly, changes in the GMVP influence the location of the whole frontier, because they characterize shifts in the frontier origin.

The importance of the GMVP for practitioners is illustrated by Michaud (1998) and Busse (1999). This portfolio avoids the estimation of the hardly predictable expected returns and exploits the predictability in the covariance

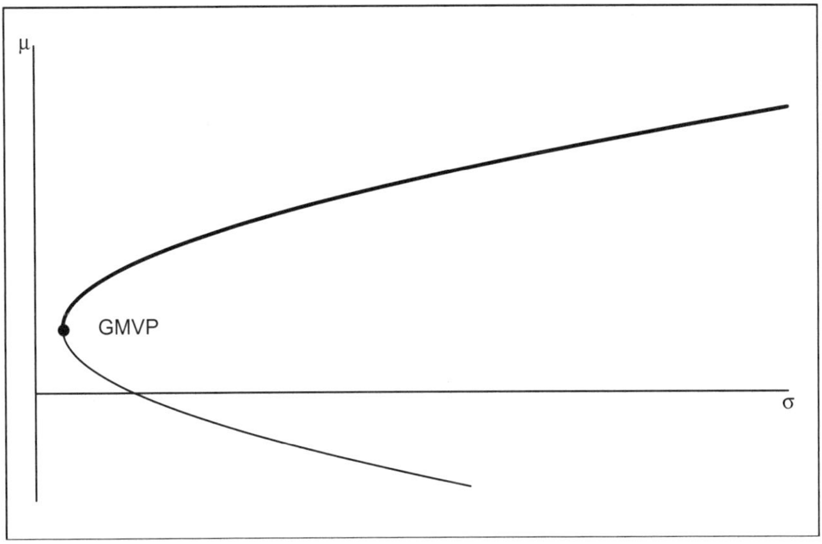

Figure 7.2 $\mu - \sigma$ space, efficiency frontier and the global minimum variance portfolio.

matrix. Numerous empirical studies, see DeMiguel *et al.* (2007) for a review, document that the GMVP often outperforms all the other portfolios from the efficient set. The usual absence of extreme short selling positions is another appealing property of the GMVP. Moreover, the GMVP is frequently used as an important benchmark portfolio for the more sophisticated portfolio rules. All these reasons stress the role of the GMVP and make its investigation important both in theoretical and applied portfolio research.

7.2.3 Distribution of the GMVP weights

The vector of the GMVP weights depends only on the covariance matrix $\mathbf{\Sigma}$, which is unknown and should be estimated. The sample covariance matrix estimate with a finite n is approved in empirical research as a way to incorporate volatility clusters in a nonparametric framework (Haerdle *et al.* 2003). Consequently, the vector of estimated GMVP weights is a function of $\hat{\mathbf{\Sigma}}_{t,n}$. Using the sample estimator (7.3) of the covariance matrix, the estimated GMVP weights are given by

$$\hat{\mathbf{w}}_{t,n} = \frac{\hat{\mathbf{\Sigma}}_{t,n}^{-1}\mathbf{1}}{\mathbf{1}'\hat{\mathbf{\Sigma}}_{t,n}^{-1}\mathbf{1}}. \tag{7.5}$$

Next we provide some results concerning the moments of the estimated GMVP weights $\hat{\mathbf{w}}_{t,n}$. Assuming that the returns $\{\mathbf{X}_t\}$ follow a stationary Gaussian process with mean $\boldsymbol{\mu}$ and covariance matrix $\boldsymbol{\Sigma}$, Okhrin and Schmid (2006) prove that the vector of the estimated optimal weights $\hat{\mathbf{w}}_{t,n}$ is asymptotically normal. Under the additional assumption that the asset returns $\{\mathbf{X}_t\}$ are independent and identically normally distributed they derive the exact distribution of $\hat{\mathbf{w}}_{t,n}$. Because the portfolio weights sum up to the unity, i.e. $\hat{\mathbf{w}}'_{t,n}\mathbf{1} = 1$, the k-dimensional weight vector has no density. Okhrin and Schmid (2006) show that the $(k-1)$-dimensional vector $\hat{\mathbf{w}}^*_{t,n}$ consisting of the first $k-1$ elements of $\hat{\mathbf{w}}_{t,n}$ follows a $k-1$-variate elliptical t-distribution. The density function is given by

$$f_{\hat{\mathbf{w}}^*_{t,n}}(\mathbf{x}) = \frac{(\mathbf{1}'\boldsymbol{\Sigma}^{-1}\mathbf{1})^{(k-1)/2}}{\pi^{(k-1)/2}} \frac{\Gamma\left(\frac{n}{2}\right)}{\Gamma\left(\frac{n-k+1}{2}\right)}$$
$$\times([\mathbf{1}'\boldsymbol{\Sigma}^{-1}\mathbf{1}\ (\mathbf{x} - E(\hat{\mathbf{w}}^*_{t,n}))]'\mathrm{Cov}(\hat{\mathbf{w}}^*_{t,n})^{-1}(\mathbf{x} - E(\hat{\mathbf{w}}^*_{t,n})) + 1)^{-n/2},$$

where $\Gamma(\cdot)$ denotes the gamma function. Consequently, the univariate marginal distribution of a single weight $\hat{w}_{i,n}$ is a scaled t-distribution with $n-k+1$ degrees of freedom:

$$\frac{\sqrt{\mathbf{1}'\boldsymbol{\Sigma}^{-1}\mathbf{1}}\sqrt{n-k+1}}{\sqrt{\mathbf{e}_i'\mathbf{Q}\mathbf{e}_j}}[\hat{w}_{i,t,n} - E(\hat{w}_{i,t,n})] \sim t_{n-k+1}.$$

The vector $\mathbf{e}_i$ has 1 on the ith position and 0 on all others. The expectation and the covariance matrix of the estimated portfolio weights $\hat{\mathbf{w}}_{t,n}$ are given by

$$E(\hat{\mathbf{w}}_{t,n}) = \mathbf{w}, \quad \mathrm{Cov}(\hat{\mathbf{w}}_{t,n}) = \boldsymbol{\Omega} = \frac{1}{n-k-1}\frac{\mathbf{Q}}{\mathbf{1}'\boldsymbol{\Sigma}^{-1}\mathbf{1}}. \tag{7.6}$$

Golosnoy and Schmid (2007) analyse the time series properties of the GMVP weights. They provide a statement about the covariance between $\hat{\mathbf{w}}_{t,n}$ and $\hat{\mathbf{w}}_{t-s,n}$ for the lag $s > 0$. Note that for Gaussian returns both quantities are independent for $|s| \geq n$ due to their construction. The calculation of the exact autocovariance seems to be very difficult. The asymptotic autocovariance with n tending to infinity for a fixed $s \geq 0$ is given by

$$\lim_{n\to\infty} n\,\mathrm{Cov}(\hat{\mathbf{w}}_{t,n}, \hat{\mathbf{w}}_{t-s,n}) = \frac{\mathbf{Q}}{\mathbf{1}'\boldsymbol{\Sigma}^{-1}\mathbf{1}}. \tag{7.7}$$

This result shows that for s fixed the limit of the covariance between $\hat{\mathbf{w}}_{t,n}$ and $\hat{\mathbf{w}}_{t-s,n}$ does not depend on s and it is equal to $\lim_{n\to\infty} n\mathrm{Cov}(\hat{\mathbf{w}}_{t,n})$. This means that asymptotically, i.e. as n tends to infinity, the process $\{\hat{\mathbf{w}}_{t,n}\}$ behaves like a unit root. This is not surprising because in that case the number of common values of $\hat{\mathbf{w}}_{t,n}$ and $\hat{\mathbf{w}}_{t-s,n}$ is $n - s$.

However, the result (7.7) is not very useful for practical purposes. For that reason, Golosnoy and Schmid (2007) propose to approximate the autocovariance for finite values of n by

$$\mathrm{Cov}(\hat{\mathbf{w}}_{t,n}, \hat{\mathbf{w}}_{t-s,n}) \approx \frac{n-s-1}{(n-1)^2} \frac{\mathbf{Q}}{\mathbf{1}'\boldsymbol{\Sigma}^{-1}\mathbf{1}}.$$

The exact expression for $s = 0$, i.e. $\mathrm{Cov}(\hat{\mathbf{w}}_{t,n})$, is given in (7.6). In order to have an equality for $s = 0$, it seems more useful to approximate the autocovariance by

$$\mathrm{Cov}(\hat{\mathbf{w}}_{t,n}, \hat{\mathbf{w}}_{t-s,n}) \approx \frac{n-s-1}{n-1} \frac{1}{n-k-1} \frac{\mathbf{Q}}{\mathbf{1}'\boldsymbol{\Sigma}^{-1}\mathbf{1}}. \tag{7.8}$$

The analysis of the goodness of the approximation (7.8) is presented in Figure 7.3. The study is conducted for a portfolio problem with $k = 4$ assets. The corresponding true covariance matrix $\boldsymbol{\Sigma}$ is chosen as in the simulation study in Section 7.5.1. Each component of $\mathrm{Cov}(\hat{\mathbf{w}}_{t,n}, \hat{\mathbf{w}}_{t-s,n})$ is divided by the corresponding component of $\frac{\mathbf{Q}}{\mathbf{1}'\boldsymbol{\Sigma}^{-1}\mathbf{1}}/(n-k-1)$. The value of $\mathrm{Cov}(\hat{\mathbf{w}}_{t,n}, \hat{\mathbf{w}}_{t-s,n})$ is estimated within a simulation study based on 10^8 repetitions, the average over all $k \times k$ ratios are presented on Figure 7.3.

Figure 7.3 illustrates that the approximation seems to be quite good even for large values of s if $n \geq 60$. It can be seen as appropriate to use $n = 40$ in order to have a sufficiently good working approximation for the autocovariance of the estimated portfolio weights. For smaller values of n the bias correction is needed; more on this issue is provided in Golosnoy *et al.* (2007).

The process of the estimated GMVP weights $\{\hat{\mathbf{w}}_{t,n}\}$ is a focus of this study. Next, using the presented above results, we formulate the task of sequential monitoring of changes in the GMVP proportions.

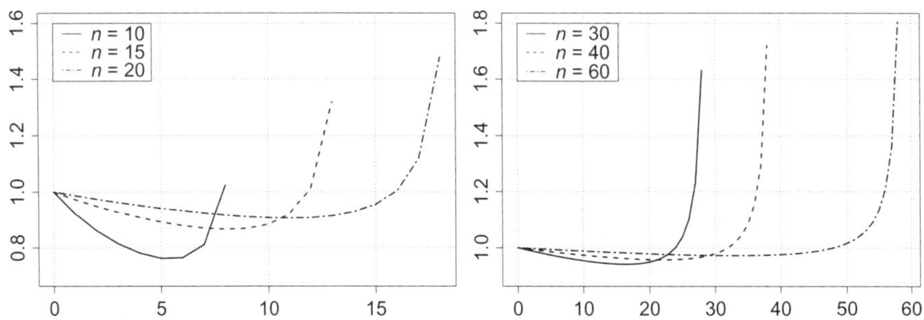

Figure 7.3 Biases in the approximation (7.8) for $\mathrm{Cov}(\hat{\mathbf{w}}_{t,n}, \hat{\mathbf{w}}_{t-s,n})$ with $n = 10, 15, 20, 30, 40, 60$ as a function of s.

7.3 Sequential methods in portfolio selection

7.3.1 Breaks in the covariance matrix

Conditional volatility modelling remains one of the mainstreams in financial econometrics since the seminal contribution of Engle (1982). However, numerous empirical studies report that even the unconditional covariance matrix of risky assets $\mathbf{\Sigma}$ is not constant over time but it is subject to sudden level changes, e.g. Lamoureux and Lastrapes (1990), Solnik *et al.* (1996), Ramchand and Susmel (1998) and Eraker *et al.* (2003). Structural breaks in the unconditional covariance matrix may cause significant alterations in the optimal portfolio weights.

The investor decides every period about wealth allocation, thus she needs to know whether the previously optimal proportions can be still considered as the optimal ones. Thus the surveillance of the optimal portfolio weights is of crucial interest for her. Monitoring the changes in the optimal weights in a sequential manner is a typical task of statistical process control (SPC). In order to introduce the monitoring approach, we consider a model of returns suggested by Hsu *et al.* (1974) and empirically investigated by Kim and Kon (1999). The returns are assumed to be independent and normally distributed with the distribution parameters subject to unpredictable changes. The time points of the changes are unknown. That means, the covariance matrix is considered to be constant up to a change point, where it suddenly changes to another level. Such an approach is a suitable alternative to explain heavy tails and volatility clusters which are frequently observed in daily financial data starting from Mandelbrot (1963) and Fama (1965).

Although changes arise in the covariance matrix, the investor is more interested in monitoring the optimal portfolio weights because she makes her investment decisions in terms of the weights. She should detect changes in the optimal portfolio weights as soon as possible in order to adjust her asset allocation and to minimize utility losses due to suboptimal portfolio holdings. A reduction of the problem dimension is an additional advantage of the direct monitoring of the weights. In particular, instead of monitoring $k(k+1)/2$ covariance matrix elements it is sufficient to control $k-1$ portfolio weights. Thus our aim is to develop a tool for the surveillance of the optimal portfolio weights.

7.3.2 Monitoring portfolio weights

Next the terminology of statistical process control is adopted in order to describe our sequential monitoring problem. In the following we distinguish between the target process $\{\mathbf{Y}_t\}$ and the actual (observed) process $\{\mathbf{X}_t\}$. This approach makes sense because the distribution of the returns may change over time. In order to determine the target process in practice it is necessary to have a suitable number of its realizations after the last change. These observations are used to calculate the parameters of the target process. Here we do not discuss the impact of estimation risk on the estimation of target parameters. Recently several papers in SPC dealt with that problem; for an overview see Jensen *et al.* (2006).

In the following it is assumed that the target process of asset returns $\mathbf{Y}_t$ is a sequence of independent and identically distributed random variables. They follow a k-dimensional normal distribution with mean $\boldsymbol{\mu}$ and covariance matrix $\boldsymbol{\Sigma}$, i.e. $\mathbf{Y}_t \sim \mathcal{N}_k(\boldsymbol{\mu}, \boldsymbol{\Sigma})$. In SPC the relationship between the observed process $\{\mathbf{X}_t\}$ and $\{\mathbf{Y}_t\}$ is usually given by

$$\mathbf{X}_t = \begin{cases} \boldsymbol{\mu} + \boldsymbol{\Delta}\,(\mathbf{Y}_t - \boldsymbol{\mu}) & \text{for} \quad t \geq \tau \\ \mathbf{Y}_t & \text{for} \quad t < \tau, \end{cases} \tag{7.9}$$

where τ denotes a date of the change. The $k \times k$ matrix $\boldsymbol{\Delta} \neq \mathbf{I}$ is unknown. It is assumed to be positive definite. The process is called in-control if $\tau = \infty$ and out-of-control if $\tau < \infty$. Hereafter the index D is used to refer to the in-control (target) process. The notation E_D, Cov_D, etc. stands for the mean, the covariance, etc. calculated with respect to the target process. The index C denotes that the process is out-of-control, consequently E_C, Cov_C, etc. are calculated with respect to the the out-of-control process assumed to be not equal to the target process. Note that (7.9) is the same model as in Chapters 5 and 6.

The considered investor makes her investment decision based on the portfolio weights. For this reason as described above the change point scheme (7.9) is not applied here. On the contrary we introduce a change point model for the optimal portfolio weights. In particular, the expected portfolio proportions are given by

$$E(\hat{\mathbf{w}}_{t,n}) = \begin{cases} \mathbf{w} & \text{for} \quad t < \tau \\ \dfrac{\boldsymbol{\Sigma}_C^{-1}\mathbf{1}}{\mathbf{1}'\boldsymbol{\Sigma}_C^{-1}\mathbf{1}} & \text{for} \quad t \geq \tau + n - 1 \end{cases}$$

with $\boldsymbol{\Sigma}_C = \boldsymbol{\Delta}\,\boldsymbol{\Sigma}\,\boldsymbol{\Delta}$. This shows that, e.g. for $\boldsymbol{\Delta} = \delta\,\mathbf{I}$ with $\delta \neq 1$ it holds that $\boldsymbol{\Sigma}_C \neq \boldsymbol{\Sigma}$ but $E(\hat{\mathbf{w}}_{t,n}) = \mathbf{w}$ for all t. If the risk behaviour of all stocks within the portfolio increases in the same way then the GMVP weights do not change at all and the investor should not adjust her portfolio. Note that $E(\hat{\mathbf{w}}_{t,n}) \neq \mathbf{w}$ implies that $\boldsymbol{\Delta} \neq \mathbf{I}$ but not necessarily vice versa.

The investor has to decide between two hypotheses at each time point $t \geq 1$:

$$H_{0,t} : E(\hat{\mathbf{w}}_{t,n}) = \mathbf{w} \qquad \text{against} \qquad H_{1,t} : E(\hat{\mathbf{w}}_{t,n}) \neq \mathbf{w}\,.$$

It is concluded that the observed process deviates from the target one if $H_{0,t}$ is rejected. We say that the weight process is in control if $H_{0,t}$ holds for all $t \geq 1$, else it is concluded to be out of control. Thus, if the weight process is out of control, the observed process is not equal to the target process. However, the fact that the weight process is in control does not imply the absence of changes in the target process of asset returns (7.9).

SPC suggests control charts as a tool for analysing whether the alternative hypothesis does or does not hold in a sequential manner. Control charts consist of a control statistic T_t and a nonrejection area $\mathcal{O}$. If the control statistic falls outside the nonrejection area at a time point t, i.e. $T_t \notin \mathcal{O}$, then it is concluded that the process is out of control. In this situation the analysis stops and the analyst has to decide how to react to the signal. Otherwise, if the control statistic is inside the nonrejection area, i.e. $T_t \in \mathcal{O}$, the procedure continues and the control statistic is evaluated at time points $t+1$, $t+2$, etc. The choices of the control statistic and of the nonrejection area determine the control chart design.

7.3.3 Choosing the nonrejection area

The considered charts and their control statistics are introduced below in Section 7.4. Here we discuss the way of choosing the nonrejection area $\mathcal{O}$. The task of monitoring the GMVP weights is of a multivariate nature. By choosing a suitable distance measure the problem is transformed into a univariate case. For this reason in our study the nonrejection area is always an interval $\mathcal{O} = [0, c]$. If the control statistic $T_t > c$ the process is concluded to be out of control.

Due to the sequential nature of our analysis it is not possible to determine the control limit c similar to the classical test theory. SPC suggests other tools for this purpose. The most popular criterion for the evaluation of control charts is the average run length (ARL). While the run length N is equal to the number of observations until a signal is given, i.e. $N = \inf\{t \in \mathbb{N} : T_t > c\}$, the average run length is defined as $E(N)$. In the in-control state $\mathrm{ARL}_D = E_D(N)$ should be large but in the out-of-control state ARL_C needs to be small. The out-of-control ARL_C is determined under the assumption that the change happens at the very beginning, i.e. at the time point $t = 1$.

In SPC the control limit c is determined in the following way. A value ξ for the desired in-control ARL_D is prespecified. In engineering the value $\xi = 500$ is frequently taken while in financial applications the number of available observations is not very large and smaller quantities like, e.g. $\xi = 60$ or $\xi = 90$ are usually used (Schipper and Schmid 2001; Severin and Schmid 1999). In our situation n previous observations are required for estimating the optimal portfolio weights. Choosing $\xi = 120$ corresponds to a half of the year without alarm and resembles the conventional frequency of portfolio revision by asset allocation decisions. Then the critical value c is obtained as the solution of the equation

$$E_D\,(N(c)) \;=\; \xi. \tag{7.10}$$

An explicit formula for the distribution of the run length and its moments is unfortunately known only for a few control procedures in the case of independent univariate random variables. If the control statistic follows a Markov process, which is the case only for a limited family of charts, the ARL can be evaluated using numerical methods; see Brook and Evans (1972) and Crowder (1987). Due to the complexity of the process of the GMVP weights the only possibility to estimate the required moments of the run length distribution is to conduct a Monte Carlo simulation study. Golosnoy and Schmid (2007) use a numerical *regula falsi* method for solving Equation (7.10). They estimated $E_D(N(\cdot))$ within a simulation study based on 10^6 repetitions in each iteration. The procedure is stopped if the absolute deviation from the prespecified in-control ARL_D is less than 0.1%.

There also exist further criteria for calibrating a control chart, see Frisén (2003) for a review. The worst case conditional expected delay performance criteria is suggested by Pollak and Siegmund (1975) as a popular alternative to the out-of-control ARL_C. The conditional expected delay for a given change point $\tau \geq 1$ is given by

$$\mathcal{D}_\tau \;=\; E_\tau(N - \tau + 1 \mid N \geq \tau),$$

where N denotes the run length and $E_\tau(\cdot)$ is the conditional expectation with respect to the change point date τ. The in-control situation is described by $\iota \to \infty$, while the worst case conditional expected delay is defined as

$$\mathcal{D}_{PS} \;=\; \sup_\tau \mathcal{D}_\tau \;=\; \sup_\tau \; E_\tau(N - \tau + 1 \mid N \geq \tau). \tag{7.11}$$

Usually it is sufficient to use the values $\tau = 1, ..., 30$ as possible change points for practical purposes.

Moreover, Golosnoy and Schmid (2007) use the median of the run length (MRL) as an alternative to the ARL. Because the distribution of the run length may be extremely right-skewed the MRL (Gan 1993) is a robust quantity with an easier interpretation compared to the ARL criterion. The critical values for the MRL-based chart are obtained in the same way as for the ARL-based chart. More discussion about the properties of these and other (Lorden 1971; Roberts 1966) performance measures can be found in Frisén (2003). Further we use both the ARL and the worst case conditional expected delay measure $\mathcal{D}_{PS}$ of Pollak and Siegmund (1975) for evaluating the goodness of the competing monitoring procedures.

7.3.4 Sequential monitoring: implementation issues

Before introducing the design of the charts, we discuss the issues of implementation of sequential methods in portfolio theory. Various control procedures have been proposed since the pathbreaking contributions of Shewhart (1931) and Hotelling (1947). Especially important schemes are the control charts with memory, such as the cumulative sum chart (CUSUM) of Page (1954) and the exponentially weighted moving average chart (EWMA) of Roberts (1959). Extensions of the CUSUM charts to multivariate processes are developed by Woodall and Ncube (1985), Crosier (1988), Pignatello and Runger (1990), and Ngai and Zhang (2001). EWMA charts for multivariate data are discussed by, among others, Lowry *et al.* (1992) and Sullivan and Jones (2002). The authors mainly concentrate on changes in the mean structure of the process.

In financial applications the underlying process is frequently autocorrelated. Alwan and Roberts (1988) show that the presence of autocorrelation dramatically influences the run length distribution compared to the independent data case. Especially in the case of a stronger autocorrelation it is necessary to determine the control limits with respect to the time series structure, otherwise the charts would be mis-specified. The monitoring schemes for dealing with autocorrelated time series were developed by, among others, Alwan and Roberts (1988), Yashchin (1993), Schmid (1995, 1997a,b) and Lu and Reynolds (1999a,b, 2001).

The task of monitoring portfolio weights is a multivariate one. Because the GMVP weights are strongly autocorrelated as it is shown in (7.8), Golosnoy and Schmid (2007) propose a number of procedures taking into account the time series structure of the process of weights. Following the line of Golosnoy and Schmid (2007), we detect changes in the GMVP weights by using the EWMA type control charts. These charts, described in Section 7.4, exploit the idea of monitoring the mean behaviour of the empirical optimal portfolio weights. Of course, changes in the parameters may also influence the covariance matrix of the estimated optimal portfolio weights in (7.6). For that reason it might be desirable to monitor the covariance matrix as well. Here, however, we exclusively focus on the mean charts, which are the most relevant tool for practical purposes. Simultaneous charts for means and variances of the multivariate process (Reynolds and Cho 2006) would be a subject for further research.

The target parameters $E_D(\hat{\mathbf{w}}_{t,n}) = \mathbf{w}$, $\text{Cov}_D(\hat{\mathbf{w}}_{t,n}) = \boldsymbol{\Omega}$ are demanded to be known in the conventional SPC. Chu *et al.* (1996) argue that in economic and financial applications this is usually not the case. In engineering there is a distinction between Phase I and Phase II (Woodall 2000). The historical data of the Phase I period is used for estimating the process parameters. The Phase

I observations are assumed to be drawn from the in-control process. Phase II, immediately following Phase I, corresponds to the monitoring period and consists of new observations that arrive sequentially. Note that in finance the distinction between Phase I and II as well as the choice and the structure of the in-control process are more complicated. By determining Phase I there exists a trade-off between the estimation risk and the usage of nonrelevant past observations in the estimation procedure (Foster and Nelson 1996). In our case the identification of a phase is explicitly treated in Section 7.6.1.

It should be ensured that the observations of Phase I really satisfy the conditions of the in-control process. In order to keep the estimation error small it is desirable to have a sufficiently large number of returns for estimating purposes. Recently, there have been several papers dealing with the influence of the estimation risk on the performance of control charts; see Albers and Kallenberg (2004a,b) and Champ *et al.* (2005). In our setting the estimation risk in the process $\{\hat{\mathbf{w}}_{t,n}\}$ is related to the number of past periods n, used for the estimation purposes. We discuss the possible implications of the choice of n on the performance of monitoring schemes in Section 7.6.3.

7.4 Control charts for the GMVP weights

Golosnoy and Schmid (2007) suggest a number of control charts for monitoring the GMVP weights. Their charts can be seen as extensions of the procedures of Kramer and Schmid (1997) and Rosolowski and Schmid (2003), adopted for portfolio monitoring purpose. Here we introduce their two most successful control charts, which refer to a class of modified charts. In the literature a control chart is called 'modified' if the underlying process is autocorrelated and the control limits are adjusted by taking into account its autocorrelation structure. In the present case the underlying returns are independent but the sample estimators of the target parameter, the GMVP weights, are strongly autocorrelated, see (7.8). Thus the charts introduced in this section are 'modified'. Another type of control charts, considered by Golosnoy and Schmid (2007), is based on differencing the portfolio weights in order to get rid of autocorrelation. However, the difference charts are not considered here because of the problems with their empirical implementation.

Now we describe the monitoring schemes. The sum over all components of $\hat{\mathbf{w}}_{t,n}$ is equal to 1 and thus its covariance matrix is not regular. For that reason we consider the random vector $\hat{\mathbf{w}}^*_{t,n}$ consisting of the first $k-1$ components of $\hat{\mathbf{w}}_{t,n}$. The vector $\mathbf{w}^*$ is defined by analogy. $\mathbf{\Omega}^*$ denotes the $(k-1) \times (k-1)$ matrix obtained by dropping the kth row and the kth column of the matrix $\mathbf{\Omega}$, see Equation (7.6). Consequently, the rank of the matrix $\mathbf{\Omega}^*$ equals to $k-1$.

The additional technical details can be found in Golosnoy and Schmid (2007) and Golosnoy *et al.* (2007).

7.4.1 Mahalanobis control chart

The distance between the estimated GMVP weights $\hat{\mathbf{w}}^*_{t,n}$ and the target weights $\mathbf{w}^* = E_D(\hat{\mathbf{w}}^*_{t,n})$ is measured by the squared Mahalanobis norm. This leads to

$$T_{t,n} = (\hat{\mathbf{w}}^*_{t,n} - \mathbf{w}^*)' \, \boldsymbol{\Omega}^{*-1} \, (\hat{\mathbf{w}}^*_{t,n} - \mathbf{w}^*), \quad t \geq 1 .$$

Then the quantities $T_{1,n}$, ..., $T_{t,n}$ are exponentially smoothed by EWMA recursion

$$Z_{t,n} = (1-\lambda)Z_{t-1,n} + \lambda T_{t,n} \quad \text{for} \quad t \geq 1.$$

The starting value $Z_{0,n}$ is set equal to its target values $Z_{0,n} = E_D(T_{t,n}) = k-1$. The fast initial response feature (Montgomery 2005, pp. 413ff) is not exploited here. The smoothing factor λ represents the memory of the chart and takes the values within (0, 1]. The smaller λ leads to the larger influence of the past observations, while $\lambda = 1$ corresponds to the no-memory Shewhart control chart.

The control chart gives a signal at time t if $Z_{t,n} > c_1$. The constant $c_1 > 0$ is determined in a such way that the ARL of the control chart is equal to a prespecified quantity in the in-control state, as in (7.10).

For the calculation of $T_{1,n}$ we make use of the previous values $\mathbf{X}_{1-n}, .., \mathbf{X}_1$, consequently the past observations are needed. It has to be ensured, however, that the observations at times $t \leq 0$ stem from the target process, i.e. refer to Phase I. Because at time $t = 1$ only one observation may be contaminated and the other $n-1$ stem from the target process the influence of a change on the estimated weight may be small. This causes inertia in detecting the changes in the optimal portfolio proportions. This undesired effect can be relaxed by choosing an appropriate number of estimation periods n. A simulation study in Section 7.5 provides evidence concerning consequences of choosing different n values for our task.

7.4.2 MEWMA control chart

This chart is more flexible compared to the upper one, because each component of $\hat{\mathbf{w}}^*_{t,n}$ is exponentially smoothed by its own smoothing factor. Using matrix notation the $(k-1)$-dimensional EWMA recursion can be written as

$$\mathbf{Z}_{t,n} = (\mathbf{I} - \mathbf{R})\mathbf{Z}_{t-1,n} + \mathbf{R}\hat{\mathbf{w}}^*_{t,n}, \qquad t \geq 1.$$

$\mathbf{I}$ is the $(k-1) \times (k-1)$ identity matrix and $\mathbf{R} = \text{diag}(r_1, ..., r_{k-1})$ is a $(k-1) \times (k-1)$ diagonal matrix with $0 < r_i \leq 1$ for $i \in \{1, ..., k-1\}$. The starting value $\mathbf{Z}_{0,n}$ is taken as $\mathbf{Z}_{0,n} = E_D(\hat{\mathbf{w}}^*_{t,n}) = \mathbf{w}^*$. The vector $\mathbf{Z}_{t,n}$ can be presented as

$$\mathbf{Z}_{t,n} = (\mathbf{I} - \mathbf{R})^t \mathbf{Z}_{0,n} + \mathbf{R} \sum_{v=0}^{t-1} (\mathbf{I} - \mathbf{R})^v \hat{\mathbf{w}}^*_{t-v,n}.$$

Consequently it holds that $E_D(\mathbf{Z}_{t,n}) = \mathbf{w}^*$. The covariance matrix of the multivariate EWMA statistic $\mathbf{Z}_{t,n}$ in the in-control state is given by (Golosnoy and Schmid 2007)

$$\text{Cov}_D(\mathbf{Z}_{t,n}) = \mathbf{R} \left(\sum_{i,j=0}^{t-1} (\mathbf{I} - \mathbf{R})^i \ \text{Cov}_D(\hat{\mathbf{w}}^*_{t-i,n}, \hat{\mathbf{w}}^*_{t-j,n}) \ (\mathbf{I} - \mathbf{R})^j \right) \mathbf{R}. \quad (7.12)$$

This formula is evaluated using the asymptotic covariance matrix, given in (7.8). The values $n \geq 60$ seem to be large enough to apply an approximation for $n \to \infty$. Golosnoy *et al.* (2007) suggest using a bias correction for calculating the covariance matrix for values of $n < 60$. For our problem setting the corresponding bias correcting factors are presented on Figure 7.3.

A control chart is constructed using the distance between $\mathbf{Z}_{t,n}$ and its target $E_D(\mathbf{Z}_{t,n}) = \mathbf{w}^*$ that is measured by the Mahalanobis statistic. This leads to the quantity

$$[\mathbf{Z}_{t,n} - E_D(\mathbf{Z}_{t,n})]' \left[\lim_{n \to \infty} \text{Cov}(\mathbf{Z}_{t,n}) \right]^{-1} [\mathbf{Z}_{t,n} - E_D(\mathbf{Z}_{t,n})] .$$

Thus the control chart gives a signal at time t if

$$(n - k - 1)\, \mathbf{1}'\mathbf{\Sigma}^{-1}\mathbf{1}\ (\mathbf{Z}_{t,n} - \mathbf{w}^*)'\ \mathbf{Q}^{*\,-1}\ (\mathbf{Z}_{t,n} - \mathbf{w}^*) > c_2.$$

The value $c_2 > 0$ is determined via the in-control ARL as described earlier. Similar to the above scheme this chart suffers from inertia in detecting changes for the large values of n.

7.5 Simulation study

The simulation study presented here follows two aims. First, it serves to determine the critical limits of the control procedures introduced above. Second, it provides an out-of-control analysis of the monitoring schemes with respect to the most important types of changes. Additionally, it investigates the consequences of choosing different values of (n, λ), which determine the design of the control chart, for monitoring purposes.

7.5.1 In-control simulations

The in-control parameters for the simulation study are estimated from the longest identified Phase I from our empirical study, namely the period 16 Jan 1997–13 Sep 1999, see Section 7.6.1. The estimated in-control covariance matrix is taken as a true matrix for the target process $\mathbf{\Sigma}_D$ and is given by

$$\mathbf{\Sigma}_D = \mathbf{S}\mathbf{\Upsilon}_D\mathbf{S} = \begin{pmatrix} 0.0243461 & 0.0090885 & 0.0018627 & 0.0023164 \\ 0.0090885 & 0.0342469 & 0.0002492 & -0.0013083 \\ 0.0018627 & 0.0002492 & 0.0162222 & 0.0028677 \\ 0.0023164 & -0.0013083 & 0.0028677 & 0.1456425 \end{pmatrix},$$

where $\mathbf{S}$ is the diagonal matrix of standard deviations and $\mathbf{\Upsilon}_D$ represents the correlation matrix. Then the corresponding optimal portfolio proportions $E_D(\hat{\mathbf{w}}_{t,n})$ and the covariance matrix $(n-k-1)\mathrm{Cov}_D(\hat{\mathbf{w}}_{t,n}) = \mathbf{Q}/\mathbf{1}'\mathbf{\Sigma}_D^{-1}\mathbf{1}$ are equal to

$$E_D(\hat{\mathbf{w}}) = \begin{pmatrix} 0.25072 \\ 0.19172 \\ 0.50869 \\ 0.04888 \end{pmatrix}, \quad \frac{\mathbf{Q}}{\mathbf{1}'\mathbf{\Sigma}_D^{-1}\mathbf{1}} = \begin{pmatrix} 0.348 & -0.157 & -0.172 & -0.019 \\ -0.157 & 0.252 & -0.090 & -0.005 \\ -0.172 & -0.090 & 0.297 & -0.035 \\ -0.019 & -0.005 & -0.035 & 0.059 \end{pmatrix}.$$

Note that we neglect here the estimation risk and consider the parameters given above as if they were the true values. This approach is justified from the practical point of view, but may lead to mis-specifications of monitoring schemes (Champ *et al.* 2005; Jensen *et al.* 2006). However, for the portfolio investor it presents no danger, because the falsely specified in-control parameters only increase the probability of an early signal, indicating possible mis-specifications. Later, in the empirical study in Section 7.6.2 we encounter such situations.

For determining the critical values c for all charts we solve Equation (7.10), written now with respect to the design of our control charts

$$\mathrm{ARL}_D(c, \lambda, n, \mathbf{\Sigma}_D) = \xi = 120. \tag{7.13}$$

The solution is obtained using the numerical Monte Carlo simulations for each specific control chart design. Additional studies show that other reasonable values of ξ, such as $\xi = 250{,}500$, do not lead to any significant changes in the results.

7.5.2 Out-of-control study

The out-of-control simulation study provides the comparison of the control charts with respect to their ability to detect economically important changes.

We consider the parameter constellations leading to the largest perturbations in the optimal portfolio composition. For this reason we concentrate exclusively on changes in the variances of the considered risky assets. Consequently, the correlation structure $\boldsymbol{\Upsilon}_D$ remains here unchanged. Following Golosnoy *et al.* (2007), the changes are modelled by a diagonal matrix $\boldsymbol{\Delta}$, given by

$$\boldsymbol{\Delta} = \begin{pmatrix} \delta_1 & 0 & 0 & 0 \\ 0 & 1 & 0 & 0 \\ 0 & 0 & 1 & 0 \\ 0 & 0 & 0 & \delta_4 \end{pmatrix}, \qquad \delta_1 \in \{0.5,\ 1,\ 1.5,\ 2,\ 3\},\ \ \delta_4 \in \{0.5,\ 1,\ 2,\ 3\}.$$

This implies that the out-of-control covariance matrix $\boldsymbol{\Sigma}_C$ is given by

$$\boldsymbol{\Sigma}_C = \boldsymbol{\Delta}\,\mathbf{S}\,\boldsymbol{\Upsilon}_D\,\mathbf{S}\,\boldsymbol{\Delta} = \boldsymbol{\Delta}\,\boldsymbol{\Sigma}_D\,\boldsymbol{\Delta}. \tag{7.14}$$

The situation with $\delta_1 = \delta_4 = 1$ corresponds to the in-control case. Note that the changes in δ_1 and δ_4 are not symmetric, the increase in δ_1 and simultaneously the decrease in δ_4 lead to the largest perturbations in the optimal portfolio proportions.

Now we report the performance of the charts with respect to different values of $n \in \{10, 15, 20, 30, 40, 60\}$ and $\lambda \in \{0.1, 0.2, ..., 1.\}$. Table 7.1 presents the smallest out-of-control ARL_C performance measures, while Table 7.2 reports the corresponding worst case conditional expected delays $\mathcal{D}_{PS}$, see (7.11).

Table 7.1 Minimal ARL_Cs of the `Mahal` and `MEWMA` control charts. The optimal parameters (n, λ) are given in parentheses.

δ_1/δ_4	0.5	1.0	2.0	3.0
0.5	19.46 (20, 0.2)	21.77 (20, 0.1)	21.63 (20, 0.2)	21.56 (30, 1.0)
	19.24 (20, 0.2)	20.70 (20, 0.1)	20.48 (20, 0.1)	20.40 (20, 0.1)
1.0	25.22 (20, 1.0)	`Mahal` (n, λ)	84.43 (60, 0.1)	67.14 (60, 0.1)
	25.27 (20, 1.0)	`MEWMA` (n, λ)	79.37 (60, 0.1)	62.83 (60, 0.1)
1.5	21.49 (20, 1.0)	43.37 (60, 0.1)	38.71 (60, 0.1)	33.44 (60, 0.1)
	21.45 (20, 1.0)	41.26 (60, 0.1)	35.76 (60, 0.1)	31.10 (60, 0.1)
2.0	20.05 (20, 1.0)	25.25 (60, 0.6)	23.54 (60, 0.9)	20.93 (60, 1.0)
	19.64 (20, 0.1)	24.48 (60, 0.1)	23.05 (60, 0.2)	20.61 (60, 0.3)
3.0	12.56 (60, 1.0)	13.61 (60, 1.0)	13.21 (60, 1.0)	12.00 (60, 1.0)
	12.53 (60, 0.9)	13.57 (60, 0.7)	13.18 (60, 0.8)	12.02 (60, 1.0)

Table 7.2 Minimal $\mathcal{D}_{PS}$ of the `Mahal` and `MEWMA` control charts. The worst case change point m^* and the optimal parameters (n, λ) are given in parentheses.

δ_1/δ_4	0.5	1.0	2.0	3.0
0.5	19.92 (20, 20, 0.2)	22.25 (14, 20, 0.2)	21.98 (15, 20, 0.2)	21.93 (25, 20, 0.2)
	20.50 (4, 20, 0.4)	22.56 (11, 20, 0.3)	22.11 (11, 20, 0.3)	22.02 (20, 20, 0.2)
1.0	26.08 (23, 20, 1.0)	123.10 (25, 20, 0.2)	92.09 (28, 60, 0.1)	73.32 (26, 60, 0.1)
	26.42 (8, 20, 1.0)	128.46 (29, 30, 1.0)	98.53 (28, 60, 0.1)	77.37 (20, 60, 0.1)
1.5	22.20 (19, 20, 0.9)	47.44 (27, 60, 0.1)	41.90 (30, 60, 0.1)	36.13 (30, 60, 0.1)
	22.38 (14, 20, 1.0)	50.21 (4, 60, 0.3)	43.99 (11, 60, 0.3)	38.09 (22, 60, 0.3)
2.0	20.63 (13, 20, 0.3)	27.59 (26, 60, 0.3)	25.64 (28, 60, 0.4)	22.75 (29, 60, 0.5)
	20.94 (13, 20, 0.9)	28.75 (7, 60, 0.6)	26.50 (8, 60, 0.8)	23.50 (9, 60, 0.7)
3.0	13.83 (29, 60, 0.8)	14.93 (29, 60, 0.7)	14.42 (29, 60, 0.8)	13.12 (30, 60, 1.0)
	14.22 (13, 60, 0.9)	15.34 (20, 60, 1.0)	14.86 (11, 60, 1.0)	13.36 (20, 60, 1.0)

The optimal values of n and λ for both performance criteria are provided in parentheses. For the worst case delay measure $\mathcal{D}_{PS}$ we additionally report the least favourable change point $m^* \in \{1, ..., 30\}$.

The ARL_C performance criterion provides the following results. The optimal estimation window length is about $n = 60$ for detecting a change implied by an increase in the asset return variances, and about $n = 20$ for detecting a change implied by a decrease in the variances. This motivates us to choose values $n \geq 20$ to reduce the estimation risk in the estimated GMVP weights $\hat{\mathbf{w}}_{t,n}$. Not surprisingly, small values of λ are more appropriate for detection of small changes, while $\lambda = 1$ is the best choice for detecting large changes. The difference between the best `MEWMA` and `Mahal` charts is marginal. On the one hand, `MEWMA` charts are more flexible due to their construction, but, on the other hand, they are more influenced by the estimation risk, due to the evaluation of the covariance matrix (7.12). These two effects seem to compensate each other.

The worst case delay $\mathcal{D}_{PS}$ provides similar results, however, in this case the `Mahal` charts are always superior. The worst case change point m^* is close to 30, indicating on the presence of the inertia property (Woodall and Mahmoud 2005). Summarizing the evidence, we can conclude that the best results are obtained for `Mahal` charts with $n = 60$ and $\lambda = 1$. This chart is superior for detecting the large changes in the GMVP weights, which have the greatest practical importance.

7.6 Empirical study

The empirical study aims to illustrate the application of the introduced sequential monitoring methodology for practical purposes. This study focuses on the ability of multivariate EWMA control charts of Golosnoy and Schmid (2007) to detect changes in the GMVP weights. Chu *et al.* (1996) indicate a significant difference in application of sequential monitoring in economics and engineering. In the case of an alarm in engineering the process could (and often should) be stopped and thoroughly investigated for the causes of the signal. After finding the reason for the signal it should be removed and the monitoring procedure can be started again. On the contrary, in economic and financial applications the process almost never stops. This forces the investor to adjust her decisions based on the knowledge that the underlying in-control assumptions might be no longer valid. For this reason this study separates the issues of monitoring and reacting on the changes. Our prime goal is to obtain a signal, while the economic effect of monitoring (Golosnoy 2007) as well as the after-signal policy recommendations remain beyond the scope of this chapter.

The considered investor is a pure volatility minimizer choosing the GMVP weights in (7.5) at each decision time point. She needs to decide whether the weights, optimal up to now, could be seen as optimal for further purposes. Here we investigate the effectiveness of control charts for detecting changes in the optimal portfolio composition. For our study we consider $k = 4$ assets, namely the US market, a non-US index, oil and gold. This portfolio dimension is typical for Markowitz-based portfolio approaches, see Fleming *et al.* (2001, 2003). The whole set of historical prices is taken for the period from January, 1984 to September, 2006 with a daily frequency. This frequency is the highest possible one, i.e. there are no intraday observations available for the selected time series. All data is obtained from `DataStream`.

The covariance matrix of returns $\mathbf{\Sigma}$ estimated over the whole sample is given by

$$\mathbf{\Sigma} = \begin{pmatrix} 0.021682 & 0.007186 & 0.002280 & -0.000551 \\ 0.007186 & 0.026583 & -0.001888 & -0.004073 \\ 0.002280 & -0.001888 & 0.018487 & 0.005392 \\ -0.000551 & -0.004073 & 0.005392 & 0.135519 \end{pmatrix}. \qquad (7.15)$$

Consequently, the expectation of the GMVP weights $E(\hat{\mathbf{w}}_{t,n})$ and their covariance matrix $(n-k-1)\mathrm{Cov}(\hat{\mathbf{w}}_{t,n}) = \mathbf{Q}/\mathbf{1}'\mathbf{\Sigma}^{-1}\mathbf{1}$ are equal to

$$E(\hat{\mathbf{w}}_{t,n}) = \begin{pmatrix} 0.2405 \\ 0.2808 \\ 0.4258 \\ 0.0528 \end{pmatrix}, \quad \frac{\mathbf{Q}}{\mathbf{1}'\mathbf{\Sigma}^{-1}\mathbf{1}} = \begin{pmatrix} 0.365 & -0.186 & -0.166 & -0.011 \\ -0.186 & 0.265 & -0.072 & -0.006 \\ -0.166 & -0.072 & 0.278 & -0.039 \\ -0.011 & -0.006 & -0.039 & 0.058 \end{pmatrix}.$$

These unconditional values may be used as a reference point of our analysis. Again, as in the simulation study, the estimation risk is neglected here.

SPC makes a clear distinction between the in-control Phase I and monitoring Phase II (Woodall 2000). For our purposes we need to identify the Phase I periods where the GMVP weights are stable from the statistical point of view. Having identified the Phase I stage, the investor can start monitoring Phase II by sequentially checking for the validity of the in-control assumptions. Note that our procedure does not exploit all available information, but is based on the historical price observations only. On the contrary, in practice any retrospective or sequential analysis should rely not only on historical asset data, but also on the whole information, available to a financial market analyst. This shows some limitations of our approach, however control charts are suggested here primarily as a supplementary analysis tool.

The rest of the empirical study is organized as follows. First we identify the in-control Phases I in Section 7.6.1. Then we provide a Phase II analysis, where the charts detect alternations in the GMVP composition. This allows us to give practical advice about the usage of the considered procedures. Finally, we evaluate the obtained results and point out possible extensions and limitations of the discussed methodology.

7.6.1 Selecting Phase I

A literature review concerning Phase I selection for sequential monitoring purposes can be found in Montgomery (2005). The basic requirement is to choose a period with the returns coming from the distribution without a structural break. In our case we are primarily interested in no changes in the GMVP weights. For this reason next we consider two alternative approaches for Phase I selection. The first one is grounded on testing the equality of the return covariance matrices, while the second one concentrates exclusively on checking the GMVP weights for changes.

For empirical purposes it is sufficient to have the observations of one year ($N = 252$) for the prerun Phase I estimation, because it allows the reduction of the estimation error in the estimated weights $\hat{\mathbf{w}}_{t,N}$ to a practically acceptable level (Okhrin and Schmid 2006). In order to identify Phase I, we divide the whole sample from January, 1984 to September, 2006, into nonoverlapping $N_i = 63$ day periods, which correspond to a quarter of a year. Then Phase I is considered to be identified, if the null hypothesis 'no change during at least four consecutive periods' is not rejected.

7.6.1.1 Testing covariance matrices

The first possibility is to test whether the whole covariance matrix $\boldsymbol{\Sigma}$ remains unchanged from one period to another. A simultaneous test for the equality of r period covariance matrices is provided by Muirhead (1982, pp. 298ff.).

The ith period $(k \times k)$-dimensional sample covariance matrix $\mathbf{S}_i$ is estimated by

$$\mathbf{S}_i = n_i^{-1} \sum_{j=1}^{N_i} (\mathbf{X}_{ij} - \overline{\mathbf{X}}_i)(\mathbf{X}_{ij} - \overline{\mathbf{X}}_i)', \qquad \overline{\mathbf{X}}_i = N_i^{-1} \sum_{j=1}^{N_i} \mathbf{X}_{ij}, \quad i = 1, ..., r.$$

Then the modified likelihood ratio statistic is given by

$$\Lambda^* = \frac{\prod_{i=1}^{r} (\det n_i \mathbf{S}_i)^{n_i/2}}{(\det \mathbf{S})^{n/2}} \frac{n^{kn/2}}{\prod_{i=1}^{r} n_i^{kn_i/2}}, \tag{7.16}$$

where $n_i = n_0 = N_0 - 1$, $n = N - r$, $\mathbf{S} = \sum_{i=1}^{r} n_i \mathbf{S}_i$. The critical values are provided by Muirhead (1982, pp. 310f.).

Taking $r = 4$, i.e. each period has $N_i = N_0 = N/r = 63$ daily observations and $N = 4N_i = 252$, then for $\alpha = 5\%$ significance level and $n_0 = 63$ the value of the statistic is approximately $c_f(\alpha) = 45.08$. The H_0 hypothesis $\boldsymbol{\Sigma}_1 = \ldots = \boldsymbol{\Sigma}_r$ is rejected, if $-2 \ln \Lambda^* > c_f(\alpha)$.

The corresponding test statistic $-2 \ln \Lambda^*$ for all consecutive four period clusters is plotted on Figure 7.4. The horizontal line denotes a value $c_f(\alpha = 0.05)$. Figure 7.4 presents evidence that hypothesis H_0 is almost always strongly rejected. However, this result is not really disappointing for our problem, because the test statistic (7.16) is based on the determinant of the covariance matrix, which is irrelevant for GMVP composition, see (7.5). For this reason we consider the choice of Phase I using the estimated GMVP weights directly.

7.6.1.2 Testing portfolio weights

In the retrospective analysis we check whether the weights, estimated in nonoverlapping periods, are statistically (significantly) different. Okhrin and Schmid (2006) show that the distribution of the estimated GMVP weights is asymptotically normal. Using their results, we undertake a sequence of tests for equality of the weights, estimated in all consequent periods. The period length is again selected to be one quarter, which corresponds to 63 daily observations. Then for all weights $i = 1, ..., k$ we test whether the GMVP weights have changed from one quarter to the previous one. This should be sufficient

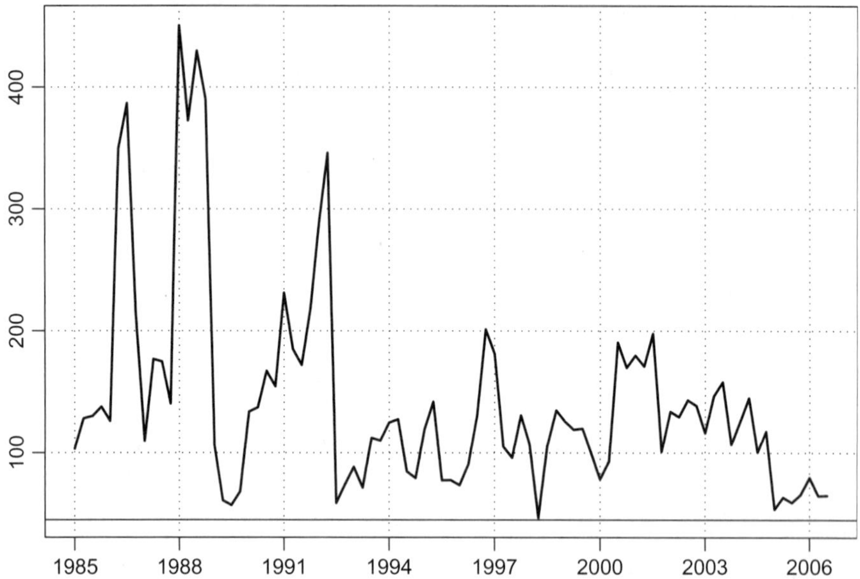

Figure 7.4 Test statistic $-2\ln\Lambda_t^*$ for detecting changes in the covariance matrix.

for our analysis, because the investor has an economic reason to monitor only the changes in the weights.

For this purpose we adopt the mean difference test with the unknown unequal variances (Lehmann and Romano 2005, pp. 159ff.), where the test statistic $T_{i,t}$ for all $\hat{w}_{i,t}$ in the vector $\hat{\mathbf{w}}_t = (\hat{w}_{1,t}, ..., \hat{w}_{k,t})'$ is distributed under H_0 as

$$T_{i,t} = \frac{\hat{w}_{i,t} - \hat{w}_{i,t-1}}{(V(\hat{w}_{i,t}) + V(\hat{w}_{i,t-1}))^{1/2}} \xrightarrow{L} \mathcal{N}(0, 1). \tag{7.17}$$

We consider the value $T_t^* = \max_i |T_{i,t}|$, which is plotted on Figure 7.5.

The horizontal line on Figure 7.5 denotes a critical value $c^*(\alpha/2) = 2.81$, which corresponds to $\alpha = 0.5\%$ significance level. Now we are able to identify all historical Phase I periods, where the H_0 hypothesis has not been rejected during at least one year. Table 7.3 reports these periods, detected based on the statistic (7.17).

The detected Phase I periods roughly correspond to the historical periods without large perturbations on the financial markets. The longest period where H_0 is not rejected, namely from 16 Jan 1997 to 13 Sep 1999, a total of 693 daily returns, is used as a Phase I in-control period for parameter estimation in the simulation study above. Having identified the Phase I periods by a retrospective

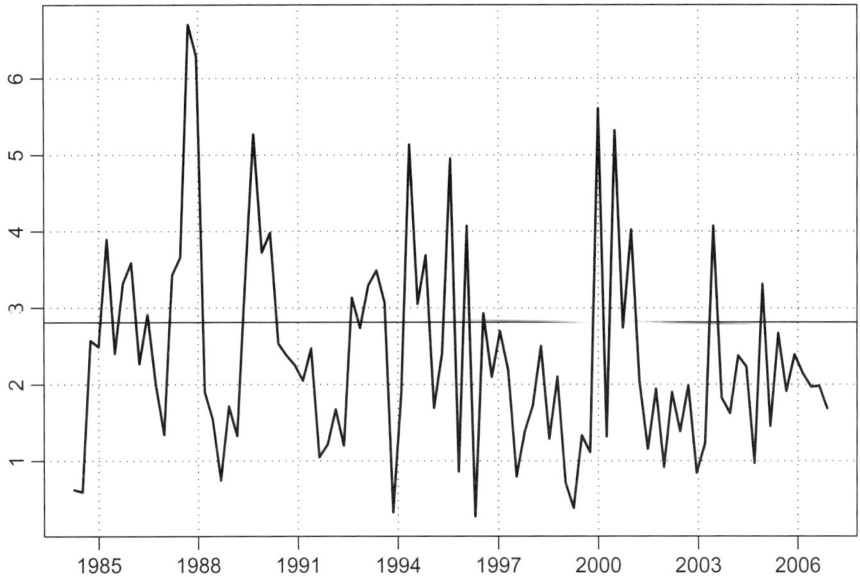

Figure 7.5 Test statistic T_t^* for detecting changes in the optimal portfolio weights.

Table 7.3 Identified Phase I periods, minimum 252 daily observations

02 Jan 1984–18 Mar 1985	06 Mar 1986–20 May 1987
10 Feb 1988–25 Apr 1989	13 Apr 1990–15 Jun 1992
16 Jan 1997–13 Sep 1999	28 Nov 2000–29 Jan 2003
29 Apr 2003–12 Jul 2004	08 Oct 2004–13 Sep 2006

analysis, we can now start with the Phase II sequential monitoring. Our task is to compare the ability of the control charts in Phase II to detect changes in the GMVP weights for each identified Phase I period.

7.6.2 Phase II monitoring analysis

The monitoring procedure is organized as follows. The control charts are constructed on a daily basis with the in-control parameters estimated from each Phase I period. The estimation risk is neglected here. The design of

Table 7.4 Dates of signals for Phase II. The number of days before signals in Phase II and λs, leading to the quickest signals, are given in parentheses.

Phase I	Chart	$n = 20$	$n = 30$	$n = 40$	$n = 60$
02.01.84–	`Mahal`	27.03.85 (7, 0.1)	19.03.85 (1, 1.0)	19.03.85 (1, 1.0)	19.03.85 (1, 1.0)
18.03.85	`MEWMA`	19.03.85 (1, λ)	19.03.85 (1, λ)	19.03.85 (1, λ)	19.03.85 (1, λ)
06.03.86–	`Mahal`	03.06.87 (10, 1.0)	22.05.87 (2, 1.0)	21.05.87 (1, 1.0)	21.05.87 (1, 1.0)
20.05.87	`MEWMA`	04.06.87 (11, 1.0)	22.05.87 (2, λ)	21.05.87 (1, λ)	21.05.87 (1, λ)
10.02.88–	`Mahal`	07.06.89 (31, 0.1)	08.06.89 (32, 0.3)	29.05.89 (24, 0.6)	25.05.89 (22, 1.0)
25.04.89	`MEWMA`	12.06.89 (33, 1.0)	13.06.89 (34, 1.0)	13.05.89 (34, 1.0)	12.05.89 (33, 1.0)
13.04.90–	`Mahal`	23.06.92 (6, λ)	16.06.92 (1, 1.0)	16.06.92 (1, 1.0)	16.06.92 (1, λ)
15.06.92	`MEWMA`	01.07.92 (12, 0.1)	18.06.92 (3, 0.1)	16.06.92 (1, λ)	16.06.92 (1, λ)
16.01.97–	`Mahal`	27.09.99 (10, 1.0)	27.09.99 (10, 1.0)	27.09.99 (10, 1.0)	24.09.99 (9, 1.0)
13.09.99	`MEWMA`	28.09.99 (11, 1.0)	27.09.99 (10, 1.0)	27.09.99 (10, 1.0)	27.09.99 (10, 1.0)
28.11.00–	`Mahal`	10.02.03 (8, 0.8)	19.02.03 (15, 1.0)	27.02.03 (21, 1.0)	18.02.03 (14, 0.7)
29.01.03	`MEWMA`	12.02.03 (10, 1.0)	20.02.03 (16, 1.0)	05.03.03 (25, 1.0)	12.03.03 (30, 1.0)
29.04.03–	`Mahal`	10.11.04 (87, 1.0)	13.07.04 (1, 1.0)	13.07.04 (1, 1.0)	13.07.04 (1, 1.0)
12.07.04	`MEWMA`	19.07.04 (5, 0.1)	13.07.04 (1, λ)	13.07.04 (1, λ)	13.07.04 (1, λ)

the applied charts is described in Section 7.4; the critical limits are chosen to guarantee the in-control $\text{ARL}_D = 120$. The considered monitoring schemes use the optimal weights $\hat{\mathbf{w}}_{t,n}$ estimated based on $n \in \{20, 30, 40, 60\}$ previous daily returns. `Mahal` and `MEWMA` charts are started for the memory parameter $\lambda \in \{0.1, 0.2, \ldots, 1.0\}$. Then the investor decides at every new date, whether the in-control parameters are still valid. If the chart gives a signal, the procedure should be stopped. The dates of signals for all in-control Phase I periods are presented in Table 7.4 for different values of n. The number of days after the start of monitoring and the parameter λ, providing the quickest signal, are given in parentheses; 'λ' in Table 7.4 denotes that all charts give the signal simultaneously.

The obtained results can be summarized as follows. The choice of the estimation window $n = 20$ is clearly not an appropriate one, because it leads to a delay compared to larger values of n. The `Mahal` charts give their signals in the majority of situations earlier or at the same time, as their `MEWMA` competitors, for this reason they should be preferred. Concerning the choice of a smoothing parameter λ, the value $\lambda = 1$ seems to provide the best performance for $n \geq 30$ and `Mahal` charts. Because the majority of the charts give the signals roughly simultaneously, we conclude that the reasons for the signals are the true changes, or, in the case of an immediate signal at $t = 1$, possible mis-specifications of Phase I. Based on the evidence we conclude that `Mahal` charts with $n = 30$ or $n = 40$, $\lambda = 1$ are the procedures to be recommended for practical sequential monitoring of the GMVP weights.

7.6.3 Empirical results: discussion and extensions

After a chart provides a signal, the investor needs to identify its reasons and consequences. This is not difficult for obvious and publicly known events, such as crashes on stock exchanges or some natural cataclysms. However, for the signals without a clear economic interpretation this task may be problematic. In the case of an alarm we suggest waiting until the number of observations after the alarm gets sufficiently large for identifying a new Phase I period. Afterwards the new Phase I (in control) parameters need to be identified and the procedure has to be started again. Unfortunately, the sequential monitoring procedures cannot differentiate between changes in the parameters and outliers. For this reason a financial analyst is required for decisions concerning further portfolio policy after each obtained signal.

The identification of Phase I is a hot issue nowadays in sequential monitoring research. In this study we have identified the Phase I periods by testing the equality of portfolio weights. An alternative approach would be to consider the whole historical period and to estimate the unconditional covariance matrix, as in (7.15). Then the sequential monitoring task can be seen as detecting deviations from the long-run GMVP composition. Golosnoy (2007) provides empirical evidence for monitoring the unconditional GMVP on the German stock market.

A practical recommendation for the choice of the parameters (n, λ) and the monitoring procedure is required for implementing the described approach. Studies of Golosnoy and Schmid (2007) and Golosnoy *et al.* (2007) provide numerous simulations for determining the best control chart design. Summarizing the evidence, it is often useful to combine some types of control charts. For example, the detecting ability can be improved by simultaneously using a chart with $\lambda = 1$, capturing large changes, and a chart with $\lambda = 0.1$, which is better for small alterations. `Mahal` charts often outperform `MEWMA` procedures, primarily because of the complicated asymptotic covariance matrix (7.12), required for the derivation of `MEWMA` charts. The value of $n = 30$ seems to be optimal for monitoring $k = 4$ asset portfolio, because it is large enough to keep the estimation risk under control, and it small enough to allow flexibility of the control statistic and to mitigate the impact of inertia.

The suggested procedure is heavily grounded on the normality assumption for asset returns. However, this assumption is not crucial for derivation of control charts and can be relaxed. As long as the second moment of asset returns exist, the covariance matrix obtained by the bootstrap methodology can be used for calculating the GMVP weights. Then the control procedures from Section 7.4 can be applied in a similar line, although the design of all charts should be newly derived. The proposed methodology is based on daily data frequency. Availability of ultra-high frequency observations requires the use of a different family of sequential monitoring methods (Andreou and Ghysels 2004, 2006).

7.7 Concluding remarks

The conventional investor holds a portfolio composition chosen according to some asset allocation rule. The composition should be revised in case of significant alterations in the optimal portfolio weights, which may happen due to the changes in the underlying model parameters. The investor is interested in the timely detection of changes in the optimal portfolio proportions, otherwise she would suffer from utility losses because of nonoptimal portfolio holdings. The control for the relevant changes can be done by means of sequential monitoring of the portfolio composition. The control charts for monitoring the optimal portfolio weights are a core issue of this chapter.

We illustrate the monitoring of the optimal portfolio proportions by considering the investor choosing the portfolio with the lowest attainable variance, known as the global minimum variance portfolio (GMVP). For the stated problem, multivariate exponentially weighted moving average charts are introduced. These charts are constructed taking into account the statistical properties of the process of the estimated optimal portfolio weights. The implementation of the suggested monitoring schemes is discussed and illustrated within a Monte Carlo simulation study. The empirical application shows the usefulness of sequential methods for a practically relevant portfolio monitoring problem. Moreover, we provide a practical recommendation for the application of the analysed schemes and point out some extensions and directions for further research in this field.

The suggested monitoring rules cannot be seen as an automatic trading system. Control charts are mainly useful for their ability to give a warning signal concerning significant (from the statistical point of view) alterations in the parameters of interest. They suggest a supplementary tool for a financial analyst, who makes a decision using the whole available information. Each obtained signal should be analysed separately for its causes. Moreover, control charts give no further recommendations about actions to be taken in response to the obtained signals. From a practical point of view, the control chart should be seen as an additional tool for a financial analyst to detect statistically significant alterations in the optimal portfolio proportions. The after-signal actions, however, remain far beyond the scope of sequential monitoring analysis.

Acknowledgements

The authors are grateful to Marianne Frisén and the participants of the Workshop on 'Financial Surveillance' in Göteborg, Sweden for valuable comments.

References

Albers, W. and Kallenberg, W. C. M. (2004a). Are estimated control charts in control? *Statistics*, **38**, 67–79.

Albers, W. and Kallenberg, W. C. M. (2004b). Estimation in Shewhart control charts: Effect and corrections. *Metrika*, **59**, 207–234.

Alwan, L. and Roberts, H. (1988). Time-series modeling for statistical process control, *Journal of Business and Economic Statistics*. **6**, 87–95.

Andreou, E. and Ghysels, E. (2004). The impact of sampling frequency and volatility estimators on change-point tests. *Journal of Financial Econometrics*, **2**, 290–318.

Andreou, E. and Ghysels, E. (2006). Monitoring disruptions in financial markets. *Journal of Econometrics*, **135**, 77–124.

Banerjee, A. and Urga, G. (2005). Modeling structural breaks, long memory and stock market volatility: an overview. *Journal of Econometrics*, **129**, 1–34.

Best, M. and Grauer, R. (1991). On the sensitivity of mean–variance-efficient portfolios to changes in asset means: Some analytical and computational results. *Review of Financial Studies*, **4**, 315–342.

Brandt, M. (2007), Portfolio choice problems. In *Handbook of Financial Econometrics*, eds. Ait-Sahalia, Y. and Hansen, L. Elsevier and North Holland, Amsterdam.

Brandt, M., Santa-Clara, P. and Valkanov, R. (2005). Parametric portfolio policies: exploiting characteristics in the cross-section of equity returns. Working paper, available at http://faculty.fuqua.duke.edu/~mbrandt/papers/working/paramport.pdf

Brook, D. and Evans, D. (1972). An approach to the probability distribution of CUSUM run length. *Biometrika*, **59**, 539–549.

Busse, J. (1999). Volatility timing in mutual funds: Evidence from daily returns. *Review of Financial Studies*, **12**, 1009–1041.

Champ, C., Jones-Farmer, L. and Rigdon, S. (2005). Properties of the T^2 control charts when parameters are estimated. *Technometrics*, **47**, 437–445.

Chu, C.-S., Stinchcombe, M. and White, H. (1996). Monitoring structural change. *Econometrica*, **64**, 1045–1065.

Crosier, R. B. (1988). Multivariate generalizations of cumulative sum quality-control schemes. *Technometrics*, **30**, 291–303.

Crowder, S. (1987). A simple method for studying run-length distributions of exponentially weighted moving average charts. *Technometrics*, **29**, 401–407.

DeMiguel, V., Garlappi, L. and Uppal, R. (2007). Optimal versus naive diversification: How inefficient is the $1/N$ portfolio policy? *Review of Financial Studies* (in press).

Engle, R. (1982). Autoregressive conditional heteroscedasticity with estimates of the variance of UK inflation. *Econometrica*, **50**, 987–1008.

Eraker, B., Johannes, M. and Polson, N. (2003). The impact of jumps in volatility and returns. *Journal of Finance*, **58**, 1269–1300.

Fama, E. F. (1965). The behavior of stock market prices. *Journal of Business*, **38**, 34–105.

Fleming, J. Kirby, C. and Ostdiek, B. (2001). The economic value of volatility timing. *Journal of Finance*, **56**, 329–352.

Fleming, J. Kirby, C. and Ostdiek, B. (2003). The economic value of volatility timing using 'Realized' volatility. *Journal of Financial Economics*, **67**, 473–509.

Foster, D. P. and Nelson, D. (1996). Continuous record asymptotics for rolling sample variance estimators. *Econometrica*, **64**, 139–174.

Frisén, M. (2003). Statistical surveillance. Optimality and methods. *International Statistical Review*, **71**, 403–434.

Frost, P. and Savarino, E. (1988). An empirical bayes approach to efficient portfolio selection. *Journal of Financial and Quantitative Analysis*, **21**, 292–305.

Gan, F. (1993). An optimal-design of EWMA control charts based on median run-length. *Journal of Statistical Computation and Simulation*, **45**, 169–184.

Garlappi, L., Uppal, R. and Wang, T. (2007). Portfolio selection with parameter and model uncertainty: a multi-prior approach. *Review of Financial Studies*, **20**, 41–81.

Golosnoy, V. (2007). Sequential monitoring of minimum variance portfolio. *Advances in Statistical Analysis*, **91**, 39–55.

Golosnoy, V. and Okhrin, Y. (2007). Multivariate shrinkage for optimal portfolio weights. *European Journal of Finance*, **13**, 441–458.

Golosnoy, V. and Schmid, W. (2007). EWMA control charts for optimal portfolio weights. *Sequential Analysis*, **26**, 195–224.

Golosnoy, V., Okhrin, I. and Schmid, W. (2007). Statistical methods for the surveillance of portfolio weights. Working paper.

Haerdle, W., Herwartz, H. and Spokoiny, V. (2003). Time inhomogeneous multiple volatility modeling. *Journal of Financial Econometrics*, **1**, 55–95.

Hotelling, H. (1947). Multivariate quality control – Illustrated by the air testing of sample bombsights. in *Techniques of Statistical Analysis*, eds. C. Eisenhart, M. W. H. and Wallis, W. A., McGraw Hill, New York, pp. 111–184.

Hsu, D., Miller, R and Wichern, D. (1974). On the stable Paretian behavior of stock market prices. *Journal of American Statistical Association*, **69**, 108–113.

Ingersoll, J. (1987). *Theory of Financial Decision Making*. Rowman & Littlefield, Maryland.

Jensen, W., Jones-Farmer, L. A., Champ, C. and Woodall, W. (2006), Effects of parameter estimation on control chart properties: a literature review. *Journal of Quality Technology*, **38**, 349–364.

Jorion, P. (1986). Bayes–Stein estimation for portfolio analysis. *Journal of Financial and Quantitative Analysis*, **21**, 279–292.

Kim, D. and Kon, S. (1999). Structural change and time dependence in models of stock returns. *Journal of Empirical Finance*, **6**, 283–308.

Klein, R. and Bawa, V. (1976). The effect of estimation risk on optimal portfolio choice. *Journal of Financial Economics*, **3**, 215–231.

Kramer, H. and Schmid, W. (1997). EWMA charts for multivariate time series. *Sequential Analysis*, **16**, 131–154.

Lamoureux, C. and Lastrapes, W. (1990). Persistence in variance, structural change and the GARCH model. *Journal of Business and Economic Statistics*, **8**, 225–234.

Lehmann, E. L. and Romano, J. P. (2005). *Testing Statistical Hypothesis*. 3rd edn. Springer-Verlag, New York.

Lorden, G. (1971). Procedures for reacting to a change in distribution, *Annals of Mathematical Statistics*, **41**, 1897–1908.

Lowry, C., Woodall, W., Champ, C. and Rigdon, S. (1992). A multivariate exponentially weighted moving average control chart. *Technometrics*, **34**, 46–53.

Lu, C.-W. and Reynolds, M. (1999a). EWMA control charts for monitoring the mean of autocorrelated processes. *Journal of Quality Technology*, **31**, 166–188.

Lu, C. W. and Reynolds, M. R. (1999b). Control charts for monitoring the mean and variance of autocorrelated processes. *Journal of Quality Technology*, **31**, 259–274.

Lu, C. W. and Reynolds, M. R. (2001). CUSUM charts for monitoring an autocorrelated process. *Journal of Quality Technology*, **33**, 316–334.

Mandelbrot, B. (1963). The variation of certain speculative prices. *Journal of Business*, **36**, 394–419.

Markowitz, H. (1952). Portfolio Selection. *Journal of Finance*, **7**, 77–91.

Merton, R. C. (1980). On estimating the expected return on the market. *Journal of Financial Economics*, **8**, 323–361.

Michaud, R. O. (1998). *Efficient Asset Management*. Harvard Business School Press, Boston, Mass.

Montgomery, D. C. (2005). *Introduction to Statistical Quality Control*, 5th edn. John Wiley & Sons, Ltd, New York.

Muirhead, R. J. (1982). *Aspects of Multivariate Statistical Theory*. John Wiley & Sons, Ltd, New York.

Ngai, H. M. and Zhang, J. (2001). Multivariate cumulative sum control charts based on projection pursuit. *Statistica Sinica*, **11**, 747–766.

Okhrin, Y. and Schmid, W. (2006). Distributional properties of optimal portfolio weights. *Journal of Econometrics*, **134**, 235–256.

Page, E. (1954). Continuous inspection schemes. *Biometrika*, **41**, 100–115.

Pignatello, J. J. and Runger, G. C. (1990). Comparison of multivariate CUSUM charts. *Journal of Quality Technology*, **22**, 173–186.

Pollak, M. and Siegmund, D. (1975). Approximations to the expected sample size of certain sequential tests. *Annals of Statistics*, **3**, 1267–1282.

Ramchand, L. and Susmel, R. (1998). Volatility and cross correlation across major stock markets. *Journal of Empirical Finance*, **5**, 397–416.

Reynolds, M. and Cho, G.-Y. (2006). Multivariate control charts for monitoring the mean vector and covariance matrix. *Journal of Quality Technology*, **38**, 230–253.

Roberts, S. (1959). Control chart tests based on geometric moving averages. *Technometrics*, **1**, 239–250.

Roberts, S. W. (1966). A comparison of some control chart procedures. *Technometrics*, **8**, 414–430.

Rosolowski, M. and Schmid, W. (2003). EWMA charts for monitoring the mean and the autocovariances of stationary processes. *Sequential Analysis*, **22**, 257–285.

Schipper, S. and Schmid, W. (2001). Sequential methods for detecting changes in the variance of economic time series. *Sequential Analysis*, **20**(4), 235–262.

Schmid, W. (1995). On the run length of a Shewhart chart for correlated data. *Statistical Papers*, **36**, 111–130.

Schmid, W. (1997a). CUSUM control schemes for Gaussian processes. *Statistical Papers*, **38**, 191–217.

Schmid, W. (1997b). On EWMA charts for time series. In *Frontiers in Statistical Quality Control*, eds. Lenz, H.-J. and Wilrich, P.-T. Physica-Valag, Heidelberg, pp. 115–137.

Severin, T. and Schmid, W. (1999). Monitoring changes in GARCH models. *Allgemeines Statistisches Archiv*, **83**, 281–307.

Shewhart, W. (1931). *Economic Control of Quality of Manufactured Product.* van Nostrand, Toronto.

Solnik, B., Bourcrelle, C. and Le Fur, Y. (1996). International market correlation and volatility. *Financial Analysts Journal*, **52**, 17–34.

Sullivan, J. H. and Jones, L. A. (2002). A self-starting control chart for multivariate individual observations. *Technometrics*, **44**, 24–33.

Wang, Z. (2005). A shrinkage approach to model uncertainty and asset allocation. *Review of Financial Studies*, **18**, 673–705.

Woodall, W. H. (2000). Controversies and contradictions in statistical process control. *Journal of Quality Technology*, **32**, 341–378.

Woodall, W. H. and Mahmoud, M. (2005). The inertial properties of quality control charts. *Technometrics*, **47**, 425–436.

Woodall, W. H. and Ncube, M. M. (1985). Multivariate CUSUM quality control procedures. *Technometrics*, **27** 285–292.

Yashchin, E. (1993). Performance of CUSUM control schemes for serially correlated observations. *Technometrics*, **35**, 37–52.

8

Likelihood-based surveillance for continuous-time processes

Helgi Tómasson

University of Iceland, Faculty of Economics and Business Administration, Oddi v/ Sturlugotu, IS-101 Reykjavik, Iceland

8.1 Introduction

In everyday life individuals and firms are exposed to a constant flow of data. An example is the news broadcast on modern television, a person is reading the news and at the bottom of the screen there is a banner with a constant flow of quotes of a financial market. Information on time of transactions, prices and volume are observed and stored electronically by systems such as Reuter and Bloomberg. It is of interest to form objective tools to do inference based on such streams. To assess the informative value of such a data stream a suitable statistical model is required together with a matching statistical inference strategy. Firms, like banks, financial supervisory authorities, central-banks and other financial institutions are facing similar data stream. Firms, buying, selling or distributing financial products need to keep track of the content of data, and update their prices accordingly. These institutions need an objective tool as a basis for their decision. The aim of this chapter is to suggest an approach of monitoring changes in a continuous-time price process.

In order to develop an objective tool some formal definitions are necessary. Due to the academic success of mathematical finance in recent years, many of the individuals making decisions are aware of pricing rules such as

Financial Surveillance Edited by Marianne Frisén

Black–Scholes, etc. The Black–Scholes rule is an example of a popular way of formalizing dynamics with help of stochastic differential equations (SDE). The mathematical pricing rules, such as Black–Scholes, are based on parameters in the SDE and the institutions decide on a change in their prices based on their inference about an eventual parameter shift in the SDE. The SDE describes a continuous process, but in practice the observed data are collected discretely, often at uneven time intervals. An example of such data flow is the quotes from financial markets.

In statistical subcultures there are many approaches for the phenomenon of a parameter shift. Broemeling and Tsurumi (1987) give the following classification for the varying parameter problem (change-of-regime, structural-break etc.). First, the case of a known break-point and an abrupt change in parameter. Second, an abrupt change in parameter at an unknown time point. Third, that there is a gradual (non-stochastic) change in the parameter over a certain period. Fourth, the parameters might follow a stochastic process. And fifth, data might be from a mixture of populations. In this chapter, only breaks of the second type are considered. The concept of parameter constancy, often in terms like structural-change, regime-shift is treated in many textbooks on econometrics. Article collections on statistical analysis of economic structural change are given in Hackl (1989) and Hackl and Westlund (1991). In this chapter the focus is on on-line detection of a deterministic (exogenous) abrupt parameter change. The methodological approach is based on ideas of statistical surveillance. The concepts of surveillance are reviewed in Frisén (2003).

In this chapter, an approach to implementing likelihood-based surveillance tools to continuous-time diffusion processes is given. In Chapter 2 some background to diffusion processes is given. The regime-shift is supposed to be a jump in a parameter. The dynamics of the process are of the form:

$$\begin{aligned} \mathrm{d}X(t) &= \mu(X(t),t)\mathrm{d}t + \sigma(X(t),t)\mathrm{d}W(t), \\ \mu(X(t),t) &= \mu_1(X(t),t)I_{[t<\tau]} + \mu_2(X(t),t)I_{[\tau<t]}, \\ \sigma(X(t),t) &= \sigma_1(X(t),t)I_{[t<\tau]} + \sigma_2(X(t),t)I_{[\tau<t]}. \end{aligned}$$

Popular continuous-time financial models for interest rates are single factor mean-reverting processes. Even though they look simple, data analysis in these simple continuous-time models is somewhat problematic because important features, such as the transition density, are not known in closed form. Recently, Aït-Sahalia (1999, 2002) has in a couple of articles suggested a Taylor approximation of the transition density and thereby opened the computational possibility of working numerically with the likelihood function. This allows the possibility of likelihood inference, i.e. maximum-likelihood estimation, Bayesian estimation and likelihood-based surveillance. In this chapter on-line monitoring a simple continuous-time model is addressed. Data are supposed to be generated by a simple mean-reverting process, the CKLS (Chan, Karolyi, Longstaff and Sanders 1992),

$$dX(t) = \kappa(\alpha - X(t))dt + \sigma X(t)^{\rho} dW(t) \tag{8.1}$$

which contains the popular CIR (Cox, Ingersoll and Ross 1985), $\rho = 1/2$, as a special case. Schmid and Tzotchev (2004) describe an approach for surveillance of a CIR model. In this chapter the CKLS model is treated as an illustrative example. The reasons for that choice are perhaps the same as for its popularity in interest rate analysis; it is a simple model that can capture several stylized facts of interest rate movements. It is always positive, it is asymmetric around its mean and it can have heavy tails. The parameters have somewhat an easy interpretation. The α parameter is the long-term mean. The κ parameter controls the speed of convergence to that mean. The σ and ρ parameters are linked to properties such as variance and tail behaviour. The CKLS family is, as most diffusions, analytically difficult. Therefore traditional statistical analysis, like maximum-likelihood estimation has to rely on numerical approaches. In Section 8.2 a brief review of the problems of calculating the likelihood function is given. The approximations of the likelihood function are done by means of a Taylor expansion. A somewhat more detailed description is given in Appendix A. For understanding the ability to detect a break it is necessary to have an idea about the precision of the estimates of the parameters and the nature of the model. In Section 8.3 some numerical properties of the CKLS model are discussed and some examples of maximum-likelihood estimates shown. Having an algorithm for calculating the likelihood functions offers the possibility of applying well known likelihood-based surveillance tools such as (exponential)-CUSUM and the Shiryaev–Roberts (SR) statistics. In this chapter the CUSUM and SR statistics for dynamic continuous-time processes are implemented in the notation and spirit of Shiryaev (2002). Even though the dynamics are specified in continuous-time, data are assumed to consist of discrete observations and thus also decisions are made in discrete time. The simulation of a continuous-time process and calculation of the likelihood therefore require some discretization tools. The implementation of the discretization tools is described in Section 8.4. An illustration of implementation and computation is given in Section 8.5. The calculations are performed using the statistical environment, R, and R-packages written by the author. A version of the R-surveillance package for the CKLS model may be downloaded from the present editions of the website of the book. A brief description of R-commands used in this article is given on the website of the book.

8.2 Likelihood approximations for diffusion processes

The diffusion process in the form of a stochastic differential equation (SDE) is written in the form:

$$\mathrm{d}X(t) = \mu(X(t), t)\mathrm{d}t + \sigma(X(t), t)\mathrm{d}W(t).$$

The concept of a diffusion process is briefly reviewed in Chapter 2 of this volume. In this case only one-dimensional time-homogeneous SDEs are considered. The drift and diffusion term are assumed to be on the form $\mu(x, \boldsymbol{\theta})$ and $\sigma(x, \boldsymbol{\theta})$ where $\boldsymbol{\theta}$ is a vector of parameters.

The log-likelihood process, $\Lambda(T)$, for a continuous-time diffusion process observed in the time interval $[0, T]$ is of the form:

$$\log(\Lambda(T)) = c + \int_0^T \frac{\mu(X(t), t)}{\sigma(X(t), t)^2}\mathrm{d}X(t) - \frac{1}{2}\int_0^T \frac{\mu(X(t), t)^2}{\sigma(X(t), t)^2}\mathrm{d}t. \quad (8.2)$$

Maximizing this expression, (8.2), with respect to the parameter vector $\boldsymbol{\theta}$ will yield the maximum-likelihood estimator, $\boldsymbol{\theta}_{ML}$, which will (in general) be a function of the entire path of $X(t)$ in the time interval $[0, T]$. For some special process analytical solutions of the maximization of (8.2) are available (Kutoyants 2004). In most cases it is not possible to write (8.2) in closed form, let alone find a simple analytical form of the maximum-likelihood estimator. In practical cases the process, $X(t)$, is only observed at discrete time points, $t_1, \ldots, t_n$. A discretizised version of (8.2) is:

$$\sum_{i=2}^{T} \frac{\mu(X(t_{i-1}), t_{i-1})}{\sigma(X(t_{i-1}), t_{i-1})^2}(X(t_i) - X(t_{i-1})) - \frac{1}{2}\sum_{i=2}^{T} \frac{\mu(X(t_{i-1}), t_{i-1})^2}{\sigma(X(t_{i-1}), t_{i-1})^2}\Delta_i,$$
$$\Delta_i = t_i - t_{i-1}. \quad (8.3)$$

Equation (8.3) represents a particular approximation of the true likelihood. For calculating the true likelihood of a discretely observed diffusion it is necessary to have a way of calculating its transition density, $f(X(t + \Delta)|X(t))$. The transition density of a diffusion process $X(t)$ is the density function of random-variable $X(t + \Delta)$ conditioned on its value $x_0 = X(t)$ at time t. For most diffusions no analytical form of the transition density is known. The idea of Aït-Sahalia (1999, 2002) is to utilize the fact that the transition density, $f(x|x_0, \Delta)$ of a diffusion satisfies the Kolmogorov forward equation,

$$\frac{\partial f(x|x_0, \Delta)}{\partial \Delta} + \frac{\partial(\mu(x, \boldsymbol{\theta}) f(x|x_0, \Delta))}{\partial x} - \frac{1}{2}\frac{\partial^2(v(x, \boldsymbol{\theta}) f(x|x_0, \Delta))}{\partial x^2} = 0, \quad (8.4)$$

where $v(x) = \sigma^2(x)$. The technique is based on doing a Taylor expansion of $f(x|x_0, \Delta)$ in Δ around 0. Instead of working with f directly it is possible to work with $l(x|x_0, \Delta)) = \log(f(x|x_0, \Delta))$. That way it is ensured that the approximation of the density will stay positive. Substituting $\exp(l(x|x_0, \Delta)$ into Equation (8.4) we see that $l(x|x_0, \Delta)$ will satisfy a new partial differential equation:

$$\frac{\partial l(x|x_0, \Delta)}{\partial \Delta} + \mu'(x) + \mu(x)\frac{\partial l(x|x_0, \Delta)}{\partial x} - \frac{1}{2}\left[\frac{\partial l(x|x_0, \Delta)}{\partial x}\right]^2 - \frac{1}{2}\frac{\partial^2 l(x|x_0, \Delta)}{\partial x^2} = 0. \tag{8.5}$$

Then $l(x|x_0, \Delta)$ is Taylor expanded around $\Delta = 0$,

$$\begin{aligned} l(x|x_0, \Delta) = & -\frac{1}{2}\log(2\pi\Delta) - \frac{1}{2}\log(v(x)) + \frac{c_{-1}(x|x_0)}{\Delta} \\ & + c_0(x|x_0) + c_1(x|x_0)\Delta + c_2(x|x_0)\frac{\Delta^2}{2!} + \cdots \\ & + c_k(x|x_0)\frac{\Delta^k}{k!} + R_k(\Delta, x|x_0). \end{aligned} \tag{8.6}$$

Substituting (8.6) into (8.5) and matching powers of Δ gives a system of differential equations. A brief review of the system is given in Appendix A.

Each set of functions $\mu(x)$ and $\sigma(x)$ have their Taylor expansion. If a set of observations of the diffusion process, $x(t_1), x(t_2), \ldots, x(t_n)$ is obtained then the log-likelihood can be approximated with l^*, and the optimization problem:

$$\max_{\boldsymbol{\theta}} l^*(\boldsymbol{\theta}|X(t_1), \ldots, X(t_n)),$$

is solved with numerical methods. The quality of the approximation depends on the number of terms in the expansion, Δ, $(x - x_0)$, and some properties of μ and σ.

Remark There exist many equivalent parameterizations of the diffusion processes. For example a Box–Cox transformation of the CKLS process has unit diffusion coefficient. If $X(t)$ is CKLS and $Y(t)$ is defined as:

$$Y(t) = \frac{1}{\sigma}\frac{X(t)^{1-\rho}}{1-\rho}.$$

Then the diffusion $Y(t)$ will have unit diffusion with drift:

$$\begin{aligned} \mu_Y(y) = & \tfrac{\kappa}{\sigma}\left(((\rho-1)\sigma y)^{\frac{1}{1-\rho}}\right)^{-\rho}\left(\alpha - ((\rho-1)\sigma y)^{\frac{1}{1-\rho}}\right) \\ & -\tfrac{1}{2}\rho\sigma\left(((\rho-1)\sigma y)^{\frac{1}{1-\rho}}\right)^{\rho-1}. \end{aligned} \tag{8.7}$$

A process with unit diffusion coefficients may be easier to deal with, e.g. with nonlinear least squares software, but the interpretation of the parameters in (8.7) is less transparent than in the standard CKLS representation. In the CKLS case the Box–Cox transformation, aimed to make the process 'more normal' is an example of a transformation that converts the diffusion function into a constant. Such a transformation is available for all one-dimensional diffusion processes. See Appendix A.

8.3 Illustration of numerical properties

Results by Jensen and Poulsen (2002), Lindström (2004) and Aït-Sahalia (1999, 2002) show that estimation methods based on the Taylor approximations described in Section 8.2 seem to perform favourably to alternative methods for many simple models used in the financial literature. A popular class of models is the class of the mean-reverting models. The idea is that there is a force that pushes the series towards an equilibrium value. Some further details are described in Chapter 2. A simple form is:

$$\mathrm{d}X(t) = \kappa(\alpha - X(t))\mathrm{d}t + \sigma(X(t))\mathrm{d}W(t),$$

which has a solution:

$$X(t) = x_0 + \int_{t_0}^{t} \kappa(\alpha - X(s))\mathrm{d}s + \int_{0}^{t} \sigma(X(s))\mathrm{d}W(s). \tag{8.8}$$

It is easy to show that the conditional expectation of $X(t)$ given by equation (8.8) is:

$$E(X(t)|X(t_0) = x_0) = \alpha + exp(-\kappa t)(x_0 - \alpha). \tag{8.9}$$

The interpretation is straightforward, α is the long-term equilibrium and κ describes the speed of convergence to this equilibrium. The half-time of a deviance from equilibrium in equation (8.9) is $\log(2)/\kappa$. The function $\sigma(x)$ controls the stochastic impact. For some simple forms of $\sigma(x)$ it is possible to derive a closed form of the conditional variance $V(X(t)|X(t_0) = x_0)$. In general for the CKLS model of equation (8.1), not much can explicitly written down except for the mean. Analytically, $\rho = 1/2$ and $\rho = 1$ are a little bit easier. For $\rho = 1/2$, the CIR model, the theoretical likelihood function is known in closed form. For $\rho = 1$ it is possible to calculate the conditional mean and variance, i.e. $E(X(t)|X(s))$ and $V(X(t)|X(s))$ for $t > s$.

The process $X(t)$ is sampled at time points $t_1, \ldots, t_n$, and generates data in the form:

$$x(t_1), \ldots, x(t_n), \quad t_1, \Delta_2 = t_2 - t_1, \ldots, \Delta_n = t_n - t_{n-1}.$$

In traditional statistical textbooks consistent estimates are obtained by letting $n \to \infty$. In the continuous-time environment things are more complicated and it is needed that both $\Delta_i \to 0$ and $t_n - t_1 \to \infty$. Letting $n \to \infty$, but say, keeping $t_n - t_1 = 1$ would give consistent estimates on $\sigma(x)$ but not on the drift, and similarly, if all the Δ_i's are large, the estimate of the diffusion function $\sigma(x)$ is likely to be poor.

To illustrate the usefulness of the Taylor approximation of the likelihood and the sensitivity to time span and sampling interval a brief simulation study was performed. The Taylor expansions mentioned in Section 8.2 were programmed in FORTRAN, and the maximization of the likelihood was performed numerically with the R-optimization routine *optim* which allows constraints on the parameters. The CKLS process of equation (8.1) was simulated using the Milstein scheme (Kloeden and Platen 1992). The parameter values were constrained to realistic values, κ, α, σ were bounded away from zero, ρ was bounded away from 1/2. The setup of the simulation was as follows:

- The time spans used are $T = 1, 10, 100$.
- $\Delta = 1$, 0.1 and 0.01 are used.
- Ten process points per observation were simulated by the Milstein scheme.
- $\kappa = 0.24$, $\alpha = 0.07$, $\sigma = 0.08838$, $\rho = 0.75$ (values from CKLS and others).

Results based on 25 replications are shown in Tables 8.1–8.4. It is clear from the tables that the quality of parameter estimates in the drift function, κ and α increases with the length of the time interval T. It is also clear from the tables that the quality of the parameters in the diffusion function, σ and ρ, increase with increased sampling frequency, and smaller Δ. A sample of size

Table 8.1 Average estimates, for $\Delta = 0.01$.

	$T = 1$	$T = 10$	$T = 100$
$\hat{\kappa}$	3.2742	0.5352	0.2709
$\hat{\alpha}$	0.0793	0.0962	0.0695
$\hat{\sigma}$	0.1115	0.0979	0.0899
$\hat{\rho}$	0.7695	0.7732	0.7570

Table 8.2 Average estimates, for $T = 100$.

	$\Delta = 1$	$\Delta = 0.1$
$\hat{\kappa}$	0.8232	0.2916
$\hat{\alpha}$	0.0644	0.0744
$\hat{\sigma}$	0.0984	0.0864
$\hat{\rho}$	0.7342	0.7299

Table 8.3 Standard deviation of simulations, for $\Delta = 0.01$.

	$T = 1$	$T = 10$	$T = 100$
s.d. $\hat{\kappa}$	1.8703	0.3995	0.0693
s.d $\hat{\alpha}$	0.0234	0.1269	0.0048
s.d. $\hat{\sigma}$	0.0542	0.0309	0.0042
s.d $\hat{\rho}$	0.2794	0.1084	0.0176

Table 8.4 Standard deviation of simulations, for $T = 100$.

	$\Delta = 1$	$\Delta = 0.1$
s.d. $\hat{\kappa}$	0.4680	0.1807
s.d $\hat{\alpha}$	0.0099	0.0248
s.d. $\hat{\sigma}$	0.0510	0.0196
s.d $\hat{\rho}$	0.2190	0.0851

100, $\Delta = 1$ and $T = 100$ is more informative about κ than an sample of size 100 with $\Delta = 0.01$ and $T = 1$.

Just as in ARMA and many other models, identifiability is an issue. In the general CKLS it will be hard to distinguish between ρ and σ if both are large. An intuitive explanation is as follows. If we have a diffusion of the above form then the longterm average is 0.07. Then the diffusion term is frequently close to:

$$\sigma 0.07^{\rho} \left(\frac{x}{0.07}\right)^{\rho}.$$

If ρ is increased and x close to its mean, the effect on the diffusion function rarely affected if σ is increased accordingly. In more complex diffusion models it seems reasonable that this issue will become more complicated.

Under certain conditions there exists an invariant stationary distribution of the diffusion. When it exists, its density is of the form:

$$f(x) \propto \frac{1}{\sigma^2(x)} \exp\left(\int_c^x \frac{2\mu(s)}{\sigma^2(s)} ds\right).$$

For the case $\rho = 1/2$, it is possible to write down the transition density. The stationary distribution will be a gamma distribution:

$$f(x) \propto x^{\frac{2\kappa\alpha}{\sigma^2}-1} \exp\left(-\frac{2\kappa}{\sigma^2}x\right),$$

with

$$\text{mean} = \alpha \text{ and variance } = \frac{\sigma^2}{2\kappa}.$$

The parameter κ controls the speed of convergence to the mean, a low value means slow convergence and therefore a large unconditional variance. If $2\kappa\alpha/\sigma^2 > 1$ the CIR process will never hit zero, the invariant distribution will be a gamma distribution and all moments of exist. The case for $\rho > 1/2$ is slightly different and it will not be possible to write down a closed form transition density. The case $\rho = 1$ is analytically convenient. For that case the invariant distribution will be:

$$f(x) \propto \underbrace{x^{-\frac{2\kappa}{\sigma^2}-2}}_{A} \underbrace{\exp\left(-\frac{2\kappa\alpha}{\sigma^2 x}\right)}_{B}. \tag{8.10}$$

The distribution in Equation (8.10) is the inverse gamma distribution. If $2\kappa\alpha/\sigma^2 > 0$ it has mean α and if $2\kappa/\sigma^2 > 1$ it will have a variance. For large values of x, part B in Equation (8.10) will be close to 1, and the behaviour of the density will be dominated by part A which is a Pareto-type tail. The existence of moments is therefore analogous to the Pareto distribution. Due to this fact, for some values of the parameters only a few moments might exist. Therefore extreme observations are likely to occur and due to the dynamic properties of the process, observations close in time to the extreme observation are likely to be extreme as well.

When $1/2 < \rho < 1$ all moments of the invariant distribution exists, but if κ is small, the behaviour of the process can be quite similar to a heavy-tail case. For example, if $\rho = 3/5$ then the invariant distribution is:

$$f(x) \propto \underbrace{\frac{1}{\sigma^2 x^{6/5}}}_{A} \underbrace{\exp\left\{-\left[\frac{5\kappa(4\alpha + x)}{2\sigma^2 x^{1/5}}\right]\right\}}_{B}. \tag{8.11}$$

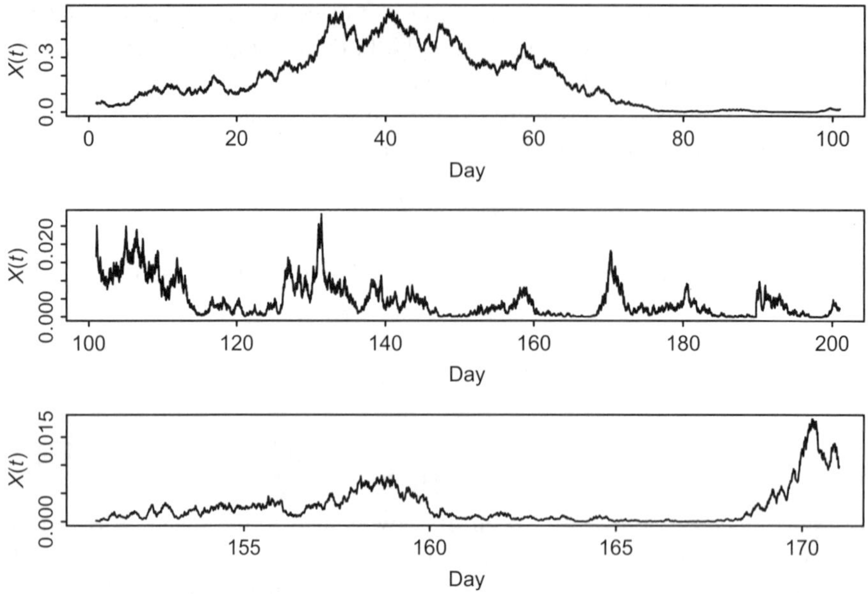

Figure 8.1 A short horizon view of a semi-heavy tail process. The top figure shows the first 100 days, the middle one days 101–200 and the bottom one shows a 20 day period from day 151–170.

Part B, the exponential part, in Equation (8.11) ensures the existence of moments. However, for a large range of x-values the impact of the exponential part is very small. Say, if $\alpha = 0.05$, then most of the time the $X(t)$ process stays in low values. If κ is low the effect of the exponential part of (8.11) is very small and part A dominates the behaviour of the invariant distribution. In this particular case the term A is similar to a (heavy-tailed) Pareto distribution. Therefore if κ is very small, extreme values are likely to be somewhat persistent. As illustration a simulation experiment with $\kappa = 0.01, \alpha = 0.05, \sigma = 0.09, \rho = 0.6$ was performed. A plausible path of 4000 days (time units) is illustrated in Figures 8.1 and 8.2. For long periods the process stays below 0.02 and even below 0.01, i.e. corresponding to interest rates of 1–2%. Occasionally it climbs up to 30–40%. In the period between day 500 and day 1000 it rockets off to several hundred percent, i.e. a factor hundred times its usual value. When a long horizon (4000 days) is viewed on the last graph on Figure 8.2, it seems that the extreme period was short. From the above, it is clear that the CKLS can generate a large variety of paths.

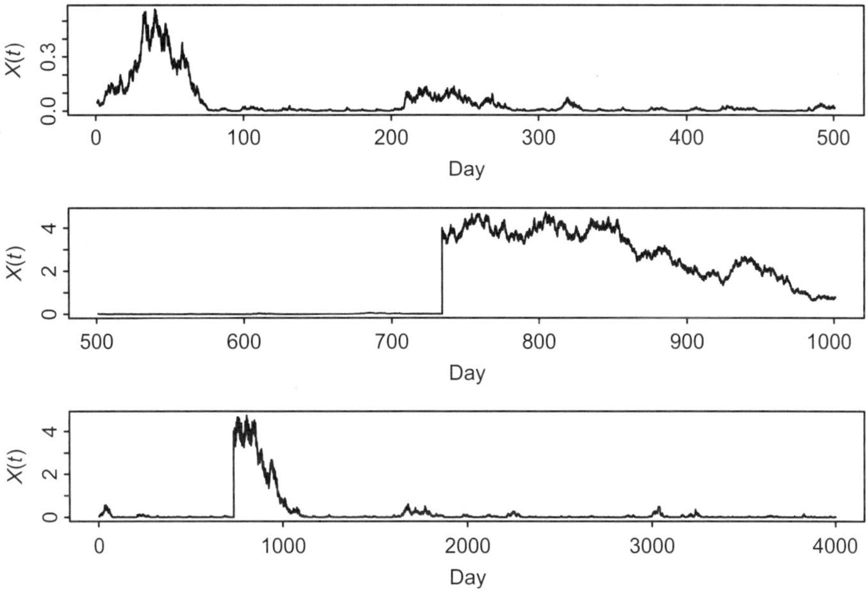

Figure 8.2 A long horizon view of a semi-heavy tail process, The top figure shows the first 500 days, the middle one days 500–1000, and the bottom one shows days 1–4000.

8.4 Implementation of surveillance tools

The tools of numerical approximation of the likelihood function of a diffusion process offer the possibility of applying standard surveillance tools, such as the CUSUM and Shiryaev–Roberts. Here the notation of Shiryaev (2002) for a continuous-time setup is used. Shiryaev (2002) and Srivastava and Wu (1993) give analytical results for the case where there is a change at time τ in the drift of a Wiener process.This turns out to be an analytically tractable case,

$$X(t) = r(t-\tau)I_{[\tau,\infty[}(t) + \sigma W(t),$$

$$\mathrm{d}X(t) = rI_{[\tau,\infty[}(t) + \sigma \mathrm{d}W(t),$$

where $W(t)$ is a standard Wiener process (Brownian motion). Following the notation of Shiryaev (2002), two probability measures, P_τ and P_∞ are to be compared. The probability measures represent a shift in the model at time τ and a model with no shift ($\tau = \infty$), respectively. The system is observed up to time t. The principal tool is the likelihood-ratio process:

$$L(t) = \frac{\mathrm{d}P_\tau}{\mathrm{d}P_\infty}[t, X(t)],$$

where $\mathrm{d}P_\tau/\mathrm{d}P_\infty$ is the Radon–Nikodym derivative. The CUSUM statistic in continuous time is:

$$\log(\gamma(t)) = \log\left(\max_{\tau\le t}\frac{L(t)}{L(\tau)}\right). \tag{8.12}$$

The Shiryaev–Roberts (SR) statistic in continuous time is:

$$\psi(t) = \int_0^t \frac{L(t)}{L(\tau)}\mathrm{d}\tau. \tag{8.13}$$

In this case it is easily derived by substituting into the log-likelihood process formula, Equation (8.2), that the log-likelihood ratio process, for shift at $\tau = 0$, is:

$$\log\left[L(t)\right] = \frac{r}{\sigma^2}X(t) - \frac{r^2}{2\sigma^2}t.$$

In general $\log\left[L(t)\right]$ will be a function of an integral involving the whole path of $X(t)$. The processes $\gamma(t)$ and $\psi(t)$ are in general dynamic processes with some dependency structure. In the case of the Wiener process, only the value $X(t)$ and some deterministic function of t enter formula of the likelihood-ratio process, which makes analytical derivations possible. By the use of Ito's formula, Srivastava and Wu (1993) give the dynamics of $\psi(t)$ in the Wiener process case:

$$\mathrm{d}\psi(t) = \mathrm{d}t + \frac{r}{\sigma^2}\psi(t)\mathrm{d}X(t).$$

Srivastava and Wu (1993) show that the distribution of the CUSUM process, $\log\left[\gamma(t)\right]$, is in the case of a shift of drift in a Wiener process, the same as for $|Z(t)|$ where:

$$\mathrm{d}Z(t) = \frac{r}{2}\mathrm{sgn}(t-\tau)\mathrm{sgn}(Z(t))dt + \mathrm{d}W(t),$$

and sgn is the sign operator.

In the Wiener process with drift case, both the CUSUM process and the SR process can therefore be described by a stochastic differential equation. To the author's knowledge dynamic forms of $\gamma(t)$ and $\psi(t)$ are not known explicitly for other cases. In a recent article by Baron and Tartakovsky (2006) it is stated that 'for continuous-time models, beyond the problem of detecting a change in the drift of a Brownian motion, little is known about the properties of CUSUM and SR procedures'. Baron and Tartakovsky (2006) also review some asymptotic optimality questions on the change-point detection problem.

In this article the focus is on empirical methods and computations. In practical cases the process, $X(t)$, is observed at discrete time points, $t_1, \ldots, t_n$. Here, no particular sampling strategy is assumed. Srivastava and Wu (1994) consider the possibility of dynamic sampling in the case of a shift in drift of a Wiener process. In financial data it is also conceivable that the sampling time-points are generated by separate trading process. Here the trading process is assumed to be exogenous. In the notation of Shiryaev (2002) the discretizesed version of equations (8.12) and (8.13), for sampling points, $t_1, t_2, \ldots, t_n$, are:

$$\log(\gamma(t_k)) = \log(f_1(x(t_1), \ldots, x(t_k))) - \min_{0 \le \tau \le t_k} (\log(f_0(x(t_1), \ldots, x(\tau)))), \tag{8.14}$$

and

$$\psi(t_k) = \sum_{j=1}^{k} \frac{f_1(x(t_1), \ldots, x(t_k))}{f_0(x(t_1), \ldots, x(t_j))}(t_k - t_{k-1}), \tag{8.15}$$

where f_0 and f_1 are the densities before and after time τ, respectively. Frisén (2003) gives some optimality properties of these statistics. The alarm rules are: alarm if $\gamma(t_k) >$ constant or $\psi(t_k) >$ constant. Alarm rules will be of the form, alarm at time τ_A such that:

$$\tau_A = \inf\{t > 0; \gamma(t) > d\} \quad \text{or} \quad \tau_A = \inf\{t > 0; \psi(t) > d^*\}.$$

The constants d and $d*$ depend on which models are compared, the duration between measurements (Δ_i's), and the false alarm strategy. Illustration of some numerical properties of the maximum-likelihood estimator given in Section 8.3 shows that the amount of information about the parameters in the drift function is primarily a function of the time span, whereas information about parameters in the diffusion function increases when density of observation increases. It is therefore, intuitively clear that the nature of increased sampling frequency will affect the characteristics of the alarm statistics $\gamma(t)$ and $\psi(t)$.

The control limits, d and d^*, have to be decided according to some alarming principles like ARL_0. A way is to use quantiles of $\gamma(t)$ and $\psi(t)$ or choose a particular length of ARL_0. These depend on the distributions of the alarm statistics, $\gamma(t)$ and $\psi(t)$, that are not available in closed form so, they have to be decided by simulations.

8.5 Calculation of alarm statistics

A nice feature of the CKLS is the ease of interpretation of parameters. κ is the speed of convergence to the mean, α, σ is (partly) a scale parameter and ρ

(partly) rules the tail behaviour of the distribution. For $\rho < 1$ all moments of the invariant distribution exist but only some in the case $\rho \geq 1$. A lot of quantitative finance focuses on estimation of variance, but also tail probabilities are of interest. As seen in a previous section, CKLS can generate some interesting heavy tail patterns.

A small simulation study for illustrating the behaviour of the SR and CUSUM surveillance tools was performed and some of the results are shown in this section. The alarm rules depend on the parameter values of the competing models as well as the sizes of the sampling intervals. To illustrate the nature of this dependence some numerical experiments were performed. A base model, M_0, was set up:

$$M_0 : \boldsymbol{\theta} = (\kappa = 0.1, \alpha = 0.05, \sigma = 0.1, \rho = 0.75)$$

versus a shift in ρ

$$M_1 : \boldsymbol{\theta} = (\kappa = 0.1, \alpha = 0.05, \sigma = 0.1, \rho = 1.5)$$

versus a shift in κ

$$M_2 : \boldsymbol{\theta} = (\kappa = 0.05, \alpha = 0.05, \sigma = 0.1, \rho = 0.75).$$

These models were chosen to contrast the behaviour due to a shift in a parameter in the drift function to a shift of parameter in the diffusion function. The shift of $\rho = 0.75$ to $\rho = 1.5$ represents a shift from a light tail invariant distribution to a heavy tailed one. The shift of κ from 0.1 to 0.05 describes the situation of that the half-life of a deviance from the long-term mean ($\alpha =$ 0.05) goes from $\log(2)/0.1 \simeq 7$ to $\log(2)/0.05 \simeq 14$, so if time is measured in days, this means that the half-length of a cycle goes from one week to two weeks. Examples of shifts from M_0 to M_1 and M_2, respectively, are shown in Figures 8.3 and 8.4. The shift from M_0 is in some sense a more drastic one. The equilibrium distribution shifts from being a light tailed one to being a heavy tailed one. Risk-managers would like to avoid the heavy tailed case. The shift from M_0 to M_2 is a more moderate case. Financial analyst sometimes claim that bubble size is increasing. Analysing cases such as M_2 is a way of formalizing such a statement.

For a practical analysis of surveillance of a continuous-time process important factors are the sampling frequency and the time-span. To illustrate the behaviour of $\gamma(t)$ and $\psi(t)$ after a break has occurred, a simulated process was generated with the models M_1 and M_2, respectively, and the hypothetical parameter shift supposed to take place at time $\tau = 0$. The Milstein scheme was used to generate 100 points per day (time-unit). Two types of sampling were compared, two observations per day, $\Delta = 0.5$, and ten observations per day $\Delta = 0.1$.

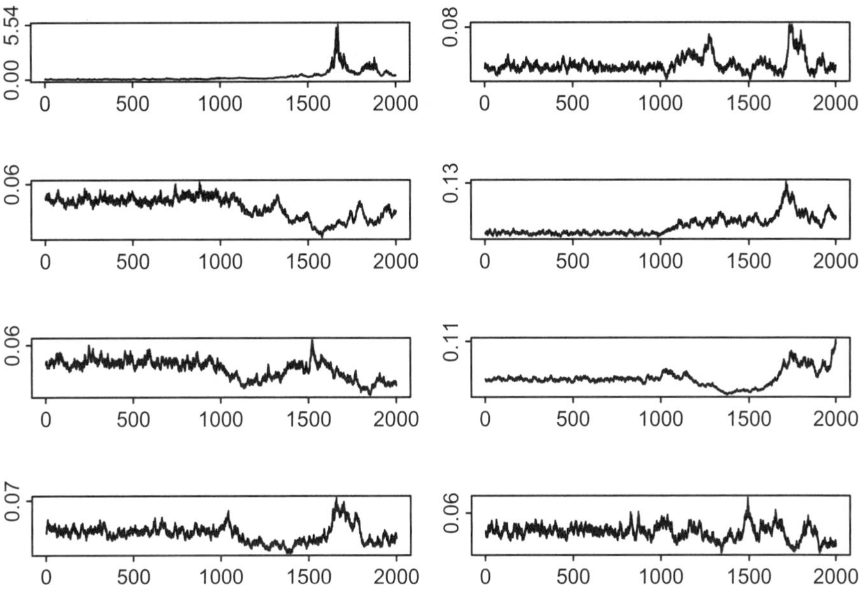

Figure 8.3 Some examples of a shift from model M_0 to M_1 at $\tau = 1000$.

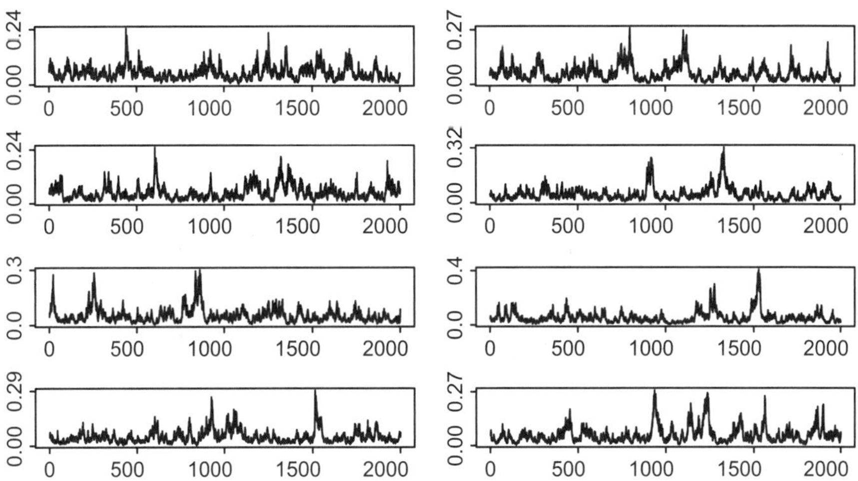

Figure 8.4 Some example of a shift from model M_0 to M_2 at time $\tau = 1000$.

Of course the true $\gamma(t)$ and $\psi(t)$ cannot be observed as they are functions of the entire sample path of $X(t)$ and $X(t)$ is only observed at discrete time points. Therefore the estimates $\hat{\gamma}(t)$ and $\hat{\psi}(t)$ are calculated based on Equations (8.14) and (8.15). In this section these estimates, and the true processes, will be denoted as $\gamma(t)$ and $\psi(t)$. The quality of the estimates will depend on Δ and the models compared. Numerically the estimate of $\psi(t)$, based on Equation (8.15) can be zero, especially when no break takes place. Therefore, the graphs illustrating the no-change case in this section $\log(\psi(t)+1)$ are plotted instead of $\log(\psi(t))$.

In Figures 8.5 and 8.6 the processes $\log(\gamma(t))$ and $\log(\psi(t)+1)$ are plotted over a period of 10 years. In Figure 8.5 the data are generated by M_0 and CUSUM and SR calculated by reference to M_1. In Figure 8.6 the same data are generated by M_0 and CUSUM and SR are calculated by reference to M_2. From the figures it is apparent that the CUSUM and SR tend to agree on the alarm. As seen in Figures 8.5 and 8.6, both $\log(\gamma(t))$ and $\log(\psi(t))+1)$ seem to be mean–variance stationary (as expected) in the case of no parameter-shift. In Figure 8.6 there are clear signs of autocorrelation in both CUSUM and SR. If the state of alarm is defined by a high value of CUSUM or SR, then the system will be in the state of alarm for days when monitoring κ, whereas when monitoring ρ the alarm periods are very short, e.g. one day. If the system has been out of alarm for a day, the waiting time for the next alarm is empirically similar to an exponential distribution for the case of ρ.

In contrast to the no-change cases described in Figures 8.5 and 8.6, there will be an upward trend in $\log\left[\gamma(t)\right]$ and $\log\left[\psi(t)\right]$, when M_0 is no longer

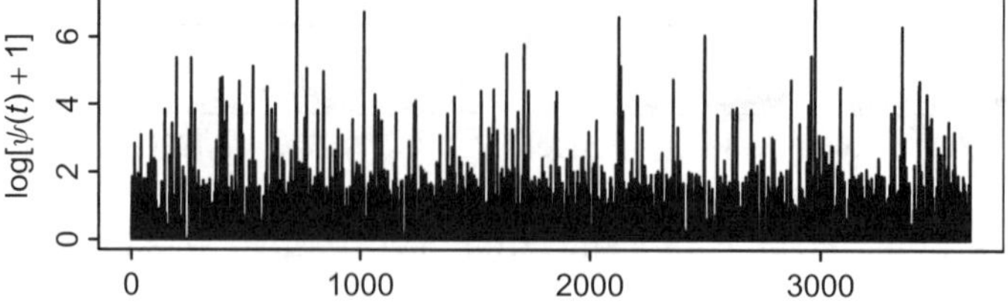

Figure 8.5 CUSUM $= \log\left[\gamma(t)\right]$ and $\log(\text{SR}) = \log\left[\psi(t)+1\right]$ for true model M_0 versus M_1.

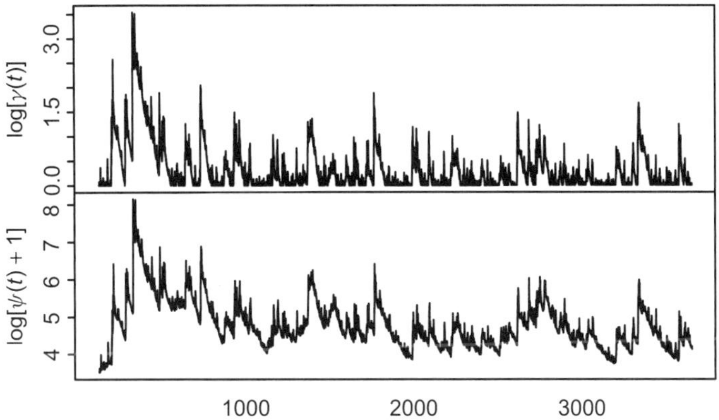

Figure 8.6 CUSUM $= \log\left[\gamma(t)\right]$ and $\log(\mathrm{SR}) = \log[\psi(t) + 1]$ for true model M_0 versus M_2.

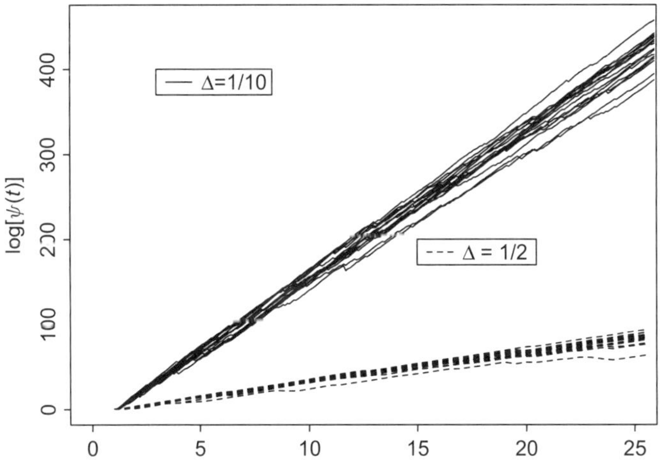

Figure 8.7 Effects of sampling density on $\log[\psi(t)]$. Sixteen replications of true model M_1 against M_0 for $\Delta = 1/2$ and $\Delta = 1/10$.

the true model. The CUSUM and SR process seem therefore to behave as they should. Except for the Wiener process with drift case, there are no closed form results for the distributional behaviour of CUSUM and SR, therefore, for exact evaluation dynamic properties, ARL_0, quantiles, etc., of $\gamma(t)$ and $\psi(t)$, extensive simulation is needed.

The results are as to be expected. In Figures 8.7 and 8.8 $\log[\psi(t)]$ and $\log\left[\gamma(t)\right]$ are plotted for 16 replications of the case when M_1 is the model

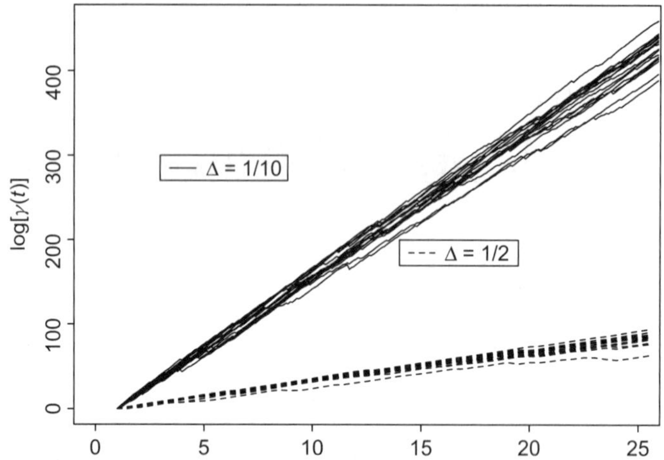

Figure 8.8 Effects of sampling density on $\log[\gamma(t)]$. Sixteen replications of true model M_1 against M_0 for $\Delta = 1/2$ and $\Delta = 1/10$.

and M_0 the reference model. In all 16 cases there is a clear upward trend and comparing to Figure 8.5 a decisive alarm is available in a few days. A small Δ will result in a quicker detection of change. For both CUSUM and SR the trend is approximately three units per day for $\Delta = 1/2$ and 15 units per day for $\Delta = 1/10$. That is, an increase of sampling frequency by a factor of 5 increases the accumulation of information by approximately a factor of 5. In contrast it will take a long time to detect a shift in κ. An increased sampling frequency adds very little information about κ. An intuitive explanation is that a value of κ of the size 0.1 can easily generate a bubble (cycle) of say 50–100 days. During such a period very little evidence is gained about an eventual change of cycle length, by increasing the sampling frequency. Therefore, to be able to detect a change in cycle (bubble) length, many cycles have to be observed, and it takes time to collect sufficiently many bubbles. In Figures 8.9 and 8.10 $\log\left[\gamma(t)\right]$ and $\log[\psi(t)]$ are plotted for a **single** replication of the case where M_2 is the true model and M_0 is the reference model. The impact of the size of Δ is virtually invisible. In this particular case Figures 8.9 and 8.10, show that it takes roughly 100–200 days before evidence against M_0 in favour of M_2 really starts to pile up. For each realization of true model M_2 versus M_0, CUSUM and SR behave somewhat like a step function indicating that there are long periods that are noninformative about κ. This behaviour was confirmed in other replications. This is natural, as change in bubble length cannot be detected if no bubbles take place.

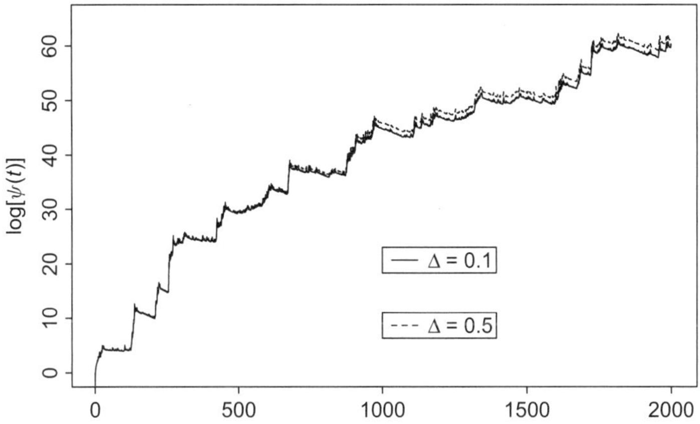

Figure 8.9 Effects of sampling frequency on $\log(\text{SR}) = \log[\psi(t)]$ for a shift from model M_0 to M_2 at $t = 0$.

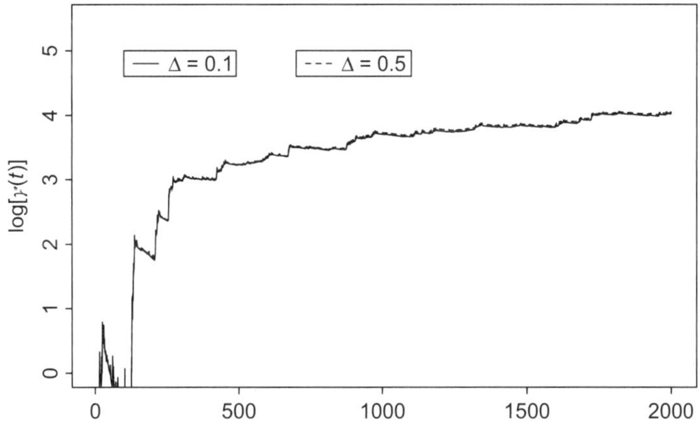

Figure 8.10 Effects of sampling frequency on CUSUM for a shift from model M_0 to M_2 at $t = 0$.

8.6 Discussion

The approximations of the likelihood function of a diffusion process offer the possibility of surveillance of parametric stability in CKLS type processes and also a large class of other univariate diffusions. Many of the popular mathematical finance models can be monitored with methods of this type. The parameters of a CKLS model are interpretable and it is important that analysts understand their nature. It is inherent in the CKLS model that it will be difficult to estimate the speed of convergence to an equilibrium, κ, when κ is a small number. It will also in some cases be difficult to separate the parameters, e.g. ρ and σ. This puts an extra responsibility on the analyst. The analyst has to make sensible restrictions on the parameters, and make preferences which features are of interest. The Taylor approximations of the likelihood together with modern computer programs on maximization and numerical integration offer a computational procedure for likelihood-based surveillance of continuous-time processes. Having a numerical approximation to the likelihood functions offers the possibility to apply standard surveillance tools such as CUSUM and SR. These tools have some optimality properties, see, Frisén (2003) and Baron and Tartakovsky (2006), but analytical results on how to calculate their dynamics have been limited. It seems evident that the properties of CUSUM and SR for CKLS models, such as worst-case scenarios, average run length, distribution of waiting times for alarms, etc., will have to be evaluated by simulation methods. The properties will depend on the parameter values as well as the sampling frequency. It seems that monitoring parameters in the drift function is difficult; nothing can replace a long period of observation, whereas monitoring parameters in the diffusion function can be improved by denser sampling of the process. That is good news, because from a financial point of view, risk management, option pricing, hedging, and the diffusion function are of vital importance. There are few analytical results available for likelihood analysis of continuous-time finance. Therefore, approximations and extensive simulations will be necessary. The popularity of mathematical finance has had the impact that there are now many individuals in the finance industry that have the knowledge and understanding of diffusion models. For them interpretation of parameters in a diffusion models is natural. A statistical tool for linking data to continuous-time modelling allows direct inference about parameters of interest in an particular model.

Appendix: Taylor expansions

The coefficient of Δ^{-1}, $c_{-1}(x|x_0)$ solves

$$
\begin{aligned}
&2c_{-1}(x|x_0) + v(x)\left[c'_{-1}(x|x_0)\right]^2 = 0 \\
&2\mu(x)c'_{-1}(x|x_0) - 2v(x)c'_0(x|x_0)c'_{-1}(x|x_0) \\
&\qquad -v'(x)c'_{-1}(x|x_0) - v(x)c''_{-1}(x|x_0) - 1 = 0.
\end{aligned}
$$

This gives:

$$c_{-1}(x|x_0) = -\frac{1}{2}\left(\int_{x_0}^{x} \frac{du}{\sqrt{v(u)}}\right)^2$$

$$c_0(x|x_0) = \int_{x_0}^{x} \left[2\mu(s)c'_{-1}(s) - c'_{-1}(s|x_0)v'(x) - v(s)c''_{-1}(s|x_0) - 1\right] /$$

$$\left[2v(s)c_{-1}(s|x_0)\right] ds.$$

The remaining equations, $j = 1, 2, \ldots$, to be solved are of the type:

$$c_j(x|x_0) + c'_j(x|x_0)k_j(x) = g_j(x)$$

where:

$$k_1(x) = -v(x)c'_{-1}(x)$$

$$k_2(x) = -\frac{1}{2}v(x)c'_{-1}(x)$$

$$k_3(x) = -\frac{1}{3}v(x)c'_{-1}(x)$$

$$k_4(x) = -\frac{1}{4}v(x)c'_{-1}(x)$$

and

$$g_1(x) = \quad \mu(x)c_0(x) \quad \mu'(x) - \frac{1}{8}\frac{v'(x)^2}{v(x)} + \frac{1}{2}v(x)c'_0(x)^2 + \frac{1}{2}$$

$$v'(x)c'_0(x) + \frac{1}{2}v(x)c''_0(x) + \frac{1}{4}v''(x) + \frac{1}{2}\frac{\mu(x)v'(x)}{v(x)}$$

$$g_2(x) = -\mu(x)c'_1(x) + v(x)c'_0(x)c'_1(x) + \frac{1}{2}v'(x)c'_1(x) + \frac{1}{2}v(x)c''_1(x)$$

$$g_3(x) = -\mu(x)c'_2(x) + v(x)c'_0(x)c'_2(x) + v(x)c'_1(x)^2$$

$$+\frac{1}{2}v'(x)c'_2(x) + \frac{1}{2}v(x)c''_2(x)$$

$$g_4(x) = -\mu(x)c'_3(x) + v(x)c'_0(x)c'_3(x) + 3v(x)c'_1(x)c'_2(x)$$

$$+\frac{1}{2}v'(x)c'_3(x) + \frac{1}{2}v(x)c''_3(x)$$

$$
\begin{aligned}
g_5(x) &= -\mu(x)c_4'(x) + v(x)c_0'(x)c_4'(x) + 3v(x)c_2'(x)^2 \\
&\quad +4v(x)c_1'(x)c_3'(x) \\
&\quad +\frac{1}{2}v'(x)c_4'(x) + \frac{1}{2}v(x)c_4''(x) \\
g_6(x) &= -\mu(x)c_5'(x) + v(x)c_0'(x)c_5'(x) \\
&\quad +5v(x)c_1'(x)c_4'(x) + 10v(x)c_2'(x)c_3'(x) \\
&\quad +\frac{1}{2}v'(x)c_5'(x) + \frac{1}{2}v(x)c_5''(x) \\
&\vdots
\end{aligned}
$$

From elementary calculus it is well known that:

$$c_j(x|x_0) = e^{-A(x)} \int_{x_0}^{x} \frac{g_j(s)}{k_j(s)} e^{A(s)} ds \tag{A.1}$$

$$A(x) = \int_{x_0}^{x} \frac{1}{k(s)} ds. \tag{A.2}$$

Solving these differential equations sequentially by integrating (A.1) can in some cases be problematic. A working solution is to Taylor expand $c_j(x|x_0)$ in x around x_0 and substitute the Taylor expansion instead of the functions c_j. A trick that is sometimes useful is to rewrite the diffusion such that $v(x) = 1$. For a one-dimensional diffusion this is always possible by using the transformation $y = h(x)$, defined by:

$$Y(t) = \int_{x_0}^{X(t)} \frac{du}{\sigma(u)}.$$

Then by Ito's lemma the dynamics of the process $Y(t)$ is described with:

$$
\begin{aligned}
\mathrm{d}Y(t) &= \mu_Y(Y(t))\mathrm{d}t + \mathrm{d}W(t) \\
\mu_Y(x) &= \frac{\mu_X(x)}{\sigma(x)} - \frac{\sigma'(x)}{2} \qquad \text{with } x = h^{-1}(y).
\end{aligned}
$$

For example if we have a CIR process:

$$
\begin{aligned}
\mathrm{d}X(t) &= \kappa\,[\alpha - X(t)]\,\mathrm{d}t + \sigma\sqrt{X(t)}\mathrm{d}W(t) \\
h(x) &= \int^{x} \frac{du}{\sigma\sqrt{u}} = \frac{2}{\sigma}\sqrt{x}, \quad h^{-1}(y) = \frac{\sigma^2 y^2}{4}
\end{aligned}
$$

$$
\begin{aligned}
\mathrm{d}Y(t) &= \mu_Y\,[Y(t)]\,\mathrm{d}t + \mathrm{d}W(t)\\
\mu_Y(y) &= \frac{4\kappa\alpha - \sigma^2}{2\sigma^2 y} - \frac{\kappa y}{2}.
\end{aligned}
$$

This transformation makes calculations of the Taylor expansion and computer programming simpler, but it can also cause numerical difficulties in some cases. In practical cases both transformed and untransformed calculations should be done for validation. For more details consult Aït-Sahalia (1999) and Aït-Sahalia (2002).

References

Aït-Sahalia, Y. (1999). Transition densities for interest rate and other nonlinear diffusion. *Journal of Finance*, **54**(4), 1361–1395.

Aït-Sahalia, Y. (2002). Maximum likelihood estimation of discretely sampled diffusions: A closed-form approximation approach. *Econometrica*, **70**(1), 223–262.

Baron, M. and Tartakovsky, A. G. (2006). Asymptotic optimality of change-point detection schemes in general continuous-time models. *Sequential Analysis*, **25**, 257–296.

Broemeling, L. D. and Tsurumi, H. (1987). *Econometrics and Structural Change*. Marcel-Dekker, New York.

Chan, K. C., Karolyi, G. A., Longstaff, F. A. and Sanders, A. B. (1992). An empirical comparison of alternative models for the short-term interest rate. *Journal of Finance*, **47**, 1209–1228.

Cox, J. C., Ingersoll, J. E. and Ross, S. A. (1985). A theory for the term structure of interest rates. *Econometrica*, **53**, 385–407.

Frisén, M. (2003). Statistical surveillance. Optimality and methods. *International Statistical Review*, **71**(2), 403–434.

Hackl, P. (Ed.) (1989). *Statistical Analysis and Forecasting of Economic Structural Change*. Springer-Verlag, Berlin.

Hackl, P. and Westlund, A. H. (Eds.) (1991). *Economic Structural Change*. Springer-Verlag, Berlin.

Jensen, B. and Poulsen, R. (2002). Transition densities of diffusion processes: Numerical comparison of approximation techniques. *Journal of Derivatives*, **9**, 18–32.

Kloeden, P. E. and Platen, E. (1992). *Numerical Solution of Stochastic Differential Equations*. Springer-Verlag, Berlin.

Kutoyants, Y. (2004). *Statistical Inference for Ergodic Diffusion Processes*. Springer-Verlag, London.

Lindström, E. (2004). Statistical modeling of diffusion processes with financial applications. PhD thesis, Lund Institute of Technology.

Schmid, W. and Tzotchev, D. (2004). Statistical surveillance of the parameters of a one-factor Cox–Ingersoll–Ross model. *Sequential Analysis*, **23**(3), 379–412.

Shiryaev, A. N. (2002). Quickest detection problems in technical analysis of financial data. In Geman, H., Madan, D., Pliska, S. and Vorst, T. (Eds.), *Mathematical Finance: Bachelier Congress 2000*. Springer-Verlag, Berlin.

Srivastava, M. S. and Wu, Y. (1993). Comparison of EWMA, CUSUM and Shiryaev–Roberts procedures for detecting a shift in the mean. *The Annals of Statistics*, **2**, 645–670.

Srivastava, M. S. and Wu, Y. (1994). Dynamic sampling plan in Shiryaev–Roberts procedure for detecting a change in the drift of Brownian motion. *Annals of Statistics*, **22**(2), 805–823.

9

Conclusions and future directions

Marianne Frisén
Statistical Reseach Unit, School of Business, Economics and Law, Göteborg University, PO Box 660, SE 405 30 Göteborg, Sweden

Financial methods have developed from mathematics and probability theory to statistical methods for decision strategies. The knowledge gained during the last decade on the stochastic properties of models for finance provides a good background to the statistical inference needed in practice. There is a need to evaluate available information in order to make timely decisions. Statistical surveillance meets this need and will hence be an important tool for financial decisions.

There is a wide range of possible models to describe financial processes. The modelling is important. Here we give an overview and some general comments on general modelling procedures in Chapter 2. Also, each chapter contains some comments on the modelling of the specific applications described in the chapter. However, the main topic in this book is not modelling *per se*, but the use of models for surveillance.

Financial surveillance is a novel area of research. From the different chapters we can nevertheless conclude that there is much knowledge on surveillance of financial time series available already. The chapters give information about how to handle a wide range of situations. In Chapter 1 some general approaches are described, and these are applied in different ways in the following chapters. In Chapter 2 the model building aspects are treated. In Chapter 3 we conclude that knowledge about the properties of methods for technical analysis can be

Financial Surveillance Edited by Marianne Frisén

gained from the theory of statistical surveillance. This also gives a tool for construction of more efficient methods. Surveillance of risk indicators is handled in Chapters 4–7. We describe methods for simple independent series in Chapters 3 and 4, methods for more complicated ones with dependencies in Chapters 5–7 and methods for continuous time in Chapter 8. In Chapters 5–7 methods for multivariate surveillance are given. For each situation described, methods are provided, and for some there are also computer programs available. Thus, the time to use surveillance in practice is already here.

Different financial series are analysed in the different chapters. In Chapter 3 the Hang Seng index is analysed with respect to optimal times for trading. In Chapter 4 a period of Standard and Poor's 500 stock market index is analysed to investigate the ability of some suggested methods to detect a change in volatility. In Chapters 5 and 6 there are several analyses of the Morgan Stanley Capital International index (MSCI) for Germany. In Chapter 5 also the MSCI indices for the US and Japan are analysed. In Chapter 7 the optimal allocations of a portfolio of the US market, a non-US index, oil and gold are determined. From the wide variety of financial data series used in the chapters we can conclude that many different financial problems can be analysed by statistical surveillance.

The powerful tool of statistical surveillance will probably be used for various kinds of financial problems in the future. Financial analysis will become more quantitative. Judging from other areas, there may be reason to expect some initial misuses where the numerical and theoretical difficulties are emphasized and the real financial problem is neglected. The financial problem, in all its complexity, should always be in focus. The results produced by a method should always be interpreted within this full context. Eventually misuses will hopefully be avoided, and financial surveillance will work as a powerful tool.

The theory of statistical surveillance developed slowly to start with, as is seen in the section on the history of surveillance in Chapter 1. The advanced theory needed for advanced applications such as those in finance is a relatively new area. The theory of statistical surveillance will be further developed by applications in various fields apart from finance. A cross-fertilization back to financial surveillance could then be expected.

Here different chapters concentrate on different issues. Much research on statistical surveillance in finance still remains to be done. This book may be used as a starting point by combining features from different chapters. Statistical approaches in one chapter can be applied to the models described in another chapter. Several models mentioned in Chapter 2 include a change or can be modified to include a change. So far, surveillance methods are available for only a few of these models. In the future, new research on the development of methods for surveillance of financial models can be expected. The general approaches described in this book can be applied to new models. The likelihood of the specific model can be compared with corresponding model with a change in order to provide the basis for likelihood-based surveillance. There are also other such comparisons which could provide the basis for new methods.

A very rapid development of practice and research on financial surveillance can be expected in the future. Efficient computers and computer programs will play an important role in this development.

Bibliography

Acosta-Mejia, C. A. (1998). Monitoring reduction in variability with the range. *IIE Transactions*, **30**, 515–523.

Acosta-Mejia, C. A., Pignatello, J. J. J. and Rao, B. V. (1999). A comparison of control charting procedures for monitoring process dispersion. *IIE Transactions*, **31**, 569–579.

Acosta-Mejia, C. A. and Pignatiello, J. J. (2000). Monitoring process dispersion without subgrouping. *Journal of Quality Technology*, **32**, 89–102.

Albers, W. and Kallenberg, W. C. M. (2004a). Are estimated control charts in control? *Statistics*, **38**, 67–79.

Albers, W. and Kallenberg, W. C. M. (2004b). Estimation in Shewhart control charts: Effect and corrections. *Metrika*, **59**, 207–234.

Alexander, S. S. (1961). Price movements in speculative markets: Trends or random walks. *Industrial Management Review*, **2**, 7–26.

Alwan, L. and Roberts, H. (1988). Time-series modeling for statistical process control. *Journal of Business and Economic Statistics*, **6**, 87–95.

An, M. Y., Christensen, B. J. and Gupta, N. D. (2003). On pensions and retirement: Bivariate mixes proportional hazard modelling of joint retirement. Working paper no 163. Center of analytical finance, University of Aarhus, Aarhus Business School.

Andersen, P. K., Borgan, Ø., Gill, R. D. and Keiding, N. (1993). *Statistical Models Based on Counting Processes*. Springer-Verlag, Berlin.

Andersson, E. (2002). Monitoring cyclical processes. A non-parametric approach. *Journal of Applied Statistics*, **29**, 973–990.

Andersson, E. (2006) Robust On-line Turning Point Detection. The Influence of Turning Point Characteristics. *Frontiers in Statistical Quality Control 8*. Eds. Lenz, H.-J. and Wilrich, P.-TH., pp. 223–248

Andersson, E. (2007). Effect of dependency in systems for multivariate surveillance. Technical Report 2007:1, Statistical Research Unit, Department of Economics, Göteborg University.

Andersson, E. and Bock, D. (2001). On seasonal filters and monotonicity. Technical Report 2001:4, Department of Statistics, Göteborg University.

Andersson, E., Bock, D. and Frisén, M. (2004). Detection of turning points in business cycles. *Journal of Business Cycle Measurement and Analysis*, **1**, 93–108.

Andersson, E., Bock, D. and Frisén, M. (2005). Statistical surveillance of cyclical processes. Detection of turning points in business cycles. *Journal of Forecasting*, **24**, 465–490.

Financial Surveillance Edited by Marianne Frisén

Andersson, E., Bock, D. and Frisén, M. (2006). Some statistical aspects on methods for detection of turning points in business cycles. *Journal of Applied Statistics*, **33**, 257–278.

Andreou, E. and Ghysels, E. (2002). Detecting multiple breaks in financial market volatility dynamics. *Journal of Applied Econometrics*, **17**, 579–600.

Andreou, E. and Ghysels, E. (2004). The impact of sampling frequency and volatility estimators on change-point tests. *Journal of Financial Econometrics*, **2**, 290–318.

Andreou, E. and Ghysels, E. (2006). Monitoring disruptions in financial markets. *Journal of Econometrics*, **135**, 77–124.

Ansley, C. F. (1979). An algorithm for the exact likelihood of a mixed autoregressive moving average process. *Biometrika*, **66**, 59–65.

Arteaga, C. and Ledolter, J. (1997). Control charts based on order-restricted tests. *Statistics & Probability Letters*, **32**, 1–10.

Aït-Sahalia, Y. (1996). Non-parametric pricing of interest rate derivative securities. *Econometrica*, **64**, 527–560.

Aït-Sahalia, Y. (1999). Transition densities for interest rate and other nonlinear diffusion. *Journal of Finance*, **54**(4), 1361–1395.

Aït-Sahalia, Y. (2002). Maximum likelihood estimation of discretely sampled diffusions: A closed-form approximation approach. *Econometrica*, **70**(1), 223–262.

Bachelier, L. (1900). Theorie de la speculation. *Annales de l'Ecole Normale Superiore*, pages 21–86.

Baille, R. T., Bollerslev, T. and Mikkelsen, H. O. (1996). Fractionally integrated generalized autoregressive conditional heteroskedacity. *Journal of Econometrics*, **74**, 3–30.

Banerjee, A. and Urga, G. (2005). Modeling structural breaks, long memory and stock market volatility: an overview. *Journal of Econometrics*, **129**, 1–34.

Barndorff-Nielsen, O. E. and Shephard, N. (2001). Non-Gaussian Ornstein-Uhlenbeck-based models and some of their uses in financial economics. *Journal of the Royal Statistical Society: Series B (Statistical Methodology)*, **64**(2), 167–241.

Baron, M. and Tartakovsky, A. G. (2006). Asymptotic optimality of change-point detection schemes in general continuous-time models. *Sequential Analysis*, **25**, 257–296.

Bartlett, M. S. (1946). On the theoretical specification and sampling properties of autocorrelated time series. *Journal of the Statistical Society Supplement*, **8**, 27–41.

Beibel, M. (2000). A note on sequential detection with exponential penalty for the delay. *The Annals of Statistics*, **28**, 1696–1701.

Beibel, M. and Lerche, H. R. (1997). A new look at optimal stopping problems related to mathematical finance. *Statistica Sinica*, **7**, 93–108.

Bell, C., Gordon, L. and Pollak, M. (1994). An efficient nonparametric detection scheme and its application to surveillance of a Bernoulli process with unknown baseline. In *Change-Point Problems*, eds. E. Carlstein, H.-G. Muller and D. Siegmund, Hayward, California: IMS Lecture Notes, Monograph Series, pp. 7–27.

Benjamini, Y. and Hochberg, Y. (1995). Controlling the false discovery rate: A practical and powerful approach to multiple testing. *Journal of the Royal Statistical Society B*, **57**, 289–300.

Beran, J. (1994). *Statistics for Long-Memory Processes*. Chapman & Hall, London.

Beran, J. (1995). Maximum-likelihood estimation of the differencing parameter for invertible short and long ARIMA models. *Journal of the Royal Statistical Society Series B*, **57**(4), 659–672.

Bergstrom, A. R. (1988). The history of continuous-time econometric models. *Econometric theory*, **4**, 365–383.

Bergstrom, A. R. (1990). *Continuous-Time Econometric Modelling*. Oxford University Press, Oxford.

Beskos, A. and Roberts, G. O. (2005). Exact simulation of diffusions. *Annals of Applied Probability*, **15**, 2422–2444.

Beskos, A., Papaspiliopoulos, O., Roberts, G. O. and Fearnhead, P. (2006). Exact and efficient likelihood-based inference for discretely observed diffusion processes with discussion. *Journal of the Royal Statistical Society, series B*, **68**, 333–382.

Best, M. and Grauer, R. (1991). On the sensitivity of mean–variance-efficient portfolios to changes in asset means: Some analytical and computational results. *Review of Financial Studies*, **4**, 315–342.

Bhardwaj, G. and Swanson, N. R. (2006). An empirical investigation of the usefulness of ARFIMA models for predicting macroecnomic and financial time series. *Journal of Econometrics*, **131**, 539–578.

Bibby, B. M., Jacobsen, M. and Sørensen, M. (2004). Estimating functions for discretely sampled diffusion-type models. Preprint 2004-4, Department of Applied Mathematics and Statistics, University of Copenhagen.

Björk, T. (2004). *Arbitrage Theory in Continuous Time*. Oxford University Press, Oxford.

Black, F. (1976). The pricing of commodity contracts. *Journal of Financial Economics*, **3**, 167–179.

Black, F. and Scholes, M. (1973). The pricing of options and corporate liabilities. *Journal of Political Economy*, pages 635–654.

Blackwell, D., Griffiths, M. and Winters, D. (2006). *Modern Financial Markets*. John Wiley & Sons, Ltd, New York.

Bladt, M. and Sørensen, M. (2006). Efficient estimation of transition rates between credit ratings from observations at discrete time points. Preprint No. 2006-2, Department of Applied Mathematics and Statistics, University of Copenhagen.

Blondell, D., Hoang, H., Powell, J. G. and Shi, J. (2002). Detection of financial time series turning points: A new approach CUSUM applied to IPO cycles. *Review of Quantitative Finance and Accounting*, **18**, 293–315.

Bock, D. (2008). Aspects on the control of false alarms in statistical surveillance and the impact on the return of financial decision systems. *Journal of Applied Statistics*, **35**.

Bodnar, O. and Schmid, W. (2007). Surveillance of the mean behaviour of multivariate time series. *Statistica Neerlandica*, (to appear).

Boehm, E. A. and Moore, G. H. (1991). Financial market forecasts and rates of return based on leading index signals. *International Journal of Forecasting*, **7**, 357–374.

Bollerslev, T. (1986). Generalized autoregressive conditional heteroscedasticity. *Journal of Econometrics*, **31**, 307–327.

Bollerster, T. (1990). Modeling the coherence in short-run nominal exchange rates: A multivariate generalized ARCH approach. *Review of Economics and Statistics*, **72**, 498–505.

Box, G. E. P. and Jenkins, G. M. (1970). *Time Series Analysis, Forecasting and Control*. Holden Day, San Fransisco.

Box, G. E. P. (1979). Robustness in the strategy of scientific model building. In Launer, R. L. and G N. Wilkinson, E. (Eds.), *Robustness in Statistics*. Academic Press, New York.

Brandt, M. (2007), Portfolio choice problems. In *Handbook of Financial Econometrics*, eds. Ait-Sahalia, Y. and Hansen, L. Elsevier and North Holland, Amsterdam.

Brandt, M., Santa-Clara, P and Valkanov, R. (2005). Parametric portfolio policies: exploiting characteristics in the cross-section of equity returns. Working paper.

Brock, W., Lakonishok, J. and LeBaron, B. (1992). Simple technical trading rules and the stochastic properties of stock returns. *Journal of Finance*, **47**, 1731–1764.

Brockwell, P. and Davis, R. (1991). *Time Series: Theory and Methods*. Springer-Verlag, New York.

Brockwell, P., Chadraa, E. and Lindner, A. (2006). Continuous-time GARCH processes. *The Annals of Probability*, **16**(2), 790–826.

Brockwell, P. J. and Davis, R. A. (1991). *Time Series: Theory and Methods*. Springer-Verlag, Berlin.

Brockwell, P. J. and Marquardt, T. (2005). Levy-driven and fractionally integrated ARMA processes with continuous time parameter. *Statistica Sinica*, **15**, 477–494.

Broemeling, L. D. and Tsurumi, H. (1987). *Econometrics and Structural Change*. Marcel-Dekker, New York.

Brook, D. and Evans, D. (1972) An approach to the probability distribution of CUSUM run length. *Biometrika*, **59**, 539–549.

Brown, R. (1827). *A brief account of microscopical observations*. Unpublished, London.

Busse, J. (1999). Volatility timing in mutual funds: Evidence from daily returns. *Review of Financial Studies*, **12**, 1009–1041.

Champ, C., Jones-Farmer, L. and Rigdon, S. (2005). Properties of the T^2 control charts when parameters are estimated. *Technometrics*, **47**, 437–445.

Chan, K. C., Karolyi, G. A., Longstaff, F. A. and Sanders, A. B. (1992). An empirical comparison of alternative models for the short-term interest rate. *Journal of Finance*, **47**, 1209–1228.

Chan, W.-S., Keung, W. and Tong, H. (Eds.). (2000). *Statistics and Finance: An Interface*. Imperial College Press, London.

Chen, G., Cheng, S. W. and Xie, H. (2004). A new EWMA control chart for monitoring both location and dispersion. *Quality Technology & Quantitative Management*, **1**, 217–231.

Cherubini, U., Luciano, E. and Vecchiato, W. (2004). *Copula Methods in Finance*. John Wiley & Sons, Ltd, New York.

Chu, C.-S. J., Stinchcombe, M. and White, H. (1996). Monitoring structural change. *Econometrica*, **64**, 1045–1065.

Cizek, P., Härdle, W. and Weron, R. (eds) (2005). *Statistical Tool for Finance and Insurance*. Springer-Verlag, Berlin.

Costa, A. F. B. and Rahim, M. A. (2004). Monitoring process mean and variability with one non-central chi-square chart. *Journal of Applied Statistics*, **31**, 1171–1183.

Cowles, A. (1933). Can stock market forecasters forecast? *Econometrica*, **1**(3), 309–324.

Cox, D., Hinkley, D. and Barndorff-Nielsen. O. (1996). *Time Series Models in Econometrics, Finance and other Fields*. Chapman and Hall, London.

Cox, J. C., Ingersoll, J. E. and Ross, S. A. (1985). A theory for the term structure of interest rates. *Econometrica*, **53**, 385–407.

Crosier, R. B. (1988). Multivariate generalizations of cumulative sum quality-control schemes. *Technometrics*, **30**, 291–303.

Crowder, S. (1987). A simple method for studying run-length distributions of exponentially weighted moving average charts. *Technometrics*, **29**, 401–407.

Crowder, S. V. (1989). Design of exponentially weighted moving average schemes. *Journal of Quality Technology*, **21**, 155–162.

Crowder, S. V. and Hamilton, M. D. (1992). An EWMA for monitoring a process standard-deviation. *Journal of Quality Technology*, **24**, 12–21.

Cvitanic, J. and Zapatero, F. (2004). *Introduction to the Economics and Mathematics of Financial Markets*. MIT Press, Cambridge.

Danilov, D. (2003). The effects of pretesting in econometrics with applications in finance. PhD thesis, Tilburg University.

DeMiguel, V., Garlappi, L. and Uppal, R. (2007). Optimal versus naive diversification: How inefficient is the $1/N$ portfolio policy? *Review of Financial Studies* (in press).

Dewachter, H. (1997). Sign predictions of exchange rate changes: Charts as proxies for bayesian Inferences. *Weltwirtschaftliches Archiv–Review of World Economics*, **133**, 39–55.

Dewachter, H. (2001). Can Markov switching models replicate chartist profits in the foreign exchange market? *Journal of International Money and Finance*, **20**, 25–41.

Dickey, D. A. and Fuller, W. A. (1979). Distribution of the estimators for autoregressive time series with unit root. *Journal of the American Statistical Association*, **74**, 427–431.

Domangue, R. and Patch, S. C. (1991). Some omnibus exponentially weighted moving average statistical process monitoring schemes. *Technometrics*, **33**, 299–313.

Draper, N. and Smith, H. (1966). *Applied Regression Analysis*. John Wiley & Sons, Ltd, New York.

Duffie, D. (1996). *Dynamic Asset Pricing Theory*. Princeton University Press, Princeton.

Dufour, A. and Engle, R. F. (2000). Time and the price impact of a trade. *Journal of Finance*, **55**, 2467–2498.

Eberlein, E. (2001). Application of generalized hyperbolic levy motions to finance. In Barndorff-Nielsen, O. E., Mikosch, T. and Resnick, S. I. (Eds.), *Lévy Processes, Theory and Applications*. Birkhäuser, Boston.

Einstein, A. (1905). On the movement of small particles suspended in a stationary liquid by the molecular-kinetic theory of heat. *Annalen der Physik*, pages 549–560.

Embrechts, P., Klüppelberg, C. and Mikosch, T. (1997). *Modelling Extremal Events for Insurance and Finance*. Springer-Verlag, Heidelberg.

Engle, R. (1982). Autoregressive conditional heteroscedasticity with estimates of the variance of UK inflation. *Econometrica*, **50**, 987–1008.

Engle, R. and Kroner, K. (1995). Multivariate simultaneous generalized ARCH. *Econometric Theory*, **11**, 122–150.

Engle, R., Ng, V. and Rothschild, M. (1990). Asset pricing with a factor arch covariance structure: Empirical estimates for treasury bills. *Journal of Econometrics*, **45**(2), 213–237.

Engle, R. F. and Bollerslev, T. (1986). Modeling the persistence of conditional variances. *Econometric Reviews*, **5**, 1–50.

Engle, R. F. and Russell, J. R. (1998). Autoregressive conditional duration a new model for irregularly-spaced transaction data. *Econometrica*, **66**(5), 1127–1162.

Eraker, B., Johannes, M. and Polson, N. (2003). The impact of jumps in volatility and returns. *Journal of Finance*, **58**, 1269–1300.

Fama, E. F. (1965). The behavior of stock market prices. *Journal of Business*, **38**, 34–105.

Fan, J. and Yao, Q. (2003). *Nonlinear Time-Series: Nonparametric and Parametric Methods*. Springer-Verlag, Berless.

Fernandes, M. and Grammig, J. (2006). A family of autoregressive conditional duration models. *Journal of Econometrics*, **130**(1), 1–23.

Fleming, J. Kirby, C. and Ostdiek, B. (2001). The economic value of volatility timing. *Journal of Finance*, **56**, 329–352.

Fleming, J. Kirby, C. and Ostdiek, B. (2003). The economic value of volatility timing using 'Realized' volatility. *Journal of Financial Economics*, **67**, 473–509.

Fleming, T. R. and Harrington, D. P. (1991). *Counting Processes and Survival Analysis*. John Wiley & Sons, Ltd, New York.

Foster, D. P. and Nelson, D. (1996). Continuous record asymptotics for rolling sample variance estimators. *Econometrica*, **64**, 139–174.

Franke, J. (1999). Nonlinear and nonparametric methods for analyzing financial time series. In *Operation Research Proceedings 98*, eds. P. Kall and H.-J. Luethi. Springer-Verlag, Heidelberg.

Franke, J., Härdle, W. and Hafner, C. (2004). *Statistics of Financial Markets. An Introduction*. Springer-Verlag, Berlin.

Frisén (2003). Statistical surveillance. Optimality and methods. *International Statistical Review*, **71**, 403–434.

Frisén, M. (1986). Unimodal regression. *The Statistician*, **35**, 479–485.

Frisén, M. (1992). Evaluations of methods for statistical surveillance. *Statistics in Medicine*, **11**, 1489–1502.

Frisén, M. (1994). Statistical surveillance of business cycles. Technical Report 1994:1 (Revised 2000), Department of Statistics, Göteborg University.

Frisén, M. (2003). Statistical surveillance. Optimality and methods. *International Statistical Review*, **71**, 403–434.

Frisén, M. and de Maré, J. (1991). Optimal Surveillance. *Biometrika*, **78**, 271–280.

Frisén, M. and Gottlow, M. (2003). Graphical evaluation of statistical surveillance. Technical Report Research Report 2003:10, Statistical Research Unit, Göteborg University.

Frisén, M. and Sonesson, C. (2006). Optimal surveillance based on exponentially weighted moving averages methods. *Sequential Analysis*, **25**, 379–403.

Frisén, M. and Wessman, P. (1999). Evaluations of likelihood ratio methods for surveillance. Differences and Robustness. *Communications in Statistics, Simulation and Computation*, **28**, 597–622.

Frost, P. and Savarino, E. (1988). An empirical Bayes approach to efficient portfolio selection. *Journal of Financial and Quantitative Analysis*, **21**, 292–305.

Föllmer, H. and Schied, A. (2002). *Stochastic Finance. An Introduction in Discrete Time*. de Gruyter, Berlin.

Gailbraith, R. F. and Gailbraith, J. I. (1974). On the inverse of some patterned matrices arising in the theory of stationary time series. *Journal of Applied Probability*, **11**, 63–71.

Gan, F. (1993). An optimal-design of EWMA control chart based on median run-length. *Journal of Statistical Computation and Simulation*, **45**, 169–184.

Garlappi, L., Uppal, R. and Wang, T. (2007). Portfolio selection with parameter and model uncertainty: a multi-prior approach. *Review of Financial Studies*, **20**, 41–81.

Gerhard, F. and Hautcsh, N. (2002). Semiparametric autoregressive conditional proportional hazard models. No 2002-W2, Economics Papers from Economics Group, Nuffield College, University of Oxford.

Golosnoy, V. (2007). Sequential monitoring of minimum variance portfolio. *Advances in Statistical Analysis*, **91**, 39–55.

Golosnoy, V. and Okhrin, Y. (2007). Multivariate shrinkage for optimal portfolio weights. *European Journal of Finance*, **13**, 441–458.

Golosnoy, V. and Schmid, W. (2007). EWMA control charts for optimal portfolio weights. *Sequential Analysis*, **26**, 195–224.

Golosnoy, V., Okhrin, I. and Schmid, W. (2007). Statistical Methods for the Surveillance of Portfolio Weights. Working paper.

Gordon, L. and Pollak, M. (1997). Average run length to false alarm for surveillance schemes designed with partially specified pre-change distribution. *The Annals of Statistics*, **25**, 1284–1310.

Gourieroux, C. (1997). *ARCH Models and Financial Applications*. Springer-Verlag, New York.

Gourieroux, C. and Jasiak, J. (2002). *Financial Econometrics: Problems, Models and Methods*. University Presses Of California, Columbia and Princeton, New Jersey.

Granger, C. W. J. and Andersen, A. P. (1978). *An Introduction to Bilinear Time Series Models*. Vandenchoeck & Ruprect, Göttingen.

Granger, C. W. J. and Joyeux, R. (1980). An introduction to long memory time series models and fractional differencing. *Journal of Time Series*, **1**, 15–29.

Granger, C. (1983). *Forecasting White Noise, in, Applied Time Series Analysis of Economic Data, Proceedings of the Conference on Applied Time Series Analysis of Economic Data (October 1981), Editor. A. Zellner*. US Government Printing Office.

Haavelmo, T. (1943). The implications of a system of simultaneous equations. *Econometrica*, **11**, 1–12.

Hackl, P. (Ed.) (1989). *Statistical Analysis and Forecasting of Economic Structural Change*. Springer-Verlag, Berlin.

Hackl, P. and Westlund, A. H. (Eds.) (1991). *Economic Structural Change*. Springer-Verlag, Berlin.

Haerdle, W., Herwartz, H. and Spokoiny, V. (2003). Time inhomogeneous multiple volatility modeling. *Journal of Financial Econometrics*, **1**, 55–95.

Hafner, C. (2003). Fourth moment structure of multivariate GARCH models. *Journal of Financial Econometrics*, **1**(1), 26–54.

Hall, P., Peng, L. and Yao, Q. (2002). Prediction and nonparametric estimation for time series with heavy tails. *Journal of Time Series*, **23**(3), 313–331.

Hamilton, J. D. (1989). A New approach to the economic analysis of nonstationary time series and the business cycle. *Econometrica*, **57**, 357–384.

Hamilton, J. D. (1994). *Time Series Analysis*. Princeton University Press, Princeton.

Hand, D. J. and Jacka, S. D. (Eds.). (1998). *Statistics in Finance*. Arnold, London.

Hannan, E. J. and Deistler, M. (1988). *The Statistical Theory of Linear Systems*. John Wiley & Sons, Ltd, New York.

Harris, T. and Ross, W. (1991). Statistical process control procedures for correlated observations. *Canadian Journal of Chemical Engineering*, **69**, 48–57.

Harrison, P. J. and Stevens, C. F. (1976). Bayesian Forecasting, with discussion. *Journal of the Royal Statistical Society B*, **38**, 205–247.

Harvey, A., Ruiz, E. and Shephard, N. (1994). Multivariate stochastic variance models. *Review of Economic Studies*, **61**(2), 247–264.

Harvey, A. C. (1989). *Forecasting, Structural Time Series Models and the Kalman Filter*. Cambridge University Press, Cambridge.

Harvey, A. C. (1993). *Time Series Models*. Harverster Wheatsheaf, London.

Harvey, C. R. and Morgenson, G. (2002). *The New York Times Dictionary of Money and Investing: The Essential A-to-Z Guide to the Language of the New Market*. Times Books, New York.

Harville, D. (1997). *Matrix Algebra from a Statistician's Perspective*. Springer-Verlag, New York.

Hawkins, D. and Olwell, D. (1998). *Cumulative Sum Charts and Charting for Quality Improvement*. Springer, New York.

Hawkins, D. L. (1992). Detecting shifts in functions of multivariate location and covariance parameters. *Journal of Statistical Planning and Inference*, **33**, 233–244.

Hawkins, D. M. (1981). A CUSUM for a scale parameter. *Journal of Quality Technology*, **13**, 228–231.

Hawkins, D. M. and Olwell, D. H. (1998). *Cumulative Sum Charts and Charting for Quality Improvement*. Springer-Verlag, New York.

Hawkins, D. M. and Zamba, K. D. (2005). A change point model for a shift in the variance. *Journal of Quality Technology*, **37**, 21–37.

Hendry, D. F. and Krolzig, H.-M. (2001). *Automatic Econometric Model Selection Using PcGets*. Timberlake Consultants Press.

Honoré, P. (1998). Pitfalls in estimating jump-diffusion models. Working paper, University of Aarhus.

Hotelling, H. (1947). Multivariate quality control – Illustrated by the air testing of sample bombsights. in *Techniques of Statistical Analysis*, eds. C. Eisenhart, M. W. H. and Wallis, W. A., McGraw Hill, New York, pp. 111–184.

Hsu, D., Miller, R and Wichern, D. (1974). On the stable Paretian behavior of stock market prices. *Journal of American Statistical Association*, **69**, 108–113.

Hull, J. (1993). *Options, Futures and other Derivatives* 2nd edn. Prentice-Hall, Englewood Cliffs.

Härdle, W., Kleinow, T. and Stahl, G. (eds) (2002). *Applied Quantitative Finance. Theory and Computational Tools*. Springer-Verlag, New York.

Ingersoll, J. (1987). *Theory of Financial Decision Making*. Rowman & Littlefield, Maryland.

Ivanova, D., Lahiri, K. and Seitz, F. (2000). Interest rate spreads as predictors of German inflation and business cycles. *International Journal of Forecasting*, **16**, 39–58.

Jensen, B. and Poulsen, R. (2002). Transition densities of diffusion processes: Numerical comparison of approximation techniques. *Journal of Derivatives*, **9**, 18–32.

Jensen, W., Jones-Farmer, L. A., Champ, C. and Woodall, W. (2006), Effects of parameter estimation on control chart properties: a literature review. *Journal of Quality Technology*, **38**, 349–364.

Jiang, W., Tsui, K.-L. and Woodall, W. (2000). A new SPC monitoring method: The ARMA chart. *Technometrics*, **42**, 399–410.

Johansson, N. C. J. (2003). Moment estimation using extreme value techniques. PhD thesis, Chalmers University of Technology and Göteborg University.

Jones, D. A. (1978). Non-linear autoregressive processes, series A. *Proc. Roy. Soc. London*, **360**, 71–95.

Jorion, P. (1986). Bayes–Stein estimation for portfolio analysis. *Journal of Financial and Quantitative Analysis*, **21**, 279–292.

Judge, G. G. and Bock, M. E. (1978). *The Statistical Implications of Pre-Test and Stein-rule Estimators in Econometrics*. North-Holland, Amsterdam.

Järpe, E. (1999). Surveillance of the interaction parameter in the Ising model. *Communications in Statistics. Theory and Methods*, **28**, 3009–3025.

Järpe, E. (2000). On univariate and spatial surveillance. PhD thesis. Göteborg University, Department of Statistics.

Järpe, E. and Wessman, P. (2000). Some power aspects of methods for detecting shifts in the mean. *Communications in Statistics. Simulations and Computations*, **29**, 633–646.

Jönsson, H., Kukush, A. and Silvestrov, D. S. (2004). Threshold structure of optimal stopping strategies for American type options. I. *Theory of Probability and Mathematical Statistics*, 82–92.

Karatzas, I. (2003). A note on Bayesian detection of change-points with an expected miss criterion. *Statistics & decisions*, **21**, 3–13.

Karatzas, I. and Shreve, S. E. (1991). *Brownian Motion and Stochastic Calculus*, 2nd edn. Springer-Verlag, Berlin.

Kendall, M. G. (1953). The analysis of economic time-series part I: Prices. *Journal of the Royal Statistical Society. Series A (General)*, pages 11–25.

Kim, D. and Kon, S. (1999). Structural change and time dependence in models of stock returns. *Journal of Empirical Finance*, **6**, 283–308.

Klein, R. and Bawa, V. (1976). The effect of estimation risk on optimal portfolio choice. *Journal of Financial Economics*, **3**, 215–231.

Klimko, L. A. and Nelson, P. I. (1978). On conditional least squares estimation for stochastic processes. *Annals of Statistics*, **6**, 629–642.

Kloeden, P. E. and Platen, E. (1992). *Numerical Solution of Stochastic Differential Equations*. Springer-Verlag, Berlin.

Knoth, S. (2004). Fast initial response features for EWMA control charts. *Statistical Papers*, **46**(1), 47–64.

Knoth, S. (ed.) (2006). *The Art of Evaluating Monitoring Schemes – How to Measure the Performance of Control Charts?* (Vol. 8), eds. H.-J. Lenz and P.-T. Wilrich, Physica Verlag, Heidelberg.

Knoth, S. and Schmid, W. (2002). Monitoring the mean and the variance of a stationary process. *Statistica Neerlandica*, **56**, 77–100.

Knoth, S. and Schmid, W. (2004). Control charts for time series: A review. In *Frontiers in Statistical Quality Control*, eds. Lenz, H.-J. and Wilrich, P.-T. Physica-Verlag, Heidelberg, vol. 7, pp. 210–236.

Kokoszka, P. and Leipus, R. (2000). Change-point estimation in ARCH models. *Bernoulli*, **6**, 513–539.

Kou, S. (2002). A jump diffusion model for option pricing. *Management Science*, **48**, 1086–1101

Kramer, H. and Schmid, W. (1997). Control charts for time series. *Nonlinear Analysis*, **30**, 4007–4016.

Kramer, H. and Schmid, W. (1997). EWMA charts for multivariate time series. *Sequential Analysis*, **16**, 131–154.

Krieger, A. M., Pollak, M. and Yakir, B. (2003). Surveillance of a simple linear regression. *Journal of the American Statistical Association*, **98**, 456–469.

Kulldorff, M. (2001). Prospective time periodic geographical disease surveillance using a scan statistic. *Journal of the Royal Statistical Society A*, **164**, 61–72.

Kutoyants, Y. (2004). *Statistical Inference for Ergodic Diffusion Processes.* Springer-Verlag, London.

Kutoyants, Y. A. (1984). *Parameter Estimation for Stochasic Processes.* Helderman, Berlin.

Lai, T. L. (1995). Sequential changepoint detection in quality-control and dynamical systems. *Journal of the Royal Statistical Society B*, **57**, 613–658.

Lai, T. L. (1998). Information bounds and quick detection of parameters in stochastic systems. *IEEE Transactions on Information Theory*, **44**, 2917–2929.

Lai, T. L. and Lim, T. W. (2005). Optimal stopping for Brownian motion with applications to sequential analysis and option pricing. *Journal of Statistical Planning and Inference*, **130**, 21–47.

Lai, T. L. and Shan, Z. (1999). Efficient recursive algorithms for detection of abrupt changes in signals and control systems. *IEEE Transactions on Automatic Control*, **44**, 952–966.

Lam, K. and Wei, L. (2004). Is the perfect timing strategy truly perfect? *Review of Quantitative Finance and Accounting*, **22**, 39–51.

Lam, K. and Yam, H. C. (1997). CUSUM techniques for technical trading in financial markets. *Financial Engineering and the Japanese Markets*, **4**, 257–274.

Lambert, P. and Lindsey, J. K. (1999). Analysing financial returns using regression models based on non-symmetric stable distributions. *Applied Statistics*, **48**, 409–424.

Lambrecth, B. M., Perraudin, W. R. M. and Satchell, S. (2003). Mortgage default and possession under recourse: A competing hazard approach. *Journal of Money, Credit and Banking*, **35**(3), 425–442.

Lamoureux, C. and Lastrapes, W. (1990). Persistence in variance, structural change and the GARCH model. *Journal of Business and Economic Statistics*, **8**, 225–234.

Lancaster, T. (1990). *The Econometric Analysis of Transition Data.* Cambridge University Press, Cambridge.

Lando, D. and Skødeberg, T. M. (2002). Analyzing rating transitions and rating drift with continuous observations. *Journal of Banking and Finance,* **26**(2-3), 423–444.

Lawson, A. (2004). Some considerations in spatial-temporal analysis of public health surveillance data. In *Monitoring the Health of Populations: Statistical Principles & Methods for Public Health Surveillance*, eds R. Brookmeyer and D. F. Stroup. Oxford University Press, Oxford, pp. 289–314.

Layton, A. P. (1996). Dating and predicting phase changes in the US business cycle. *International Journal of Forecasting*, **12**, 417–428.

Layton, A. P. and Katsuura, M. (2001). A new turning point signalling system using the Markov switching model with application to Japan, the USA and Australia. *Applied Economics*, **33**, 59–70.

Lehmann, E. L. and Romano, J. P. (2005). *Testing Statistical Hypothesis.* 3rd edn. Springer-Verlag, New York.

Lindeboom, M. and Van den Berg, G. J. (1994). Heterogeneity in bivariate duration models: The importance of mixing distribution. *Journal of the Royal Statistical Society; B*, **56**, 49–60.

Lindström, E. (2004). Statistical modeling of diffusion processes with financial applications. PhD thesis, Lund Institute of Technology.

Lo, A. W. (2000). Foundations of technical analysis: Computational algorithms, statistical inference, and empirical implementation. *Journal of Finance*, **55**, 1705–1770.

Lorden, G. (1971). Procedures for reacting to a change in distribution, *Annals of Mathematical Statistics*, **41**, 1897–1908.

Lowry, C. A. and Montgomery, D. C. (1995). A review of multivariate control charts. *IIE Transactions*, **27**, 800 810.

Lowry, C. A. Woodall, W. H., Champ, C. W. and Rigdon, S. E. (1992). A multivariate exponentially weighted moving average control chart. *Technometrics*, **34**, 46–53.

Lu, C.-W. and Reynolds, Jr, M. (1999). EWMA control charts for monitoring the mean of autocorrelated processes. *Journal of Quality Technology*, **31**, 166–188.

Lu, C. W. and Reynolds, M. R. (1999). Control charts for monitoring the mean and variance of autocorrelated processes. *Journal of Quality Technology*, **31**, 259–274.

Lu, C. W. and Reynolds, M. R. (2001). CUSUM charts for monitoring an autocorrelated process. *Journal of Quality Technology*, **33**, 316–334.

Lucas, J. M. and Saccucci, M. S. (1990). Exponentially weighted moving average control schemes: Properties and enhancements. *Technometrics*, **32**, 1–12.

Lund, R. and Basawa, I. (2000). Recursive prediction and likelihood evaluation for periodic ARMA models. *Journal of Time Series Analysis*, **21**, 75–93.

Lütkepol, H. (1991). *Introduction to Multiple Time Series Analysis*. Springer-Verlag, Berlin.

MacGregor, J. F. and Harris, T. J. (1993). The exponentially weighted moving variance. *Journal of Quality Technology*, **25**, 106–118.

Maddala, G. S. and Rao, C. R. (Eds.) (1996). *Handbook of Statistics, Volume 14: Statistical Methods in Finance*. Elsevier, Amsterdam.

Magnus, J. and Neudecker, H. (1999). *Matrix Differential Calculus with Applications in Statistics and Econometrics*. John Wiley & Sons Ltd, New York.

Maheu, J. M. and McCurdy, T. H. (2004). News arrival, jump dynamics, and volatility components for individual stock returns. *Journal of Finance*, **59**, 755–793.

Mandelbrot, B. (1963). The variation of certain speculative prices. *Journal of Business*, **36**, 394–419.

Mandelbrot, B. (1963a). New methods in statistical economics. *The Journal of Political Economy*, **71**, 421–440.

Mandelbrot, B. (1963b). The variation of certain speculative prices. *Journal of Business*, **36**, 394–419.

Mao, X., Yuan, C. and Yin, G. (2005). Numerical method for stationary distribution of stochastic differential equations with Markovian switching. *J. Comput. Appl. Math.*, **174**(1), 1–27.

Markowitz, H. (1952). Portfolio Selection. *Journal of Finance*, **7**, 77–91.

Marquardt, T. and Stelzer, R. (2007). Multivariate CARMA processes. *Stochastic Processes and their Applications*, **117**, 96–120.

Marsh, I. W. (2000). High-frequency Markov switching models in the foreign exchange market. *Journal of Forecasting*, **19**, 123–134.

Marshall, C., Best, N., Bottle, A. and Aylin, P. (2004). Statistical issues in the prospective monitoring of health outcomes across multiple units. *Journal of the Royal Statistical Society A*, **167**, 541–559.

McLeod, A. I. and Li, W. K. (1983). Diagnostic checking ARMA time series models using squared-residual autocorrelations. *Journal of Time Series Analysis*, **4**, 269–273.

McLeod, A. (1993). Parsimony, model adequacy and periodic correlation in forecasting time series. *International Statistical Review*, **61**, 387–393.

Mcneil, A. J. (1997). Estimating the tails of loss severity distributions using extreme value theory. *Astin Bulletin*, **27**(1), 117–137.

Melard, G. (1983). A fast algorithm for the exact likelihood of moving average models. *Applied Statistics*, **33**, 104–114.

Merton, R. C. (1980). On estimating the expected return on the market. *Journal of Financial Economics*, **8**, 323–361.

Merton, R. C. (1976). Option pricing when underlying stock returns are discontinuous. *Journal of Financial Economics*, **3**, 125–144.

Merton, R. C. (1990). *Continuous-Time Finance*. Blackwell, Oxford.

Michaud, R. O. (1998). *Efficient Asset Management*. Harvard Business School Press, Boston, Mass.

Mikosch, T. (2004). How to model multivariate extremes if one must? Maphysto, Research Report no. 21.

Mikosch, T. (2005). Copulas tales and facts. Discussion paper, International Conference on Extreme Value Analysis in Gothenburg.

Mills, T. (2004). *The Econometric Modelling of Financial Time Series*. Cambridge University Press, Cambridge.

Montgomery, D. and Mastrangelo, C. (1991). Some statistical process control methods for autocorrelated data. *Journal of Quality Technology*, **23**, 179–204.

Montgomery, D. C. (2005). *Introduction to Statistical Quality Control*, 5th edn. John Wiley & Sons, Ltd, New York.

Mood, A. M., Graybill, F. A. and Boes, D. C. (1974). *Introduction to the Theory of Statistics*, 3rd edn. McGraw-Hill, New York.

Moore, G. H., Boehm, E. A. and Banerji, A. (1994). Using economic indicators to reduce risk in stock-market investments. *International Journal of Forecasting*, **10**, 405–417.

Morais, M. (2001). Stochastic ordering in the performance analysis of quality control schemes. Ph.D. thesis, Universidade Técnica de Lisboa, Lisbon, Portugal.

Moustakides, G. V. (1986). Optimal stopping times for detecting changes in distributions. *The Annals of Statistics*, **14**, 1379–1387.

Muirhead, R. J. (1982). *Aspects of Multivariate Statistical Theory*. John Wiley & Sons, Ltd, New York.

Ncube, M. and Li, K. (1999). An EWMA–CUSCORE quality control procedure for process variability. *Mathematical and Computer Modelling*, **29**, 73–79.

Neely, C. J. (1997). Technical analysis in the foreign exchange market: A layman's guide. *The Federal Reserve Bank of St. Louis Review*, **79**, 23–38.

Neely, C. J. and Weller, P. A. (2003). Intraday technical trading in the foreign exchange market. *Journal of International Money and Finance*, **22**, 223–237.

Neftci, S. N. (1991). Naive trading rules in financial-markets and Wiener-Kolmogorov prediction-theory – A study of technical analysis. *Journal of Business*, **64**, 549–571.

Neftci, S. N. (1996). *An Introduction to the Mathematics of Financial Derivatives*. Academic Press, San Diego.

Nelson, D. (1991). Conditional heteroskedacity in asset returns. *Econometrica*, **59**, 347–370.

Nelson, D. B. (1990). Stationarity and persistence in the GARCH(1,1) model. *Econometric theory*, **6**, 318–344.

Ngai, H. M. and Zhang, J. (2001). Multivariate cumulative sum control charts based on projection pursuit. *Statistica Sinica*, **11**, 747–766.

Nikiforov, I. (1975). Sequential analysis applied to autoregression processes. *Automation and Remote Control*, **36**, 1365–1368.

Øksendal, B. (1998). *Stochastic Differential Equations: An Introduction with Applications*, 5th edn. Springer-Verlag, Berlin.

Ozaki, T. (1982). The statistical analysis of perturbed limit cycle processes using nonlinear time series models. *Journal of Time Series*, **3**, 29–41.

Ozaki, T. (1985). Nonlinear time series models and dynamical systems. In Hannan, E. and Krishnaiah, P. (Eds.), *Handbook of Statistics*, volume 5. North-Holland, Amsterdam.

Pagano, M. (1978). On the periodic and multiple autoregressions. *Annals of Statistics*, **6**, 1310–1317.

Page, E. S. (1954). Continuous inspection schemes. *Biometrika*, **41**, 100–115.

Page, E. S. (1963). Controlling the standard deviation by CUSUMS and warning lines. *Technometrics*, **5**, 307–315.

Perez-Amaral, T., Gallo, G. and White, H. (2003). A flexible tool for model building: The relevant transformation of the inputs network approach (RETINA). *Oxford Bulletin of Economics and Statistics*, pages 821–838.

Pettersson, M. (1998). Evaluation of some methods for statistical surveillance of an autoregressive Process. Technical Report 1998:4, Department of Statistics, Göteborg University.

Pettersson, M. (1998). Monitoring a freshwater fish population: Statistical surveillance of biodiversity. *Environmetrics*, **9**, 139–150.

Petzold, M., Sonesson, C., Bergman, E. and Kieler, H. (2004). Surveillance in longitudinal models. Detection of intrauterine growth retardation. *Biometrics*, **60**, 1025–1033.

Philips, T. K., Yashchin, E. and Stein, D. M. (2003). Using statistical process control to monitor active managers, *Journal of Portfolio Management*, **30**, 86–95.

Pignatiello, J. J. and Runger, G. C. (1990). Comparisons of multivariate CUSUM charts. *Journal of Quality Technology*, **22**, 173–186.

Pollak, M. and Siegmund, D. (1975). Approximations to the expected sample size of certain sequential tests. *Annals of Statistics*, **3**, 1267–1282.

Pollak, M. and Siegmund, D. (1985). A diffusion process and its applications to detecting a change in the drift of Brownian motion. *Biometrika*, **72**, 267–280.

Poor, V. H. (1998). Quickest detection with exponential penalty for delay. *The Annals of Statistics*, **26**, 2179–2205.

Pourahmadi, M. (2001). *Foundations of Time Series Analysis and Prediction Theory*. John Wiley & sons, Ltd, New York.

Priestley, M. B. (1991). *Non-Linear and Non-Stationary Time Series Analysis.* Academic Press, New York.

Quoreshi, A. M. M. S. (2006). Bivariate time series modeling of financial count data. *Communications in Statistics: Theory and Methods*, **35**(7), 1343–1358.

R Development Core Team (2005). *R: A language and environment for statistical computing.* R Foundation for Statistical Computing, Vienna, Austria.

Ramchand, L. and Susmel, R. (1998). Volatility and cross correlation across major stock markets. *Journal of Empirical Finance*, **5**, 397–416.

Resnick, S. I. (1987). *Extreme Values Regular Variations, and Point Processes.* Springer-Verlag, New York.

Resnick, S. I. (1997). Discussion on the data on large fire insurance losses. *Astin Bulletin*, **27**(1), 139–151.

Revuz, D. and Yor, M. (1999). *Continuous Martingales and Brownian Motion* 3rd edn. Springer-Verlag, Berlin.

Reynolds, M. and Cho, G.-Y. (2006). Multivariate control charts for monitoring the mean vector and covariance matrix. *Journal of Quality Technology*, **38**, 230–253.

Reynolds, M. R. and Soumbos, Z. G. (2001). Monitoring the process mean and variance using individual observations and variable sampling intervals. *Journal of Quality Technology*, **33**, 181–205.

Rigdon, S. E., Cruthis, E. N. and Champ, C. W. (1994). Design strategies for individuals and moving range control charts. *Journal of Quality Technology*, **26**, 274–287.

Roberts, S. W. (1959). Control chart tests based on geometric moving averages. *Technometrics*, **1**, 239–250.

Roberts, S. W. (1966). A comparison of some control chart procedures. *Technometrics*, **8**, 411–430.

Rogerson, P. A. and Yamada, I. (2004). Monitoring change in spatial patterns of disease: Comparing univariate and multivariate cumulative sum approaches. *Statistics in Medicine*, **23**, 2195–2214.

Rosolowski, M. and Schmid, W. (2003). EWMA charts for monitoring the mean and the autocovariances of stationary processes. *Sequential Analysis*, **22**, 257–285.

Rukhin, A. L. (2002). Asymptotic behavior of posterior distribution of the change-point parameter. *Journal of Statistical Planning and Inference*, **105**, 327–345.

Ruppert, D. (2004). *Statistics in Finance: An Introduction.* Springer-Verlag, New York

Ryan, T. P. (2000). *Statistical Methods for Quality Improvement* (2nd edn). John Wiley & Sons, Ltd, New York.

Saleh, A. K. M. E. (2006). *Theory of Preliminary Test and Stein-Type Estimation with Applications.* John Wiley & Sons, Ltd, New York.

Sato, K. (2001). Basic results on Lévy processes. In Barndorff-Nielsen, O. E., Mikosch, T. and Resnick, S. I. (Eds.), *Lévy processes: Theory and Application.* Birkhäuser, Boston.

Scherer, B. and Martin, R. D. (2005). *Introduction to Modern Portfolio Optimization with Nuopt^{tm}, S-Plus and S^+ Bayes^{tm}.* Springer-Verlag, New York.

Schipper, S. and Schmid, W. (2001). Sequential methods for detecting changes in the variance of economic time series. *Sequential Analysis*, **20**(4), 235–262.

Schipper, S. and Schmid, W. (2001a). Control charts for GARCH processes. *Nonlinear Analysis*, **47**, 2049–2060.

Schmid (1997a). CUSUM control schemes for Gaussian processes. *Statistical Papers*, **38**, 191–217.

Schmid (1997b). On EWMA charts for time series In *Frontiers in Statistical Quality Control*, eds. Lenz, H.-J. and Wilrich, P.-T. Physica-Verlag, Heidelberg, pp. 115–137.

Schmid, W. (1995). On the run length of a Shewhart chart for correlated data. *Statistical Papers*, **36**, 111–130.

Schmid, W. and Okhrin, Y. (2003). Tail behaviour of a general family of control charts. *Statistics & Decisions*, **21**, 79–92.

Schmid, W. and Schöne, A. (1997). Some properties of the EWMA control chart in the presence of autocorrelation. *The Annals of Statistics*, **25**, 1277–1283.

Schmid, W. and Tzotchev, D. (2004). Statistical surveillance of the parameters of a one-factor Cox–Ingersoll–Ross model. *Sequential Analysis*, **23**(3), 379–412.

Schöne, A., Schmid, W. and Knoth, S. (1999). On the run length of the EWMA chart: A monotonicity result for normal variables. *Journal of Statistical Planning and Inferences*, **79**, 289–297.

Severin, T. and Schmid, W. (1998). Statistical process control and its application in finance. In *Contributions to Economics: Risk Measurement, Econometrics and Neural Networks*. Physica-Verlag, Hevdelbag. pp. 83–104.

Severin, T. and Schmid, w. (1999). Monitoring changes in GARCH models. *Journal of the German Statistical Society*, **83**, 281–307.

Shewhart, W. A. (1931). *Economic Control of Quality of Manufactured Product*. MacMillan and Co., London.

Shiryaev, A. N. (1963). On optimum methods in quickest detection problems. *Theory of Probability and Its Applications*, **8**, 22–46.

Shiryaev, A. N. (1999). *Essentials of Stochastic Finance: Facts, Models, Theory*. World Scientific, Singapore.

Shiryaev, A. N. (2002). Quickest detection problems in the technical analysis of financial data. In *Mathematical Finance – Bachelier Congress 2000*, eds. H. Geman, D. Madan, S. Pliska and T. Vorst. Springer-Verlag, Berlin.

Shiryaev, A. N. (2004). A remark on the quickest detection problems. *Statistics & Decisions*, **22**, 79–82.

Shiryaev, A. N., Kabanov, Y. M., Kramkov, O. D. and Melnikov, A. V. (1994). Toward the theory of pricing of options of both European and American types. I. Discrete time. *Theory of Probability and its Applications*, **39**, 14–60.

Siegmund, D. and Venkatraman, E. S. (1995). Using the generalized likelihood ratio statistic for sequential detection of a change-point. *The Annals of Statistics*, **23**, 255–271.

Singleton, K. J. (2001). Estimation of affine asset pricing models using the empirical characteristic function. *Journal of Econometrics*, **102**(2), 111–141.

Sliwa, P. and Schmid, W. (2005a). Monitoring the cross-covariances of a multivariate time series. *Metrika*, **61**, 89–115.

Sliwa, P. and Schmid, W. (2005b). Surveillance of the covariance matrix of multivariate nonlinear time series. *Statistics*, **39**, 221–246.

Smith, A. F. and West, M. (1983). Monitoring renal transplants: An application of the multiprocess Kalman filter. *Biometrics*, **39**, 867–878.

Solnik, B., Bourcrelle, C. and Le Fur, Y. (1996). International market correlation and volatility. *Financial Analysts Journal*, **52**, 17–34.

Sonesson, C. (2003). Evaluations of some exponentially weighted moving average methods. *Journal of Applied Statistics*, **30**, 1115–1133.

Sonesson, C. and Bock, D. (2003). A Review and discussion of prospective statistical surveillance in public health. *Journal of the Royal Statistical Society A*, **166**, 5–21.

Sonesson, C. and Frisén, M. (2005). Multivariate surveillance. In *Spatial Surveillance for Public Health*, eds. A. Lawson and K. Kleinman. John Wiley & Sons, Ltd, New York. pp. 169–186.

Sowell, F. (1992). Maximum-likelihood estimation of stationary univariate fractionally integrated time series models. *Journal of Econometrics*, **53**, 165–188.

Srivastava, M. S. (1997). CUSUM procedures for monitoring variability. *Communications in Statistics. Theory and Methods*, **26**, 2905–2926.

Srivastava, M. S. (1994). Comparison of CUSUM and EWMA procedures for detecting a shift in the mean or an increase in the variance. *Journal of Applied Statistical Science*, **1**, 445–468.

Srivastava, M. S. and Chow, W. (1992). Comparison of the CUSUM procedure with other procedures that detect an increase in the variance and a fast accurate approximation for the ARL of the CUSUM procedure. Technical Report 9122, Department of Statistics, University of Toronto.

Srivastava, M. S. and Wu, Y. (1993). Comparison of EWMA, CUSUM and Shiryaev–Roberts procedures for detecting a shift in the mean. *The Annals of Statistics*, **21**, 645–670.

Srivastava, M. S. and Wu, Y. (1994). Dynamic sampling plan in Shiryaev–Roberts procedure for detecting a change in the drift of Browninan motion. *Annals of Statistics*, **22**(2), 805–823.

Starica, C. (1999). Multivariate extremes for models with constant conditional correlations. *Journal of Empirical Finance*, **6**, 515–553.

Starica, C. (2004). Is GARCH(1,1) as good a model as the Nobel Prize accolades would imply? Econometrics 0411015, EconWPA. available at http://ideas.repec.org/p/wpa/wuwpem/0411015.html.

Steland, A. (2002). Nonparametric monitoring of financial time series by jump-preserving control charts. *Statistical Papers, Berlin*, **43**, 401–422.

Steland, A. (2003). Jump-preserving monitoring of dependent time series using pilot estimators. *Statistics & Decisions*, **21**, 343–366.

Steland, A. (2005). Optimal sequential kernel detection for dependent processes. *Journal of Statistical Planning and Inference*, **132**, 131–147.

Stoumbos, Z. G., Reynolds Jr, M. R., Ryan, T. P. and Woodall, W. H. (2000). The state of statistical process control as we proceed into the 21st century. *Journal of the American Statistical Association*, **95**, 992–998.

Sullivan, J. H. and Jones, L. A. (2002). A self-starting control chart for multivariate individual observations. *Technometrics*, **44**, 24–33.

Sweeney, R. J. (1986). Beating the foreign exchange market. *The Journal of Finance*, **41**, 163–182.

Taylor, S. (1986). *Modelling Financial Time Series*, Vol. 1., John Wiley & Sons, Ltd, Chichester.

Tong, H. (1983). *Threshold Models in Non-Linear Time-Series Analysis*. Springer-Verlag, New York.

Tsai, H. and Chan, K. S. (2005). A note on non-negative continuous time processes. *Journal of the Royal Statistical Society: Series B (Statistical Methodology)*, **67**(4), 589–597.

Tsay, R. S. (2002). *Analysis of Financial Time Series*. John Wiley & Sons, Ltd, New York.

Tómasson, H. (1986). Prediction and estimation in ARMA models. PhD thesis, University of Gothenburg.

Vasilopoulos, A. and Stamboulis, A. (1978). Modification of control chart limits in the presence of data correlation. *Journal of Quality Technology*, **10**, 20–30.

Wang, Z. (2005). A shrinkage approach to model uncertainty and asset allocation. *Review of Financial Studies*, **18**, 673–705.

Wardell, D., Moskowitz, H. and Plante, R. (1994a). Run length distributions of residual control charts for autocorrelated processes. *Journal of Quality Technology*, **26**, 308–317.

Wardell, D., Moskowitz, H. and Plante, R. (1994b). Run-length distributions of special-cause control charts for correlated processes. (With discussion). *Technometrics*, **36**, 3–27.

Wei, W. W. S. (1990). *Time Series Analysis–Univariate and Multivariate Methods*. Addison-Wesley, Reading.

Wessman, P. (1998). Some principles for surveillance adopted for multivariate processes with a common change point. *Communications in Statistics. Theory and Methods*, **27**, 1143–1161.

Wessman, P. (1999). Studies on the surveillance of univariate and multivariate processes. Doctoral Thesis, Göteborg University, Sweden, Department of Statistics.

Wiener, N. (1923). Differential space. *Journal of Mathematical Physics*, **2**, 131–174.

Wilmott, P., Howison, S. and Dewynne, J. (1995). *The Mathematics of Financial Derivatives*. Cambridge University Press, Cambridge.

Wong, H. Y. and Li, C. P. (2006). Estimating jump diffusion structural credit risk models. Working Paper, Chinese University, Hong Kong.

Wong, W. K., Moore, A., Cooper, G. and Wagner, M. (2003). WSARE: Whats Strange About Recent Events? *Journal of Urban Health*, **80**, I66–I75.

Woodall, W. H. (1997). Control charts based on attribute data. bibliography and review. *Journal of Quality Technology*, **29**, 172–183.

Woodall, W. H. (2000). Controversies and contradictions in statistical process control. *Journal of Quality Technology*, **32**, 341–378.

Woodall, W. H. and Mahmoud, M. (2005). The inertial properties of quality control charts. *Technometrics*, **47**, 425–436.

Woodall, W. H. and Ncube, M. M. (1985). Multivariate CUSUM quality control procedures. *Technometrics*, **27** 285–292.

Yang, W. and Allen, D. E. (2005). Multivariate GARCH hedge ratios and hedging effectiveness in Australian futures markets. *Accounting and Finance*, **45**(2), 301–321.

Yashchin, E. (1993). Statistical control schemes – methods, applications and generalizations. *International Statistical Review*, **61**, 41–66.

Yashchin, E. (1993). Performance of CUSUM control schemes for serially correlated observations. *Technometrics*, **35**, 37–52.

Yashchin, E., Philips, T. K. and Stein, D. M. (1997). Monitoring active portfolios using statistical process control. In *Computational Approaches to Economic Problems. Selected Papers from the 1st Conference of the Society for Computational Economics* (Vol. 193–205), ed. H. e. a. Amman. Kluwer Academic, Dordrecht.

Zarnowitz, V. and Moore, G. H. (1982). Sequential signals of recessions and recovery. *Journal of Business*, **55**, 57–85.

Zhang, M. Y., Russell, J. R. and Tsay, R. S. (2001). A nonlinear autoregressive conditional duration model with applications to financial transaction data. *Journal of Econometrics*, **104**, 179–207.

Zivot, E. and Wang, J. (2003). *Modeling Financial Time Series with S-plus*. Springer-Verlag, New York.

Index

Page locators in **bold** indicate tables and those in *italics* indicate figures.

STATISTICS IN PRACTICE

Human and Biological Sciences

Berger – Selection Bias and Covariate Imbalances in Randomized Clinical Trials
Brown and Prescott – Applied Mixed Models in Medicine, Second Edition
Chevret (Ed) – Statistical Methods for Dose-Finding Experiments
Ellenberg, Fleming and DeMets – Data Monitoring Committees in Clinical Trials: A Practical Perspective
Hauschke, Steinijans and Pigeot – Bioequivalence Studies in Drug Development: Methods and Applications
Lawson, Browne and Vidal Rodeiro – Disease Mapping with WinBUGS and MLwiN
Lui – Statistical Estimation of Epidemiological Risk
Marubini and Valsecchi – Analysing Survival Data from Clinical Trials and Observation Studies
Molenberghs and Kenward – Missing Data in Clinical Studies
O'Hagan, Buck, Daneshkhah, Eiser, Garthwaite, Jenkinson, Oakley and Rakow – Uncertain Judgements: Eliciting Expert's Probabilities
Parmigiani – Modeling in Medical Decision Making: A Bayesian Approach
Pintilie – Competing Risks: A Practical Perspective
Senn – Cross-over Trials in Clinical Research, Second Edition
Senn – Statistical Issues in Drug Development, Second Edition
Spiegelhalter, Abrams and Myles – Bayesian Approaches to Clinical Trials and Health-Care Evaluation
Whitehead – Design and Analysis of Sequential Clinical Trials, Revised Second Edition
Whitehead – Meta-Analysis of Controlled Clinical Trials
Willan and Briggs – Statistical Analysis of Cost Effectiveness Data
Winkel and Zhang – Statistical Development of Quality in Medicine

Earth and Environmental Sciences

Buck, Cavanagh and Litton – Bayesian Approach to Interpreting Archaeological Data
Glasbey and Horgan – Image Analysis in the Biological Sciences
Helsel – Nondetects and Data Analysis: Statistics for Censored Environmental Data

Illian, Penttinen, Stoyan, H and Stoyan D–Statistical Analysis and Modelling of Spatial Point Patterns
McBride – Using Statistical Methods for Water Quality Management
Webster and Oliver – Geostatistics for Environmental Scientists, Second Edition
Wymer (Ed) – Statistical Framework for Recreational Quality Criteria and Monitoring

Industry, Commerce and Finance

Aitken – Statistics and the Evaluation of Evidence for Forensic Scientists, Second Edition
Balding – Weight-of-evidence for Forensic DNA Profiles
Brandimarte – Numerical Methods in Finance and Economics: A MATLAB-Based Introduction, Second Edition
Brandimarte and Zotteri – Introduction to Distribution Logistics
Chan – Simulation Techniques in Financial Risk Management
Coleman, Greenfield, Stewardson and Montgomery (Eds) – Statistical Practice in Business and Industry
Frisén (Ed) – Financial Surveillance
Lehtonen and Pahkinen – Practical Methods for Design and Analysis of Complex Surveys, Second Edition
Ohser and Mücklich – Statistical Analysis of Microstructures in Materials Science
Taroni, Aitken, Garbolino and Biedermann – Bayesian Networks and Probabilistic Inference in Forensic Science